AURORA

GEOPHYSICS AND ASTROPHYSICS MONOGRAPHS

AN INTERNATIONAL SERIES OF FUNDAMENTAL TEXTBOOKS

VOLUME 9

AURORA

by

ALISTER VALLANCE JONES, F.R.S.C.
Astrophysics Branch, National Research Council of Canada,
Ottawa, Canada K1A 0R8

D. REIDEL PUBLISHING COMPANY
DORDRECHT-HOLLAND/BOSTON-U.S.A.

Library of Congress Cataloging in Publication Data

Jones, Alister Vallance.
Aurora.

(Geophysics and astrophysics monographs ; v. 9)
Bibliography: p.
Includes index.
1. Auroras. I. Title. II. Series.

QC971.J66 538′.768 74–26994
ISBN 90–277–0272–1
ISBN 90–277–0273–X pbk.

First printing: December 1974

Published by D. Reidel Publishing Company,
P.O. Box 17, Dordrecht, Holland

Sold and distributed in the U.S.A., Canada and Mexico
by D. Reidel Publishing Company, Inc.
306 Dartmouth Street, Boston,
Mass. 02116, U.S.A.

Printed in Belgium

To Catherine
whose constant encouragement and support
have made this book possible

TABLE OF CONTENTS

CHAPTER 5. AURORA AND THE IONOSPHERE

CHAPTER 6. MECHANISMS OF PRECIPITATION OF AURORAL PARTICLES

APPENDICES

PREFACE

Over the past two decades auroral science has developed from a somewhat mysterious and imprecise specialty into a discipline central in the study of the ionosphere and magnetosphere. The investigation of aurora unites scientists with very different backgrounds and interests so that it is difficult to write a self-contained account of the field in a book of reasonable length. In this work I have attempted to include those aspects of theory which I have found valuable in predicting the effects on the atmosphere of auroral particle precipitation. In addition I have attempted to describe the techniques of observation with particular emphasis on optical methods which have been useful.

While the aeronomy of aurora has been regarded as central, the mechanisms by which particles are accelerated and precipitated into the atmosphere is of no less interest. This aspect of the subject has however been treated in a briefer fashion since it is a part of the immense and rapidly developing field of magnetospheric science.

Generally I have attempted to provide a coherent introduction to auroral science with an emphasis on relatively simple physical interpretations and models. References are given to enable the reader to find more extensive or rigorous discussions of particular topics. A fairly complete, quantitative atlas of the auroral spectrum is included.

The question of units is particularly difficult. In conformity with the practice of most workers in the field c.g.s. units have been generally used except for electrical quantities for which I have used S.I. units. Equations have been expressed in S.I. unit form when this would be different as is often the case for electrical relations. The conversion of the c.g.s. units employed to S.I. form is in any case very simple.

ACKNOWLEDGEMENTS

It is a pleasure to acknowledge the assistance of many who have helped in the preparation of this book. Drs D.R. McDiarmid, R.L. Gattinger, B.A. Whalen, D.C. Cartwright and G. Atkinson were very helpful in reading and commenting on sections of the manuscript. Mrs G.M. Ouellette, Mrs V. Powell, Mrs C. Chénier, Miss L. Sontheim, Mr D.E. Phinney, Mr A.E. Castellano, and the Stenographic Services of the National Research Council of Canada provided invaluable help in the preparation of the figures and manuscript. Special thanks are due to my colleague Dr R.L. Gattinger for innumerable discussions on the interpretation of the auroral spectrum and for making available the synthetic and observed spectral plots.

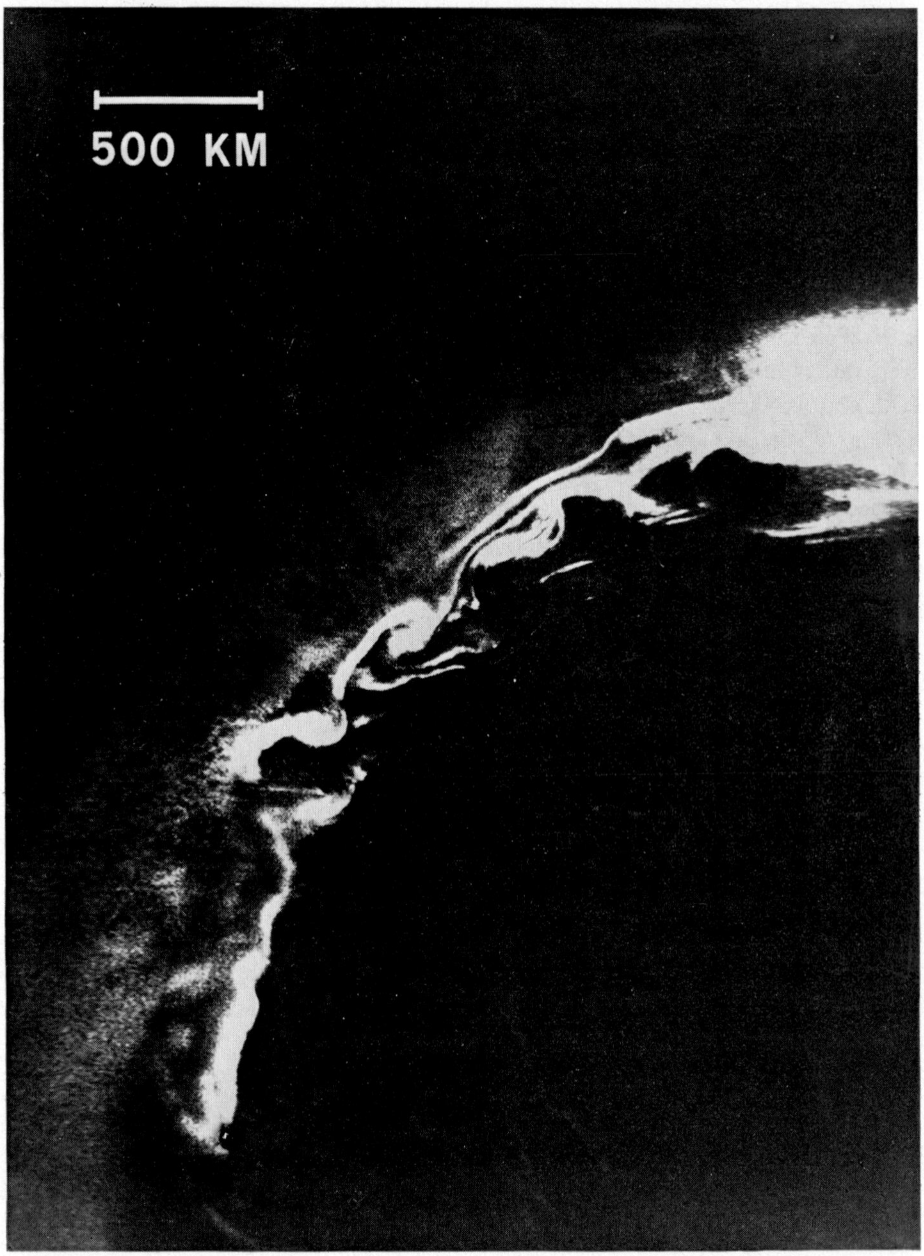

A satellite photograph of the evening-midnight quadrant of the southern auroral oval. This probably corresponds to the expansive phase of a substorm and shows several spirals (western travelling surges). This photograph was obtained from a U.S. Air Force Weather satellite according to the technique described by Rogers *et al.* (1973). I am very much indebted to Dr E. H. Rogers for this photograph.

CHAPTER 1

INTRODUCTION

This chapter is a short introduction to the phenomena which are usually described by the term aurora. Here the basic ideas will be outlined with some references to historical matters and an attempt made to relate aurora to the broader framework in which it belongs. History apart, the ideas discussed in this chapter will be taken up in more detail in later chapters.

1.1. Basic Ideas

The mysterious and beautiful luminosity which we call aurora has excited the curiosity and wonder of men since the dawn of history. Unlike some other impressive phenomena of nature such as the sunset, the rainbow or the thunder storm, it is both physically and conceptually remote since it occurs well above the lower layers of the atmosphere and its explanation involves some of the more recent disciplines of physics. When we see aurora we have visible evidence for the operation of the still unfamiliar mechanisms of the magnetosphere. In many ways the observation and measurement of the aurora has increased its mystery.

Aurora consists of a luminosity, usually distributed in regular or irregular horizontal bands which sometimes have vertically oriented striations or rays. The main luminous elements are called *auroral forms* and their classification constitutes auroral *morphology*. Most commonly seen forms have their lower limits between 100 and 120 km above the Earth's surface and have a vertical extent of less than 50 km.

Auroral bands are usually oriented very approximately in a magnetic E–W direction and have a horizontal extent of 10 to 20 km perpendicular to their length which can be 100 km or much more. The general E–W orientation may be perturbed by large or small scale folding. Much and perhaps most aurora is too low in intensity to cross the color threshold of the eye so that it is seen by scotopic vision (see Elliott and Dickson, 1951) and hence appears without definite color. Most observers report the sensation as being white. More intense forms appear green or more rarely red; purple, violet and blue are seen very occasionally.

The overall probability of seeing aurora reaches a maximum in an approximately circular zone of radius $\sim 22°$ centered on the geomagnetic poles. This is called the *auroral zone*. However if local time is taken into consideration the locus of maximum occurrence is an oval aligned along the Sun–Earth direction with radii from the geomagnetic pole of about 12° and 22°, respectively, in the solar and anti-solar directions. This is the *auroral oval*. The most intense aurora is seen in the midnight region of the oval.

The intensity of auroral forms is variable from a just detectable level comparable to that of the Milky Way, to 1000 times brighter. Displays of aurora tend to follow a pattern called the *auroral substorm* of which the most spectacular feature is an explosive increase in intensity in the midnight region of the oval followed by a spreading of the brightening poleward, westwards and eastwards. Rapid motions of the auroral forms take place at the height of each substorm and strong local magnetic perturbations are observed below on the ground due presumably to large electric currents at auroral heights.

The study of the spectra of aurora and more recently direct measurements by rockets and satellites have shown that most bright aurora is excited by energetic electrons guided downwards along the lines of the Earth's magnetic field. The average energy of the electrons is of the order of 5 to 10 keV. In addition to exciting the atoms and molecules of the atmosphere to emit optically, the bombarding electrons produce ionization which gives rise to various radio wave absorption and reflection effects. Radio aurora is a reflection of VHF and UHF waves from ionization more or less closely associated with visual aurora. It appears however that plasma instabilities associated with the presence of electric fields are necessary to generate the structure from which the radio waves are reflected. Electric fields also give rise to electric currents in the ionized regions and these currents generate strong magnetic field disturbances on the ground beneath many auroral forms.

Higher energy components of auroral electron fluxes can generate X-ray emission detectable at balloon altitudes below aurora as well as VLF radio emission. Such higher energy electrons also cause radio wave absorption due largely to the ionization they produce below 100 km. Both X-ray and particularly absorption observations have been widely used in synoptic studies since they give a measure of the entry of the higher energy component of auroral electrons independent of atmosphere conditions or time of day.

Very low frequency sound waves are generated by auroral forms under certain conditions. The time of propagation of such waves to the ground is of course of the order of 10 min while their frequency is below the threshold of audibility. Consequently it is unlikely that these waves can have any relation to reports of sounds from aurora.

In addition to electrons, there are influxes of energetic protons capable of producing a low level of luminosity in the same height range as electron aurora. This is called *proton aurora*. Sometimes it is spatially separate from electron aurora and sometimes intermingled with it. Because of charge exchange processes, incoming protons spend part of the trajectories as hydrogen atoms, thus escaping from the control of the Earth's magnetic field. This permits the proton beam to spread laterally and consequently a proton aurora never shows the detailed structure of electron aurora but rather takes the form of a broad diffuse emitting surface. Proton aurora was discovered by means of optical spectrographs and has been extensively studied by photometric and spectroscopic techniques. It is usually only faintly visible to the eye although it can appear as a definite band when seen obliquely near the horizon. Viewed in the magnetic zenith (i.e. parallel to the Earth's field lines), the hydrogen Balmer lines of proton aurora show a Doppler shift which historically provided the first direct evidence that auroral phenomena are caused by particle bombardments.

1.2. Brief History of Auroral Studies

Many authors have given excellent accounts of the early history of the knowledge of aurora and little purpose would be served by recopying their work. The main original sources are the books of Mairan (1733), Capron (1879), Fritz (1881), Tromholt (1885), Lemström (1886), and Angot (1896). Full references to these works and a list of more recent books and reviews may be found in an appendix in Chamberlain (1961). A more recent collection of interesting historical material has been compiled by Petrie (1963) while the monograph of Störmer (1955) provides a rich and accessible source of fascinating accounts of notable auroral displays. Two reviews by Chapman (1968, 1969) give a correct perspective to more recent progress in this century. Some of the most important points of the early history will be noted briefly. References can be found in the reviews listed above.

The name *aurora borealis* (northern dawn) was given by Gassendi in a book describing observations in southern France before 1621. The corresponding *aurora australis* was first seen by Cook in 1773. Cavendish made a reasonable height measurement in 1784. By 1873 the northern auroral zone had been mapped as a result of the work of Muncke, Loomis and Fritz. The connection between aurora and magnetic storms was recognized by Celsius in 1741. The sunspot cycle was discovered by Schwabe about 1844 and within 10 yr the connection between sunspot maxima and magnetic storms was recognized. The auroral spectrum was first observed by Ångström about 1868.

The discovery of the electron around 1880 paved the way for the proposal of Birkeland in 1896 that aurora was caused by electrons ejected from the Sun and guided to the auroral zones by the Earth's field. Birkeland showed by means of laboratory experiments that an electron beam would behave qualitatively as proposed. By 1910 Störmer began to calculate mathematically such particle orbits, so that quantitative testing of the theory could begin. Subsequently Störmer carried out definitive studies of auroral heights and extent by photographic triangulation.

The most prominent feature of the auroral spectrum, the green line at 5577Å, had been a mystery but was finally identified as a forbidden transition of O I by McLennan and Shrum in 1925. The hydrogen lines and their Doppler shift were identified and measured by 1950 in the work of Vegard, Gartlein and Meinel. In the following decade the exploration of the auroral spectrum was virtually complete as detailed by Chamberlain (1961).

By 1950 the theory of aurora and magnetic storms had been significantly advanced by Chapman and Ferraro who, beginning in 1930, had explored the consequences of the advance of a slab of solar plasma towards the Earth's magnetic field. The necessity of considering a plasma arose from the demonstration by Schuster in 1911 that a stream of particles of one sign would disperse by electrostatic repulsion.

Lindemann suggested that the solar particles could be in the form of a plasma, i.e., an ionized but neutral gas. Chapman and Ferraro concluded that an advancing plasma could push back the terrestrial field which would then be confined to a cavity in the plasma. The first effect is to produce an increase in the terrestrial surface field at the equator and so the initial positive phase of magnetic storms can be explained. The Chapman-Ferraro theory was extended by Martyn in 1951 but it was another decade before some of the main ideas of the modern theory of the magnetosphere were proposed by Axford and Hines (1961). Gold (1959) had suggested the name *magnetosphere* for the ionized distant regions of the Earth's magnetic field and provided some of the ideas on which the Axford-Hines theory was based. From this time until the present, development of an understanding of the magnetosphere and its behavior has been rapid. Satellites in highly elliptical orbits such as Explorer 12 defined the boundaries of the magnetosphere and showed that it did indeed lie within a cavity of the type proposed by Chapman and Ferraro. Likewise the hypothetical plasma reaching the Earth from the Sun was observed directly and named the *solar wind* since it was found to resemble a wind blowing continuously away from the Sun. Moreover this solar wind was found to carry an interplanetary magnetic field as envisaged by Alfvén in another pioneering theory of aurora. Dungey (1961) showed that the interplanetary field could play a very important role in driving the magnetospheric convection proposed by Axford and Hines. Satellite observations within the magnetosphere have given the fluxes of protons and electrons and mapped out the magnetic field configuration. Measurements have been made directly of the precipitating particles above the auroral oval. In brief, many different sorts of direct observations have contributed to an increasingly complete knowledge of the magnetosphere and the overall process of which aurora is a part. It is impossible to give an adequate history of these developments in this brief introduction. A little more discussion of magnetosphere processes will be given in the final section of this chapter and a more detailed account in Chapter 6.

1.3. Aurora and the Magnetosphere

Recent ideas about the structure of the magnetosphere are summarized in Figure 1.1. The terrestrial field is almost entirely confined inside the magnetopause, HHH, by the pressure of the solar wind, *K*. This effect can be produced simply by the advance of the solar wind plasma regarded as a perfect conductor. Eddy currents (not shown) circulate near the surface of the magnetopause to produce the modified field configuration. To maintain open lines in the tail and to account for satellite field observations a current sheet, *G*, crossing the center of the plasma sheet, *C*, is necessary. The solar wind plasma, *K*, has a supersonic velocity and thus leads to the formation of a shock front, *I*, around the nose of the magnetopause. Between the shock and the magnetopause is a region of stagnant plasma, the magnetosheath, *E*.

According to the proposal of Dungey (1961) there is an important interaction (depicted in Figure 1.2) between the lines of the interplanetary magnetic field when

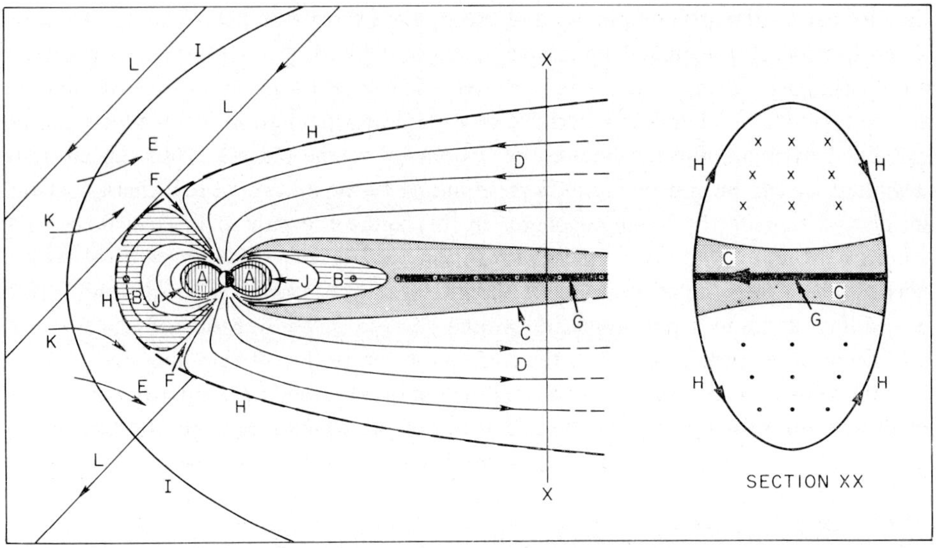

Fig. 1.1. Schematic diagram of magnetosphere. *A* : Plasmasphere. Filled with trapped cold plasma, $n \simeq 10^3$ cm^{-3}. *B* : Trapped and quasi-trapped particles 10–100 keV. Protons (electrons) drift westward (eastward) as a result of the curvature and gradient of the Earth's magnetic field. *C* : Plasma sheet particles[1]. Protons ~5 keV. Electrons ~1 keV. $n \sim 0.5$ cm^{-3}. *D* : N and S lobes[1] of terrestrial field. $n \sim 10^{-2}$ cm^{-3} $\langle E_e \rangle \sim 300$ eV; $\langle E_n \rangle \sim 1.5$ keV. *E* : Magnetosheath region. Contains stagnated solar wind plasma. *F* : Dayside cusps or clefts[2]. *G* : Current sheet[3]. Current flows west across boundary between N and S lobes. Circuit is completed in figure eight configuration around lobes. *H* : Magnetopause. *I* : Bow shock. *J* : Plasmapause. *K* : Solar wind. *L* : Interplanetary field lines. (The superscripts refer to the special set of references given in Section 3.7.)

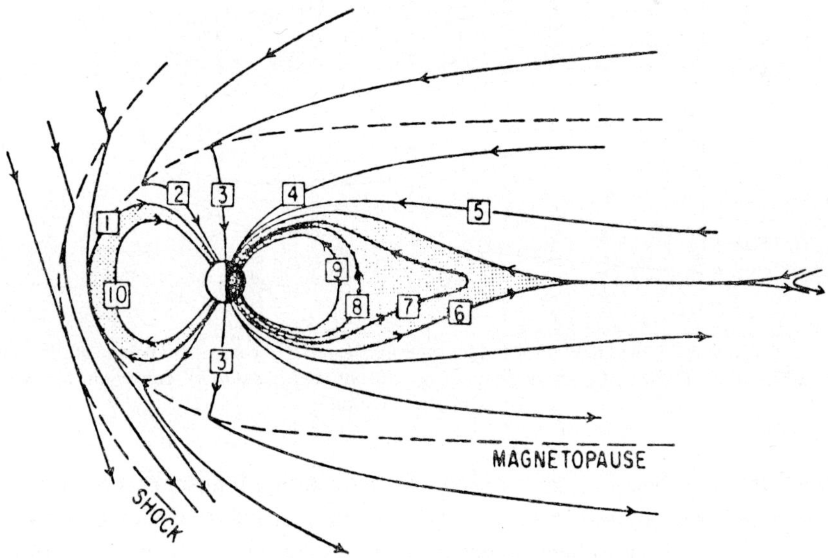

Fig. 1.2. Schematic diagram illustrating the reconnection hypothesis of Dungey from Levy *et al.* (1964). The region of reconnection may be much further down the tail. (Courtesy *AIAA Journal.*)

they have a southward component and those of the terrestrial field. The figures show schematically the postulated successive positions 1 to 10 of a field line as it connects and disconnects from a line of the terrestrial field. Field lines in an ionized medium can be taken to be identifiable because to a good approximation many effects can be explained by imagining the lines to be 'frozen in' to the plasma. Thus the motions in Figure 1.2 can be considered as corresponding to the motion of low energy plasma inside and outside the magnetosphere. In this figure the pair of lines in position 1 connect at the nose and disconnect in position 6 in the tail. According to the reconnection hypothesis the auroral oval is the intersection at the Earth of the limit of the closed lines at 1 and 6. The open lines connect to the Earth over the polar caps which are within the auroral ovals. Here the convection (or motion of field lines and plasma) is in the anti-solar direction while within the region of closed lines equatorward of the oval the motion is sunward. In the equatorial plane, which is crossed at intermediate distances by closed lines, the flow pattern is believed to be as shown by Brice (1967) and reproduced in Figure 1.3. The central region around the Earth where the flow

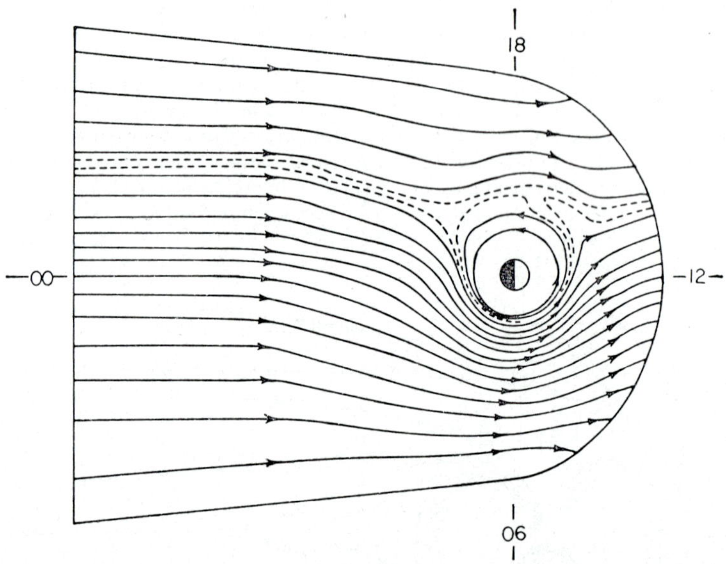

Fig. 1.3. Convection flow of field lines and low energy plasma in the equatorial plane of the magnetosphere after Brice (1967). (Courtesy *Journal of Geophysical Research*; copyright by American Geophysical Union.)

is closed (the lines being dragged around by the rotating ionosphere) corresponds to the plasmasphere, *A*, in Figure 1.1. The location of the plasmapause, the outer boundary of the plasmasphere, is believed to be determined by processes discussed in Chapter 6. The plasma sheet appears to contain solar plasma entering along the open lines and trapped by the reconnection of the lines like 6 in Figure 1.2. (The exact mode of entry is still under discussion.) This plasma is believed to convect inward, gaining energy by adiabatic compression until it is finally precipitated in the auroral oval region either by pitch angle scattering or through some discontinuous instability process.

The auroral substorm is believed to be initiated when the interplanetary field develops a sustained southward component so that the processes depicted in Figure 1.2 may occur. According to the hypothesis reviewed by Axford (1969) and elaborated, for example, by Coroniti and Kennel (1973) extra lines are transferred to the tail but their return to the nose is delayed by resistive effects in the ionosphere and back-pressure of plasma in the plasma sheet until some instability effect causes the extensive precipitation in the oval associated with an auroral substorm. These ideas will be discussed more fully in Chapter 6.

In conclusion it can be said that we are much closer to a satisfactory theory of aurora than two decades ago. It must be admitted that we still do not know exactly how auroral forms and displays arise and this is particularly true regarding the details of the fine spatial structure. It is also true that we may not yet have identified all the mechanisms which accelerate the solar wind plasma up to auroral energies, particularly the higher energies implied by the lower level types of auroras and X-ray emissions. Nevertheless, guided by direct observation, it appears that the broad outlines of the magnetospheric process in substorms are becoming clear.

CHAPTER 2

TECHNIQUES OF OBSERVATION

In this chapter some description is given of most of the techniques which are of importance in the observation of aurora. Since aurora is primarily an optical phenomenon more detail has been devoted to the description of optical methods with an emphasis on points which have been found valuable in choosing the best technique for particular studies.

2.1. Observations of Total Intensity

A large part of our knowledge of the behavior of auroral displays has been gained by observations with detectors sensitive to a wide range of wavelengths. Visual and direct photographic observations are of this type. To improve discrimination against moonlight and artificial lights, narrow spectral band passes have been used in auroral photometers designed for synoptic studies of auroral occurrence. In each case the response obtained is assumed to be proportional to the total intensity of the aurora. Here the emphasis is on the spatial and temporal variations of the emission.

The human eye is a valuable detector in qualitative observations of auroral forms. In many cases reasonably accurate intensity estimates can be made visually. However, when there is a high contrast between aurora and background the intensity may be overestimated and vice versa.

Photographic emulsions are useable from the short wavelength UV cut off of the atmosphere near 3100 Å to 8700 Å. A few spectra have been obtain photographically out to nearly 11000 Å and with rockets down to 1300 Å but the results are not very satisfactory.

Photoelectric image converters and intensifiers have been used successfully to extend the spectrographic technique particularly in the infrared to 11000 Å. Image orthicons, as discussed below, have been used in direct imaging and spectroscopy of aurora.

For the visible, UV and near IR regions the most useful detector has been the photomultiplier phototube. Because of their importance and as an introduction to later sections some of characteristics of photomultipliers will be outlined. For many purposes it is sufficient to consider only the cathode quantum efficiency for photoelectron production and the rate of emission of dark current electrons from the cathode, i_D. Peak quantum efficiencies can be as high as 20% and dark currents as low as 1 s^{-1}. As will be shown later these two parameters are sufficient for an approximate description of the signal-to-noise ratio in spectrometer or photometer systems in the pulse counting mode. The dark count is strongly temperature dependent and consequently operation at reduced temperatures is beneficial with many tubes particularly those with sensitivities extending to the red and IR. Usually there is an optimum working temperature. At present, tubes with good quantum efficiency (i.e., better than 1%) are available out to a little beyond 9000 Å while other photocathodes are sensitive to 11 000 Å at lower quantum efficiency and higher dark currents. In the UV CsI and CsTe photocathodes with long wave limits at about 3000 Å and 2000 Å are very useful because of their lack of response to longer wavelengths. Open channel multipliers may be used in the UV particularly at the very shortest wavelengths.

In the IR beyond 11 000 Å intrinsic Ge, PbS, InSb, HgCdTe and various varieties of doped Ge give response to at least 40 μm although at the cost of lower operating temperatures at the longer wavelengths.

2.1.1. Visual observations

Up to about 1950 almost all the synoptic observations of auroral occurrence had been made visually. The use of visual reports of aurora from large networks of observers was attempted during the I.G.Y. (International Geophysical Year) (1957–58) and various schemes were devised to collect the data. Aurora was reported by sketching its position on special forms together with notations as to type, intensity, and color. Reports were in many cases made at regular intervals by meteorological stations but also by large numbers of volunteer observers who might report particular events. The results were correlated and plotted on maps. The techniques and their application have been described and referenced by Millman (1968).

The visual technique is effective in acquiring information on the occurrence of aurora and historically the discovery of the auroral zones depends on such observations. The more recent visual programs have provided a body of data which may be useful particularly with reference to low latitude and red auroral displays.

2.1.2. All-sky camera observations

Several designs of camera systems capable of photographing the 180° field of the sky appeared in the years before the I.G.Y. Meinel and Schulte (1953) reported auroral photography with a 140° camera developed by Henyey and Greenstein for astronomical purposes (Struve, 1951). For auroral photography a satisfactory system is achieved by using a fast (f/1.0 to f/2.0) cine- or still-camera lens to photograph the sky image formed by a horizontal convex spherical or parabolic mirror 30 to 60 cm in diameter. Gartlein (1947) pioneered the use of such systems. Their history is discussed by Chamberlain (1961).

The systems developed for the I.G.Y. photograph the sky, usually on fast 16 mm or 35 mm film, with exposure times varying from several to several tens of seconds.

Since such cameras operate automatically, photographing the sky at perhaps one frame per minute and record also the time and date, they have proved the most valuable tool for studies of the occurrence and motions of aurora particularly of discrete forms. The use of color film, pioneered by Sandford and Heiser (1959), is becoming common since it provides additional information on the spectral distribution of the auroral emissions.

There has also been some limited application of 'fish-eye' lenses for auroral photography. Recently an f/1.4 Nikon lens with a 180° field of view, linear in zenith distance across an 18 mm field has been announced. For some applications, such as in observations from aircraft, these lenses, although costly, have advantages.

2.1.3. Television and Image Intensifier Camera Systems

For the study of fine detail and rapid movements in auroral forms a more sensitive system than direct photography is valuable. The use of image orthicon television camera systems for this purpose was first exploited extensively by Davis and Hicks (1964). Details of the method have been discussed by Davis (1966, 1967b, 1969a). The output of the camera can be recorded on film or on video tape at rates up to 60 frame s^{-1}. IBC 1 aurora can be detected readily. The gain over direct photography is of the order of 100.

Scourfield and Parsons (1969) have discussed the use of an EMI 9694 image intensifier tube (S-20 cathode) in conjunction with a vidicon tube for auroral cinematography. This also appears to be an effective tool in the study of detail in rapidly moving aurora and of pulsations. The sensitivity of these systems makes it possible to use several cameras each with a different filter and so to record time-constant effects in the excitation processes. This will be discussed in Section 3.3.7.

2.1.4. Ground Based, Rocket and Satellite Auroral Photometry

Photoelectric auroral recorders were developed from about 1948 with the object of compiling statistics of auroral occurrence with better intensity information than is possible from photographic data. An early instrument of this type was described by Penn and Currie (1949). Narrow band filters for the 5577 Å [O I] line and for a background wavelength were employed to provide discrimination against moonlight. Later instruments of this type, such as the I.G.Y. recorder described by Hunten (1956) scanned the meridian and provided automatic compensation for non-auroral light. The later instrument also pioneered the use of digital recording on paper tape to facilitate later computer processing.

In practice synoptic studies of aurora using such recorders have not been as rewarding as the all-sky camera studies because of the much greater difficulties of keeping a network of such recorders operating properly. On the other hand limited operations with a small number of stations have been very useful particularly with the multiwavelength photometers described below. For particular events a high-speed all-sky scanner has been described by Sawchuk and Anger (1972).

Simple filter photometers have been extensively employed in rockets mainly in the hope of measuring the height distributions of various spectral features. This is discussed further in the next section. A very powerful system is the scanner developed by Robertson *et al.* (1974) for obtaining medium resolution pictures of auroral forms at two wavelengths from a spinning rocket. After computer processing this system can give views of the displays as seen from the rocket or (from near apogee) as projected at the 100 km level.

Photometers have also been vital in exploratory measurements of the extreme UV region below 1000 Å by Paresce *et al.* (1972) with channel multlpliers as detectors and LiF, Al and Sn as filters to separate broad bands down to 300 Å.

Photometers have frequently been used on satellites (O'Brien, 1967; Warnecke *et al.*, 1969; Deehr and Egeland, 1972) to measure optical emissions along the path for comparison with particle measurements. This is particularly valuable in UV studies below 3100 Å as in the work of Chubb and Hicks (1970).

A new era of auroral satellite photometry has recently begun with the λ6300, λ5557 and λ3914 scanning systems of ISIS B (Anger *et al.*, 1973b; Shepherd *et al.*, 1973a, b). These photometers use satellite spin to provide a scan (usually traverse to the satellite motion) in a polar orbit. They can consequently make instantaneous maps of auroral emission over a major portion of the dark side of the Earth on each orbit. Preliminary results of these studies are discussed in Section 3.2.

2.1.5. Height Finding Systems

The classical technique for measuring the height of auroral forms by taking simultaneous photographs from two stations has been well described by Störmer (1955). His meticulous and sustained researches on this topic provide the main body of knowledge as to the height distribution of aurora. In brief the method is to measure the location of a point on an auroral form by triangulation from two ground stations about 30 km apart by using the stars photographed behind as reference directions. The accuracy is limited by the difficulty of identifying a definite point on which to base the triangulation. Moreover there is a tendency for quiet aurora to be faint and for bright structured aurora to be in rapid motion. A method of measuring heights from all sky camera photographs has been described by Brandy and Hill (1964).

It is even possible to determine the height of regular arcs from one station following a method most recently discussed by Rees (1963a). Alternatively the height distribution can be obtained as by Evans *et al.* (1972) from one station if the field lines on which it lies can be identified from particle measurements.

Another technique of determining heights and also intensity distributions within aurora has been developed by Romick and Belon (1967a, b). Two meridian scanning photometers are used. The ideal geometry is to have a simple form in the magnetic zenith above one station and to scan it from the side from the other station. In this way, with certain assumptions, the distribution in space of the emission can be determined.

The use of rocket photometers to measure height profiles has also been practiced over the last decade. The best technique is to use an up-down photometer as described by Dick and Fastie (1969); otherwise temporal intensity changes during the ascent can seriously disturb the height profile (obtained in the simplest approximation by differentiating the signal from the rocket photometer). In practice the horizontal structure of aurora, the horizontal velocity of the rocket and the changes in its orientation can make rocket height determinations very difficult except under ideal conditions or with spatially diffuse emissions.

2.2. Spectroscopic Techniques

Spectroscopic analysis of auroral emissions gives a large amount of information as to processes taking place within auroral forms. Modern spectroscopic techniques involve several complementary systems to be discussed in turn.

2.2.1. Grating spectrographs

The best photographic spectrographs were those developed by A.B. Meinel about 1950. These had 20 × 20 cm, 600 line mm^{-1}, plane diffraction gratings used in conjunction with field-flattened f/0.8, 9″ Schmidt cameras. Such instruments are capable of photographing at least 2500 Å of the near IR spectrum in the 1st order in one exposure with a resolution of 5 Å or better. In the violet, the third order would be used with correspondingly increased resolution and decreased wavelength range. The low quantum efficiency of the photographic emulsion (compared to the photomultiplier of a scanning spectrometer) is largely compensated by the fact that at least 500 spectral elements are recorded simultaneously rather than sequentially. Consequently grating spectrographs were the instruments *par excellence* for the semi-quantitative exploration of the spectrum in the photographic region at high resolution. In practice fast grating spectrographs can record lines with exposures varying from 100 kR-s at 4000 Å to perhaps 1 MR-s at 8000 to 9000 Å. For accurate intensity measurements and for studies of particular features with higher time resolution, they have been displaced by the photoelectric systems to be described.

Lower resolution, automatic spectrographs recording on film (so-called patrol spectrographs) were extensively employed for the 10-yr period beginning in 1955 (Devlin *et al.*, 1964; Clark and Romick, 1959; Vallance Jones, 1971). They are generally used as meridian spectrographs with the complete sky focussed on the entrance slit so that the intensity variation along the spectral lines corresponds to the distribution along a N–S meridian. Like the all-sky camera, these instruments provided a method of obtaining and storing vast amounts of information. Properly used they were very useful in synoptic observations of the main auroral emission and particularly of the Balmer hydrogen lines.

Spectrographs with electronic image converters or TV type cameras have been used successfully either to obtain IR response to 1.1 μ (Fedorova, 1961) or to get enhanced sensitivity in the visible region as referenced by Vallance Jones (1971).

2.2.2. Grating Spectrometers

Grating spectrometers are at present the most versatile and widely used instruments for studying the auroral spectrum. The fundamental reasons may be discussed in terms of the equation which determines the photon flux which the instrument can deliver to its detector for a given spectral band pass. This relation (derived originally in a slightly different form by Jacquinot) is

$$\mathscr{L}\mathscr{R} = A(\sin i + \sin \theta)\,\beta\tau. \tag{2.2a}$$

$\mathscr{L}$ is the luminosity defined as the flux, Φ, passed per unit monochromatic source brightness, B, in units of (photon s^{-1})/(photon cm^{-2} sr^{-1} s^{-1}). $\mathscr{R}$ is the resolution $\lambda/\Delta\lambda$ as determined by the geometrical slit widths, A is the grating area, i and θ are the angles of incidence and diffraction of the transmitted beam at the grating, β is the angular length of the slits and τ is the maximum transmission of the system for the transmitted wavelength. Equation (2.2a) shows that the performance of grating spectrometers is subject to a fundamental limit, closely approached by modern instruments. The term $(\sin i + \sin\theta)$ can be little greater than unity while τ can be little greater than 0.5. β is limited by aberrations to values of about $\frac{1}{7}$ or less. The grating area can be increased up to a limit determined by availability, cost and physical size of the spectrometer.

The other factor determining a spectrometer's performance is the characteristics of the detector. For simplicity we assume operation in the pulse count mode so that each photoelectron or dark current electron produces a count. This mode is preferable in practice to the older DC one because (i) it can be made to discriminate against the noise and offset produced by spurious pulses, too large or small to come from the photocathode, (ii) time constant effects are eliminated so that it is possible to add data from many rapid wavelength scans to minimize the effect of intensity variations and (iii) direct coupling into digital recording systems is facilitated. This in turn makes it easy to carry out the averaging noted in (ii) as well as to apply calibration factors, to subtract backgrounds and to plot the final corrected spectra. The factors involved in the choice of recording modes for photomultipliers are rather complex and even debatable. Robben (1971) and Jones *et al.* (1971) have analyzed the problem and their work should be consulted for a serious discussion.

A fundamental index of the system is the count rate per rayleigh* when the instrument is centered on a narrow emission line. Since $\mathscr{L} = \Phi/B$ where Φ is the photon flux reaching the detector and B is the source brightness as defined above, then

* The rayleigh (Hunten *et al.*, 1956) is a unit of photon brightness defined by the relation

$$B(\text{rayleigh}) = \frac{4\pi}{10^6}\,B(\text{photon cm}^{-2}\text{ sr}^{-1}\text{ s}^{-1}).$$

In the case of an optically thin emitter $4\pi B$ gives the total emission from a column of unit cross section along the line of sight. Moreover when measured in the zenith direction the brightness in rayleighs is often roughly the average emission rate in photons cm^{-3} since many auroral forms have a vertical thickness of the order of 10 km.

$$\Phi \simeq 10^6 \times \frac{B}{4\pi} \frac{A}{\mathscr{R}} \beta \tag{2.2b}$$

since $(\sin i + \sin \theta)\,\tau \simeq 1$. B is the brightness in rayleighs. Consequently the count rate per rayleigh, e_1 is

$$e_1 = \frac{\mathrm{C}}{\mathrm{B}} \simeq \frac{10^6}{4\pi} \cdot \frac{A}{\mathscr{R}} \varepsilon\beta. \tag{2.2c}$$

If we adopt reasonable values, taking $A = 10^2$ cm^2, $\varepsilon = 0.1$, $\beta = 0.1$ and $\mathscr{R} = 10^3$ ($\Delta\lambda = 5$ Å at 5000 Å) then

$$e_1 \simeq 10^2 \text{ count s}^{-1}\,\text{R}^{-1}.$$

This value is for a somewhat high quantum efficiency. Values of ε down to 1% or less are normal, so count rates/rayleigh between 5 and 100 are typical.

Since a spectrometer is almost always used to scan a finite wavelength range, suppose that a total time T is to be used to scan N spectral elements, with the wavelength setting fixed for T/N s at each wavelength observed. This total time could be divided into shorter scans the outputs of which are added. Then the signal count on a line of intensity B rayleighs will be

$$e_1 BT/N$$

while the total count for this wavelength channel will be

$$(e_1 B + i_\mathrm{D})\, T/N.$$

Since the statistical fluctuation in the sample will be equal to the square root of the total count, the signal to noise ratio, r, observed will be

$$r = \frac{\sqrt{e_1 BT/N}}{\sqrt{1 + \dfrac{i_\mathrm{D}}{e_1 B}}} \tag{2.2d}$$

This expression has been evaluated for the range of interest in Figure 2.1 for a total scan time of 10^3 s for 500 samples. A typical value of $e_1 = 10$ has been assumed. Several conclusions may be drawn if it assumed that $r \geq 30$ is required for satisfactory observations:

(1) Features of brightness down to 100 R can be measured under the standard conditions assumed for dark count rates up to 10^3 s^{-1}.

(2) An increase of quantum efficiency or grating area by a factor n would permit the abscissa to be shifted by an amount corresponding to n.

(3) An increase in T/N by a factor of n would give a shift in the ordinate scale by a factor of $n^{\frac{1}{2}}$. Thus increasing the sampling time to 20 s per sample would give a satisfactory r at 10 R if the dark current remains less than 10^2 s^{-1}. Thus dark current becomes important when the system is used to scan short wavelength regions slowly in an attempt to gain sensitivity.

(4) Increasing the resolving power $\mathscr{R}$ by a factor of n shifts the abscissa scale to the left by the same factor. This could be compensated by a $\sim n\times$ increase in sampling time if i_D is sufficiently small. However, if the same range is to be scanned a further $n\times$ increase in total observing time is required.

From the point of view of signal to noise it would appear that the spectrometer will be superior to the spectrograph when the number of spectral elements to be scanned is below a limit related to the quantum efficiences of photomultipliers and

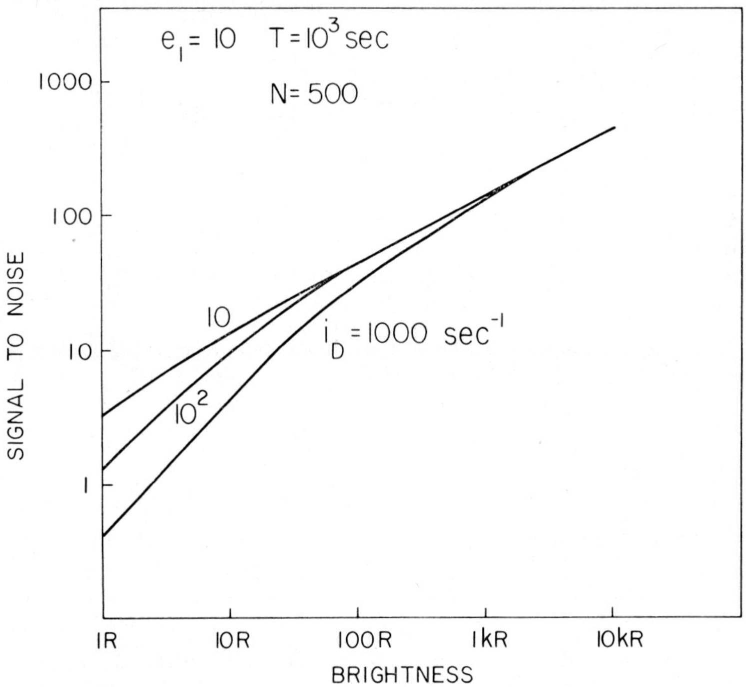

Fig. 2.1. Calculated signal to noise ratio for typical auroral grating spectrometer as described in text with assumed sensitivity, e_1, of 10 count R^{-1} and scan time of 2s/spectral element for different dark count rates, i_D.

photographic emulsions as well as other factors in the case of the latter. Bird *et al.* (1969) give a valuable discussion of the theory of the photographic emulsion as a photon detector.

For quantitative measurements spectrometers are also superior to conventional spectrographs because of the linear relation between input flux and output. Nevertheless it seems in principle possible that a spectrograph with an image intensifier or image tube of higher quantum efficiency could give better high resolution performance.

2.2.3. Fabry-Perot Spectrometers

The modern plane Fabry-Perot spectrometer has an *aperture limited* resolution $\mathscr{R}$ and luminosity $\mathscr{L}$ (see Section 2.2.2) satisfying the relation

$$\mathscr{L}\mathscr{R} = 2\pi A n^2 \tau_{av}, \tag{2.2e}$$

where A is the plate area, and n is the refractive index of the medium between the plates (Jacquinot, 1958). The basic arrangement is shown in Figure 2.2. Such a system,

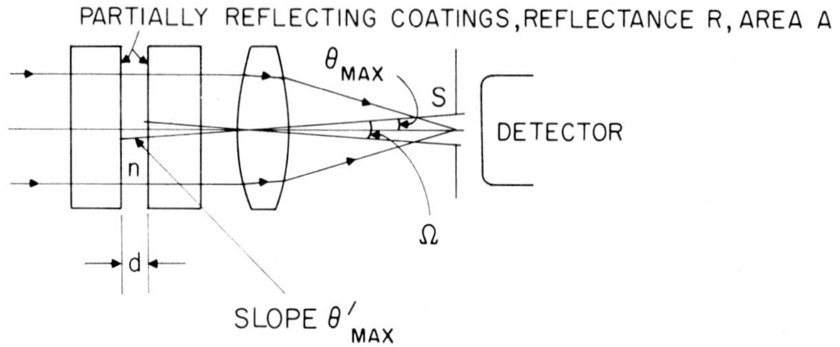

Fig. 2.2 Schematic diagram of Fabry-Perot spectrometer. The symbols are defined in the text.

with an aperture, S, of finite size, satisfies the interference condition for maximum transmission from a wavenumber $\sigma_0 = k/2nd$ (where the integer k is the order of interference) to $\sigma_0 + \Delta\sigma_a$ where $\Delta\sigma_a = \sigma_0/\mathscr{R}$ and $\mathscr{R}$ is given by the expression

$$\mathscr{R}\Omega = 2\pi n^2. \tag{2.2f}$$

Ω is the angular size of the exit aperture.

For radiation of a given wavenumber falling within the pass band $\Delta\sigma_a$, Equation (2.2e) is satisfied, with τ_{av} being the average transmission for this wavenumber across the pass band. τ_{av} is comparable to the τ of the spectrometer Equation (2.2a); it will be discussed further below.

By comparing Equations (2.2a) and (2.2e) one can see that if $n = 1$ and the grating and etalon areas are the same, the $\mathscr{L}\mathscr{R}$ product of the FP exceeds that for the grating spectrometer by a factor of about $2\pi/\beta$ ($\simeq 10^2$) which is often called the Jacquinot advantage. This ratio gives a direct measure of the relative ability of the two systems to deliver flux to the detector when set to transmit monochromatic radiation of a given wavenumber with a given aperture or slit-width limited resolution. Some further points must be considered before the two systems can be fully compared. The fundamental source of most of the equations and concepts considered is Chabbal (1953); a recent discussion is that of Hays and Roble (1971).

The transmission τ of a FP system such as shown in Figure 2.2 varies for radiation traveling parallel to its axis (and thus traversing the center of the aperture S) as shown by the solid curves in Figure 2.3. Also shown is the transmission for radiation

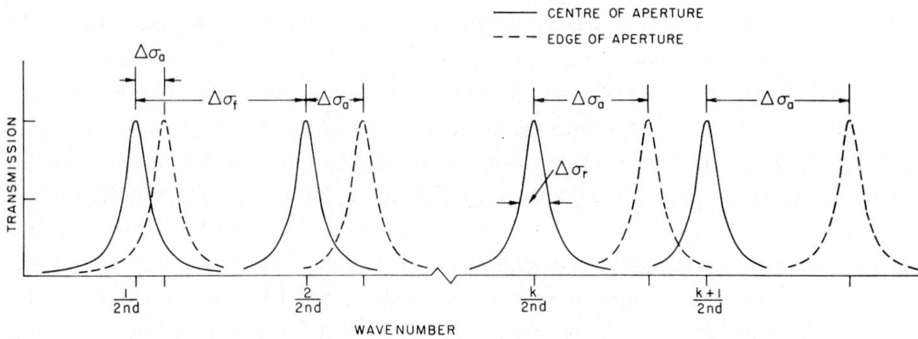

Fig. 2.3. Transmission functions for Fabry-Perot spectrometer with optical spacing *nd* between plates. Full curve shows transmission at center of aperture *S* (Figure 2.2) while dashed curve is transmission at edge of aperture.

traversing the system at such a slope angle as to pass by the outer edge of the aperture. The form of the transmission function of Figure 2.3 (called the Airy function) is determined by the reflectance of the etalon coating R so that the width at half peak, $\Delta\sigma_r$ is given by the expression

$$N_R = \Delta\sigma_f/\Delta\sigma_r = \Pi R^{\frac{1}{2}}/(1-R). \tag{2.2g}$$

$\Delta\sigma_f = 1/2nd$, the free wavenumber interval, is defined in Figure 2.3 and N_R is called the *reflective finesse*. The ratio $\Delta\sigma_f/\Delta\sigma_a$ is the *aperture limited finesse* N_A. The resultant instrumental function (i.e., the transmission curve vs σ for a fixed value of nd) may be seen from Figure 2.3 to be a convolution of the rectangular aperture functions of width $\Delta\sigma_a$ and the Airy function of width $\Delta\sigma_r$, (together with a further convolution by a plate defect function expressible by a finesse $N_D \simeq \lambda/2\Delta d$ where Δd is an average variation in d). τ_{av} in Equation (2.2e) is the average value, across the interval σ_0 to $\sigma_0 + \Delta\sigma_a$, of the Airy function having its peak at the wavenumber for which the etalon is set, i.e., $\sigma_0 + \Delta\sigma_a/2$ in the ideal case. If $N_A \simeq N_R$ the two functions have comparable widths so τ_{av} will be only a little less than τ_{max} for the Airy function. This is the favorable condition since then the overall resolution is only a little smaller than the aperture limited value $\mathscr{R}$. Further increases in N_R would not increase the resolution very much but would reduce τ_{max} and particularly τ_{av}. Smaller values of N_R would degrade the resolution severely with little increase in $\mathscr{L}$.

As the optical path, *nd*, between the plates is varied the pass bands of Figure 2.3 shift in wavenumber so that an unknown spectrum may be explored. The large $\mathscr{L}\mathscr{R}$ gain is achieved at the cost of certain difficulties which will be discussed in turn.

(i) *Multiple Pass Bands*

In order to obtain unambiguous spectral information it is usually necessary to suppress all but one of the multiple pass bands of Figure 2.3. These lie at intervals of

$$\Delta\sigma_f = 1/2nd.$$

The free wavelength range can be increased by decreasing *nd* which implies a lower order of interference k. For fixed resolution $\Delta\sigma_r$, an increase in $\Delta\sigma_f$ must be accompanied by a proportional increase in N_R. But N_R is limited by absorption losses as discussed above, to values of about 50 or less, so that larger free wavenumber ranges are only possible for low resolutions. Even for a resolution of 5 Å (a convenient value for a grating spectrometer) a free wavelength range of only 250 Å could be attained. (This would correspond to a plate spacing of about 5 μm for 5000 Å radiation.) At higher resolutions the free spectral range is proportionally smaller. The unwanted pass bands must be suppressed in some way. At low resolutions an interference filter may be sufficient but at high resolutions a second or third FP in series may be needed. The additional FPs can be of lower resolution with wider free spectral ranges (Bens *et al.*, 1965) or high resolution ones with incommensurate ratios between their free spectral ranges (Mack *et al.*, 1963).

(ii) *Wavelength Scanning Modes*

The means of scanning the FP may be restrictive. A popular method has been to change the pressure (and hence the refractive index) of the gas in the etalon. This works well for large plate spacings since a change of *nd* of several wavelengths can be easily attained. It is obviously not effective for low resolution. A second method is that of mechanical change in plate separation. This can now be achieved piezoelectrically, a method which is very promising for both high and low resolutions (e.g., Hernandez, 1970). Third, there is the method of spatial scanning (Shepherd *et al.*, 1965) which is essentially the exploration of the variation in optical path difference by means of an annular aperture of constant area. This is again limited in wavelength range but can be very stable since the FP may be a single solid plate.

One can thus conclude that the FP spectrometer is very useful for scanning a limited wavelength range at high to very high resolution with a luminosity higher than can be obtained with a grating spectrometer. Some typical applications of FP systems are listed in Table 2.1. This list contains only a selection of the systems reported in the literature but gives an idea of the range of resolutions and their applications.

2.2.4. Interference Filter Photometers

These are very widely employed in auroral studies because of their simplicity reliability, and good wavelength resolution. Interference filters with half peak transmission widths down to several angstroms are available. In principle such a filter is a fixed low order FP with a high (~ 2) refractive index spacer material so that according to Equation (2.2e) the $\mathscr{L}\mathscr{R}$ product may be $4\times$ larger than an air-spaced FP.

Analogously to Equation (2.2b) one has

$$\Phi = \frac{10^6 \times B}{4\pi} \cdot \frac{\pi A n^2}{\mathscr{R}} \quad \text{since} \quad \tau \simeq \tfrac{1}{2}$$

so that the count rate per rayleigh, e_1 is

TABLE 2.1
Applications of Fabry-Perot spectrometers

Use	Diam.	Instrumental width	Free range	Spacer	Scan mode	Pass band isolation	Reference
Hβ profile	6 cm	2 Å	40 Å	0.06 mm	Mechanical	Tilting interference filter	Zwick and Shepherd (1963)
He I 1.08 μ	12 cm	1.2 Å	~25 Å	0.19 mm	Spatial	10 Å	Neo and Rundle (1969)
[N I] λ5200 measurement	6.7 cm	0.5 Å	3 Å	0.4 mm	Piezo-electric	Interference filter	Hernandez (1970)
[O I] λ5577, λ6300 temperature measurement	10 cm	0.01 Å	0.18 Å	5–10 mm	Pressure	10 Å interference filter	Feibelman *et al.* (1972)
[O I] λ6300 dayglow	10 cm	0.02–0.07 Å	0.2–0.6 Å		Pressure	1.0–0.6 Å FP + 10 Å interference filter	Bens *et al.* (1965)

$$e_1 = \frac{C}{B} = \frac{10^6}{4\pi} \frac{\pi A n^2}{\mathscr{R}} \varepsilon. \tag{2.2h}$$

For a modest value for $A \simeq 5$ (1 in. diameter) and $n \simeq 2$, $\varepsilon = 0.1$ and $R = 10^3$ then

$$e_1 = 5 \times 10^2 \text{ count } s^{-1}\text{R}^{-1},$$

it is clear that a small sized filter photometer can give an output superior to that of a spectrometer. The disadvantage of observation at a single wavelength can be offset in several ways.

2.2.4.1. *Multiple Channel Systems*

In many applications the intensity at a limited number of wavelength points is sufficient, so that several independent photometers can be used or several filters can be used in succession with a single detector. Since the statistical noise varies as the square root of the sample count, multiplexing 10 channels would reduce the signal to noise ratio by less than a factor of 5×. Such systems have been described by Omholt and Harang (1955), Meriwether *et al.* (1970), and Gattinger and Vallance Jones (1972). Modern types have up to 11 channels, band widths of 25 to 45 Å and sampling rates of 20 s^{-1} to follow fast auroral pulsations.

2.2.4.2. *Use of Tilting-Filter Systems*

If an interference filter is rotated by an angle θ, a wavelength shift $\Delta\lambda$ arises for the rays parallel to the photometer axis according to the relation

$$\Delta\lambda = \lambda\theta^2/2n^2. \tag{2.2i}$$

Thus if $n = 2$ a rotation of 20° at $\lambda = 5000$ Å would produce a shift of about 75 Å. There is also a broadening and skewing of the aperture function. The limited wavelength scan produced in this fashion is extremely valuable for measuring and detecting line emissions in the presence of background continuum and for such specialized tasks as determining Hβ line profiles or even OH rotational temperatures.

Photometers of this type for Hβ observations have been described by Meredith and Montalbetti (1960) and by Eather and Jacka (1966b). In this application band widths of 3 to 10 Å have been used with scan times of about 1 min. A general discussion is given by Eather and Reasoner (1969) while Schaeffer *et al.* (1972) have applied them in rocket studies of the dayglow.

2.2.4.3. *Spatial Scanning Techniques*

A variant of the tilting-filter is to use an annular aperture of fixed area but variable radius. The wavelength shift is then still given by 2.2.4.1 but the aperture function is unchanged. Fiber optics or some other means can be used to pick off a set of adjacent wavelength channels by means of the same principle.

An example of a static system has been described by Miiller and Shepherd (1968). This gives two 5 to 10 Å wide channels separated by about 20 Å and has been applied successfully to rocket studies of Hβ and the λ4278 N_2^+ band in aurora.

2.2.4.4. *Wedge Interference Filters*

The filter can be made with a variable band pass by using a spacer layer of varying thickness disposed as a linear or circular wedge. Such a system has been described by Holm *et al.* (1972). The resolution is about 40 Å.

Rotating circular wedge filters have also proved useful in the 2–15 μm IR region for low resolution studies of emissions associated with auroral activity.

2.2.4.5. *High Order Interference Filter Systems*

In principle the spatial scanning or filter tilting technique can be applied very advantageously with filters of higher order and correspondingly small free spectral ranges. Thus a filter with a free range of 10 Å and a resolution of the order of 0.5 Å is realizable, with a solid spacer having a thickness of the order of 0.3 mm. Such filters have been constructed with mica or fused silica solid spacers (Dobrowolski, 1959; Title, 1971) and appear promising for some high resolution auroral studies (McCormac, private communication, 1974).

The range of possible variations of filter photometer systems is very wide. Such photometers are particularly valuable in rocket, satellite, and automatic ground-based operations where reliability and stability are of primary importance.

Photometer calibration is discussed in Appendix 2A.

2.2.5. Fourier Spectroscopy

The technique of Fourier spectroscopy as conceived by Fellgett (1951) was first put to effective use in the study of atmospheric spectra by Connes and Gush (1960). The principle is simple and illustrated by Figure 2.4a. As the mirror M_1 is moved a

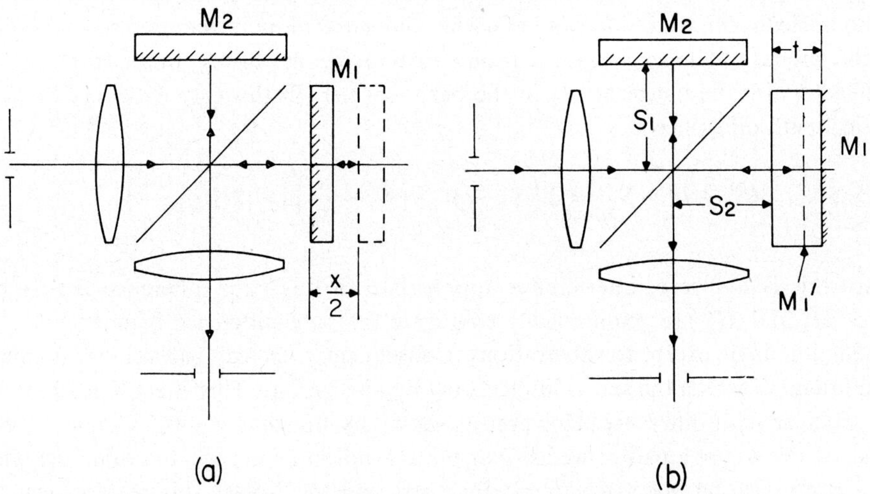

Fig. 24. (a) Schematic diagram of Michelson interferometer showing the moveable mirror in two positions differing by optical path difference x (for $n = 1$). (b) Schematic diagram of field-widened Michelson. M_1' is the virtual image of the reflecting surface of M_1 seen from the front.

sinusoidal intensity variation occurs in the output so that for an increase in the path difference of λ there is one cycle in the intensity variation of the component with that wavelength. The spectrum is recovered by a Fourier analysis of the output. The resolution depends on the total path difference scanned, x_{max}; the minimum wavenumber difference resolved is of the order of $1/x_{max}$. The aperture size is limited by the need to maintain the modulation out to x_{max} by accepting only the central fringe of the Haidinger interference pattern. This imposes a limitation on the luminosity $\mathscr{L}$ giving in fact exactly the same equation for $\mathscr{L}\mathscr{R}$ as for the FP (Equation (2.2e)) with $n \simeq 1$. The Michelson thus shares the Jacquinot advantage over the grating spectrometer.

In addition each wavelength component of the input spectrum reaches the detector during the total observing time T rather than for T/N as when N spectral elements are observed in succession. Thus the total signal 'count' associated with a spectral element is $N\times$ larger than for a serially scanning system so the signal to noise increases by $\sqrt{N}$. This is the so called Fellgett advantage. If the detector is an ideal photomultiplier with negligible dark current, the noise from the count associated with all N spectral elements contributes to the noise associated with any one so that the Fellgett advantage is reduced. In fact it then becomes equal to $\sqrt{B/B_{av}}$ where B is the brightness in a given spectral element and B_{av} is the average brightness over the spectral band passed to the detector. Thus the strong lines enjoy an advantage while the weaker features suffer a loss in signal-to-noise ratio. For detectors, such as photoconductive ones, the noise is largely independent of the total signal and the full Fellgett advantage is realized.

2.2.5.1. *Field Compensated Michelsons*

The limitation on diaphragm size mentioned above arises because of the variation in optical path difference between rays parallel to the system axis and those inclined at an angle θ. This variation can in fact be compensated by including a block of glass in the optical system such that the geometrical image, M_1' of M_1, in the front surface coincides with the image of M_2 in the beam splitter. In this case $S_1 = S_2 + t/n$. The optical path difference x is

$$2(S_2 + nt - S_1) = 2\left(S_2 + nt - \left(S_2 + \frac{t}{n}\right)\right) = 2t\left(n - \frac{1}{n}\right).$$

Since this is not zero, interference appropriate to this path difference occurs but since M_2 and M_1' are geometrically conjugate the path difference is independent of inclination angle except for aberrations. Consequently one can use a larger diameter diaphragm of which the size is limited only by aberrations. This effect holds only for a particular path difference. However as shown by Bouchareine and Connes (1963), it is possible to use a pair of wedge shaped back-reflecting mirrors to obtain the effect of a compensating plate of variable thickness and to displace this plate in just the right direction to cause the reflecting surface M_1 to move so that compensation is maintained over a range of path differences. One can thus obtain a system with a value of $\mathscr{LR}$ perhaps 50× greater than a conventional Michelson. Ring and Schofield (1972) have reviewed field compensated Michelsons.

To date the application of Fourier techniques to auroral spectroscopy has developed slowly. The time variation of auroral emissions is a problem, especially in the photoconductive region beyond 1.1 μm where there is an intense and *constant* nightglow emission. It should be possible to overcome this problem by rapid scanning and signal averaging. In the photomultiplier region the absence of the Fellgett advantage has discouraged attempts to use classical Michelson systems. The success of the field widened system reported by Despain *et al.* (1971) in nightglow studies in the PM IR indicates that the application to auroral studies should be profitable. An initial attempt by O. Harang with a system of the Bouchareine and Connes type was not successful probably because of the simultaneous presence of the varying auroral and constant nightglow components.

A different application by Hilliard and Shepherd (1966a, b) of the field widened principle to the measurement of 5577 Å [OI] profiles in aurora was very successful. In this case the scan was restricted to fringes near the compensated path difference so that the line broadening could be determined from the fringe contrast. Temperature measurements were possible in times of about 1 s compared to scans of the order of minutes for the FP method.

2.2.6. Birefringent Filter Photometers

Birefringent systems related to the Lyot filter have been widely employed in airglow photometry. A description with references has been given by Hunten *et al.* (1967). McCormac (private communication, 1974) reports success in the production of tunable Lyot filters with band widths less than 1 Å and transmissions as good as those of interference filters.

2.3. Particle Measurements

The use of rocket- and satellite-borne particle detectors for direct measurements of the characteristics of the particles responsible for the excitation of aurora has made possible a much greater understanding of auroral excitation processes. Particle detectors carried on satellites in polar orbits at heights of several hundred kilometers provide direct observations of the particle fluxes originally inferred from optical observations. Rocket borne detectors have made possible more detailed studies of individual auroral forms. The direct particle measurements can provide very complete information on the angular and energy distributions of auroral primary and secondary particles as well as some information on their time and space variations.

Particle detector systems should ideally give measurements of the distribution in energy, angle and position for electrons and ions in the 0.1 to 100 keV range. For secondary electrons, measurements to still lower energies are desirable.

First, various particle detectors will be briefly described followed by a description of energy analyzing systems.

2.3.1. Particle detectors

(a) *Geiger Counters*

Geiger counters with thin windows can detect electrons with energies greater than 20 keV and protons greater than 300 keV. They have been extensively employed in magnetospheric measurements of particle fluxes mainly as an indicator of geomagnetic field topology. This energy range is, however, above the energies of typical auroral primaries.

(b) *Scintillation Detectors*

Plastic or inorganic scintillating crystals followed by photomultipliers have been much used as detectors of energetic charged particles. Threshold energies down to a few keV for electrons and protons may be achieved. Since the amplitude of the light pulse produced in the scintillator depends linearly on particle energy these detectors may also be used directly for energy distribution measurements.

(c) *Solid State Detectors*

Various solid state detectors such as Li drift silicon junction, Si p-n junction, and surface barrier type detectors can be used for electrons and protons in the energy range down to 15 keV for electrons and 25 keV for protons. As with scintillators, pulse height analysis is possible to obtain the energy spectrum. Such detectors are sensitive to light which must either be absent or excluded by a surface coating which often consists of a layer of aluminum because of its low specific energy loss for charged particles. In the latter case the energy threshold is increased.

(d) *Electron Multipliers*

The most common sensor used for single lower energy electron or proton detection is the electron multiplier. Originally open multiplier structures with dynodes similar to those in photomultiplier tubes were used. A most important development has been that of the channel electron multiplier (channeltron). This consists typically of a helical tube semiconductor with an internal diameter about 0.5 mm and a length about 120 mm. Ruggedness and insensitivity to exposure to the atmosphere has made the channeltron the most popular low energy particle sensor. The tube has a resistance of the order of 10^9 Ω and with 3 to 4 kV applied between the ends the interior of the tube acts as a continuous set of dynodes. The entrance may be held at various potentials depending on the mode of operation. An energetic particle entering the negative end strikes the interior semiconductor surface; secondaries produced in its collisions with the walls are accelerated down the tube and produce additional secondaries when they strike the walls. The resulting avalanche produces a pulse at the positive end of the tube of over 10^8 electrons. The output pulses may be detected like those from a photomultiplier.

(e) *Faraday Cups*

Low energy electrons (<200 eV) may simply be collected in a Faraday cup and measured as a current. The normally low particle intensities in the magnetosphere make such current measuring devices less useful than the single particle detectors described above.

2.3.2. Measurement of Particle Energy Distribution

Particle intensities and fluxes may be expressed in several ways.

(i) The particle intensity is $N(E, \theta, \phi)$ where $N(E, \theta, \phi)\,\mathrm{d}E\,\mathrm{d}\omega\,\mathrm{d}S$ gives the number of particles in the energy range E to $E+\mathrm{d}E$ in the solid angle $\mathrm{d}\omega$ in the direction defined by the polar coordinates θ and ϕ and crossing the area element $\mathrm{d}S$ perpendicular to this direction.

(ii) The omnidirectional flux,

$$\Phi(E) = \int N\,\mathrm{d}\omega. \tag{2.3a}$$

(The omnidirectional flux multiplied by the cross section of a target particle for a given process gives the number of such processes due to the flux in the energy range $\mathrm{d}E$.)

(iii) The flux per unit solid angle in the direction θ, through a unit area perpendicular to a *fixed* direction (such as the magnetic field or vertical), is $N\cos\theta$, if the fixed direction is along the polar axis of the coordinates.

(iv) The total flux per unit area defined as in (iii) is

$$F(E) = \int N\cos\theta\,\mathrm{d}\omega. \tag{2.3b}$$

$F(E)$ is used *inter alia* in giving particle fluxes incident on the atmosphere from above.

An *isotropic* flux is one in which $N(E, \theta, \phi)$ is independent of θ and ϕ. In this case $\phi(E) = 4\pi N(E)$ while $F(E) = 0$ since the upgong cancels the downgong component. Sometimes the flux is isotropic but confined to the downward direction so that $F(E) = \pi N(E)$.

As in the optical case spectral information may be obtained by an analyzer capable of transmitting particles in a limited range of energies or in the simplest case by means of a fixed or variable energy threshold. Some ways of achieving this will be mentioned.

(a) *Threshold Systems Using Windows*

Geiger counters detect all particles above a threshold determined by window thickness. Likewise Al foil windows were used with scintillation counters to give thresholds to 2 keV for electrons and 4 keV for protons in the pioneering satellite measurements of auroral particles by Sharp and Johnson (1968). Solid state detectors have thresholds determined by their physical structure or by an external window.

(b) *Threshold Systems Using Electrostatic Effects*

A repelling voltage on a grid at the entrance aperture has been used to provide a lower threshold variable over a wide range of energies. Thus Sharp and Johnson (1968) varied the threshold between 80 eV and 1.6 keV ahead of a scintillation type detector. (The selected particles can be accelerated to increase the detection efficiency before they reach the detector.) At lower energies multiple grid systems as described by Maier and Rao (1970) can measure fluxes with thresholds from 5 to 200 eV.

(c) *Electrostatic Band Pass Analyzers*

For electrons and protons in the range important in exciting optical aurora (1 to 40 keV) various types of electrostatic analyzers have been extensively employed. The performance of such systems is commonly specified in terms of a *geometrical factor*, $G(E)$ analogous to the luminosity of an optical spectrometer. $G(E)$ is the rate at which particles reach the detector per unit particle flux (in particles $\text{cm}^{-2}\ \text{sr}^{-1}\ \text{s}^{-1}$) for particles of energy E. This would normally equal $A\Omega$ where A is the area of the input aperture and Ω the angular field of view. Sometimes the total energy geometric factor, $\mathcal{G} = \int G(E)\ \text{d}E$, integrated over the detector pass band is given. This gives the ratio of count rate to particle intensity N defined above when N may be assumed constant over the angular field and energy band pass of the detector.

A typical electrostatic analyzer described by Rearwin (1971), is shown in Figure 2.5. This instrument scans the electron energy spectrum in the range 1 to 35 keV by varying the voltage across the cylindrical concentric plates. The resolution is about 8 and the geometrical factor, $\mathcal{G}$, at energy E keV, about $3 \times 10^{-4}\ E\ \text{cm}^2$ sr keV. The efficiency of channeltrons in detecting single electrons or protons traversing the analyzer is crucial to the practicability of such systems.

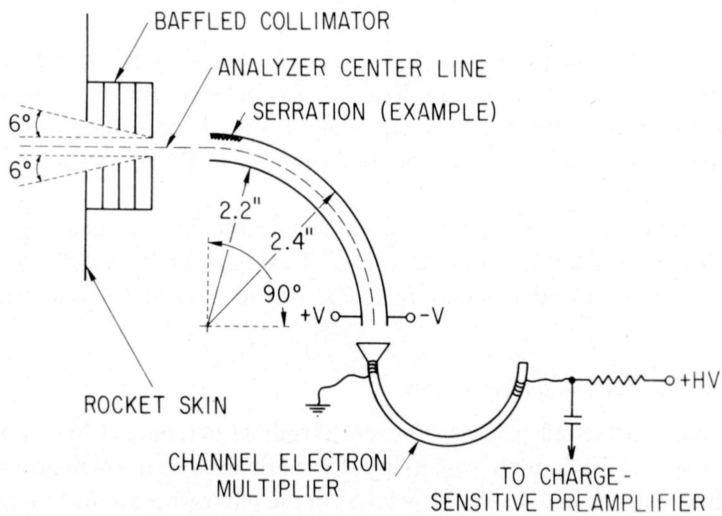

Fig. 2.5. Schematic diagram of typical electron energy spectrometer (Rearwin, 1971). (Courtesy *Journal of Geophysical Resarch*; copyright by American Geophysical Union.)

Bernstein *et al.* (1969) used a similar system to measure the energy spectrum and flux of H atoms. Incoming ions were first removed by electrostatic deflection and then the incoming H atoms ionized by passage through a carbon foil.

Somewhat similar electrostatic analyzer systems have been used at lower energies with special precautions to ensure the rejection of particles intercepting the analyzer plates. A hemispherical analyzer for the 6 to 50 eV range has been described by Doering *et al.* (1970). This instrument uses a grid-type electron multiplier with 250 V post-acceleration.

(d) *Magnetic Band Pass Analyzers*

For satellite observations Reed *et al.* (1969) have described a system which includes analysis of incoming beams of electrons and protons by means of a magnetic field which selects electrons in energy ranges from 1 to 27 keV and protons from 10 to 500 keV. The magnetic field system should be very reliable which is important in satellite applications but has the undesirable effect of producing spurious magnetic fields. Also electrical power considerations require the field be produced by permanent magnets which do not allow a variable energy band pass.

(e) *Combined Electrostatic Magnetic Systems*

For a study of H^+, He^+ and He^{++} auroral particle fluxes Whalen *et al.* (1971a) have employed a double spectrometer. With the combination of electrostatic analyzer which defines the energy/charge of an ion beam and magnetic analysis which defines the momentum/charge it is possible to determine the mass/charge and energy/charge spectrum of the beam.

2.4. Radio Reflection Techniques

Ionization produced by auroral bombardment is capable of reflecting VHF and UHF radio waves under favorable conditions. The techniques for obtaining and studying such reflections may be classified briefly as follows.

2.4.1. Pulse Radar Systems

Most frequently pulses are transmitted and received by the same antenna although in some cases the receiving antenna is separate. Frequencies from 40 to 100 MHz have been used in most studies. Important work has also been done in the 400 to 3000 MHz range. For the 40 to 100 MHz region the radar beam widths vary from 5° to 60° whereas for the 400 to 3000 MHz large parabolic reflectors have been used with beam widths as small as 0.4°. Pulse lengths from 16 to 500 μs have been used with peak transmitted powers of 1 to 180 kW.

Such systems give the range and amplitude of the reflected signals. The VHF systems generally give imprecise information as to the azimuth and elevation of the reflecting region while the UHF systems give precise values.

In some radars it is possible to measure the Doppler shift of the frequency in the reflected pulse by beating it against an unshifted signal derived from the transmitter.

The polarization characteristics of the reflected signal can also be examined by using receiving antennas of different polarization properties.

HF radars operating from 4 to 64 MHz have been employed by Bates *et al.* (1969). At these frequencies reflection via the ground and *F* region becomes common and observations in a stepping mode over the frequency range become necessary to interpret the results which have some similarities to those of ionospheric sounders.

2.4.2. Continuous wave reflection techniques

This technique offers the possibility of following rapid changes in the nature of the reflections as well as making observable reflections in several directions from a scattering region. It is also particularly suitable for Doppler-shift measurements and some systems have been designed to operate in both modes. Sofko and Kavadas (1971) employed a CW system in one of the most complete studies to date of the polarization state of the reflected signal. By means of a receiving system consisting of two dipoles inclined at 45° to the vertical together with appropriate circuitry they were able to resolve the scattered wave into its vertical- and horizontal-linear and right- and left-circularly polarized components. They were also able to detect fluctuations in the reflected signal with frequencies at least up to 10 Hz.

Egeland (1962), Czechowsky and Lange-Hesse (1971), and Forsyth (1969) have reported extensive studies obtained using the multiple-station continuous wave reflection technique. This method is simple and reliable but the reflecting region is defined only approximately by the intersection of the transmitter and receiver beams.

2.4.3. Ionosphere sounders

The standard ionospheric sounding technique as employed from the ground produces distinctive ionograms in the presence of aurora. Such a system gives a plot of the apparent vertical height of ionospheric reflections as a function of frequencies in a range of the order of 1 to 20 MHz. Early studies of such reflections were made by Meek (1953, 1954), Heppner *et al.* (1952), Omholt (1955), and Knecht (1956).

A detailed presentation of typical results, with ionograms and corresponding all-sky camera photographs well reproduced, has been given by Hunsucker and Owren (1962). They show that various kinds of sporadic-*E* type reflections are typical of auroral displays. Thus *Esa* (so-called auroral), *Esf* (flat type), *Esr* (retardation type) and *Ess* (slant type) all occur in association with auroral displays. Omholt found that aurora which was uniform over a large area in the zenith gives a regular ionogram from which the electron density can be determined in the same way as for the normal *E*-layer. Reddy *et al.* (1969) have discussed the nature of auroral ionograms in the light of simultaneous rocket measurements of the electron density profile.

term is of the same order of magnitude as the fields near auroras. This technique has been applied in rockets by Aggson (1969) and Kelley *et al.* (1971a) and on satellites by Cauffman and Gurnett (1971).

The technique has been extended to balloon heights by Mozer and Serlin (1969) who argue that horizontal electric fields measured by horizontal probes represent ionospheric fields averaged over several hundred kilometers. The reliability of the balloon method has not been entirely established; it would have advantages in giving continuous results over a longer period than is possible with rocket- and satellite-borne probes.

(c) *Ion Drift Measurements*

The drift velocity of ions is directly measurable by using ion spectrometers of the type described in Section 2.3.2c. In a field of 100 mV m^{-1} the ion drift velocity is about 2 km s^{-1} which is easily measurable. Green and Whalen (1974) have described the theory and accuracy while its application to a particular aurora has been described by Whalen *et al.* (1974).

2.6.4. Detection of infrasonic emissions from aurora

Under certain conditions infrasonic waves with frequencies of the order of 0.03 Hz are generated by auroral forms. These are detectable by special low frequency condenser microphones. These consist of 4 in. diameter brass diaphragms forming one plate of an oscillator. The microphone chamber on one side of the diaphragm is usually connected to a pipe about 300 m in length with open side tubes at about 1.5 m intervals. The pipe serves to average out local random pressure fluctuations (Burridge, 1971; Grover, 1971). A ballast chamber and leak serve as an acoustic filter for the 0.1 to 0.01 Hz range. As employed in recent work by C.R. Wilson (private communication, 1970) four such microphones were set out on a 36 km square. Infrasonic waves are identified by correlating signals obtained from all four receivers to eliminate local disturbances. The phase shifts necessary to achieve the correlation give the angle of arrival of the infrasonic wave front.

Wilson (1969) has noted some aspects of the technique. A more general discussion of infrasonic techniques is to be found in pp. 1–425 of *Geophys. J.* (*Roy. Astron. Soc.*) **26** (1971).

2.6.5. Detection of X-rays from aurora

The first work in the detection of X-rays under aurora was done using Geiger and proportional counters. These are not very sensitive and give only a crude indication of the spectral distribution of the X-rays. Scintillation counter systems are used in modern work. These generally consist of a NaI (Tl) crystal a few centimeters in height and diameter, optically coupled to a photomultiplier from which the pulses are fed into a pulse height analyzer. This can provide discrimination into as many as

100 energy channels in the 15 to 150 keV range. Detailed descriptions of such systems have been given by Maral (1970a, b) and Pilkington (1970). A relatively cheap commercially produced detector package for balloon studies has been used in much of the European work (see e.g., Jentsch and Kremser, 1972).

X-rays are absorbed by the lower atmosphere and are observable only from balloon heights. The transmission of X-rays is discussed in Section 5.6.2.

CHAPTER 3

OCCURRENCE AND MORPHOLOGY

3.0. Introduction

In this chapter we attempt to summarize the stages through which our knowledge of auroral displays has progressed, from the first good descriptions of its geographical distribution to the detailed accounts now emerging. The morphology of electric currents in aurora and their magnetic effects is discussed. Finally auroral displays are related to the particle fluxes and magnetospheric processes which are their immediate cause.

3.1. Occurrence of Visible Aurora

General historical references have already been given in Chapter 1.

3.1.1. The auroral zones

The first step in a systematic study of auroral occurrence was the compilation of statistics on the frequency of observation of aurora as a function of geographical coordinates. This was first done by Fritz (1881) who correlated all the available visual reports. It was not an easy task because reliable observations over a wide and difficult geographical region are required. Chamberlain (1961) gives a figure from Vestine (1944) showing an improved contour plot of the results in terms of the fraction of clear nights on which aurora can be observed from a given point. It has been shown by Bond and Jacka (1962) and Hultqvist (1961) that the maximum of the distribution lies along a line of IN Lat or corrected dipole latitude* close to 67.5°. The distribution vs corrected dipole colatitude found by Bond and Jacka is shown in Figure 3.1. The auroral region has been defined in the International Auroral Atlas (1963) as the region from 60° to 75° IN Lat. No strict definition of the *auroral* '*zone*' exists although it is usually taken to be the long-term mean position of the region with a high incidence of aurora. Stringer and Belon (1967) show how the distribution varies over the solar cycle. The maximum does not shift but the curve is skewed more to higher latitudes at sunspot minimum.

3.1.2. The auroral oval

Diurnal variations in the geomagnetic latitude of aurora had long been recognized and as noted, for example, by Currie (1947), for stations equatorwards of the zone,

* See Appendix 3A for notes on geomagnetic latitude and time.

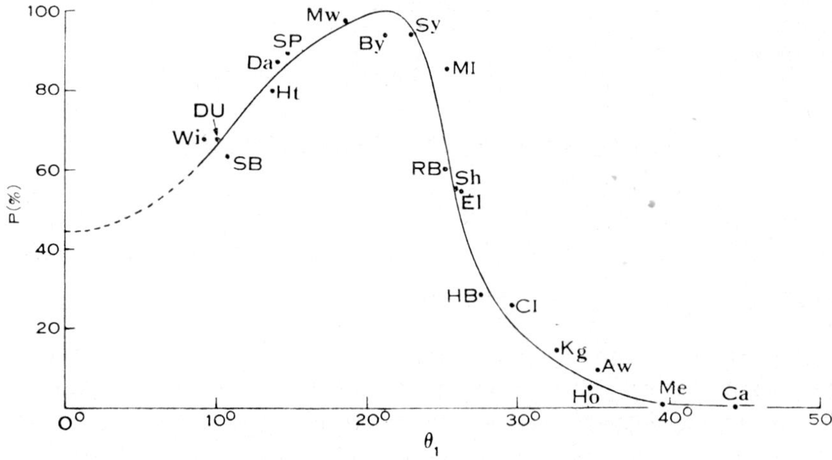

Fig. 3.1. Distribution of probability of occurrence of aurora vs corrected geomagnetic colatitude from Antarctic IGY observations after Bond and Jacka (1962). The symbols refer to the observing stations. (Courtesy *Australian Journal of Physics*.)

the aurora moves equatorwards until near midnight and then retreats. Attempts to characterize this variation in detail were hampered by lack of sufficient auroral observations although the pattern advocated by Meek (1955) from analysis of magnetic records came very close to that distribution in geomagnetic latitude and time which is now known as the *auroral oval*. On the basis of I.G.Y. all-sky camera observations this concept was introduced by Feldstein (1964) who found a distribution similar to that shown in Figure 3.2b. The analysis of Davis (1962) revealed a similar pattern.

The *auroral oval* might be taken to be either the long-term overhead auroral frequency distribution in geomagnetic time-latitude coordinates as in Figure 3.2b or distributions of subsets of the data, for example, for various ranges of magnetic disturbance indices*. Such distributions are shown in Figure 3.2 where plots for ranges of the magnetic index* Q are reproduced as given by Feldstein and Starkov (1967).

Finally the term *instantaneous auroral oval* is sometimes used to denote the location of the quiet arc system associated with the development stage of a substorm (see Section 3.2). The instantaneous oval appears often, especially for more disturbed conditions, to be continuous (Buchau *et al.*, 1972; Anger and Lui, 1973) but for less active conditions this may not be so. Feldstein (1973), Buchau *et al.* (1972), Lassen (1970), and Eather and Mende (1972) have discussed the possibility that the instantaneous oval may sometimes be discontinuous in the evening and morning regions; this is reasonable because the midday and night injection mechanisms are different.

* See Appendix 3B for brief definitions of these indices.

The term auroral oval as used above has been implicitly taken to refer to aurora *as detected by all-sky cameras*, i.e., discrete forms with a minimum intensity of 0.5 to 1.0 kR for the 5577 Å [O I] line for normal green aurora or perhaps 1 to 2 kR for the 6300–64 Å [O I] lines for red aurora. There is often considerable diffuse emission spread across the whole sky during auroral displays and it is interesting to enquire whether the same distributions would be obtained if the mean intensity of λ5577 were plotted rather than the ASC data. The time-latitude distributions of λ5577, λ3914 and λ6300 were studied by Sandford (1964, 1968) from patrol spectrograph exposures. The results for λ5577 are shown in Figure 3.3. This distribution is similar to that of the oval in the night sector but deviates in the daytime. Sandford argued from a careful analysis of a large body of data that in addition to the normal discrete aurora there was a diffuse type called '*mantle aurora*' which followed a

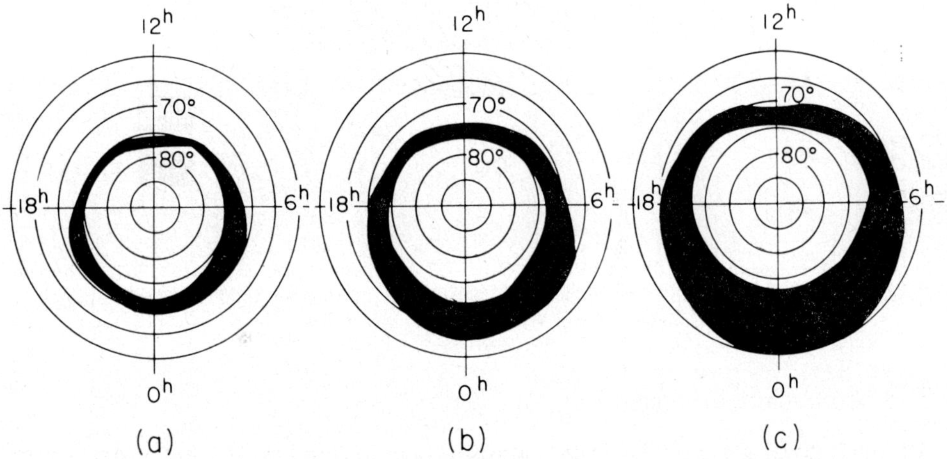

Fig. 3.2. Auroral oval for various levels of magnetic disturbance index Q after Starkov and Feldstein (1968). (a), (b), and (c) are for Q equal to 0, 3, and 7 respectively. The mean oval is similar to (b).

circular zone 20° in radius on the time-latitude plot. This is further discussed in Section 3.3.6c.

Much work has been done on the orientations of quiet arcs as a function of time and latitude. In general it appears that these forms are mostly oriented parallel to the direction of the auroral oval (Gustafsson *et al.*, 1969; Bond and Thomas, 1971; Akasofu *et al.*, 1972). Arcs within the polar cap are usually oriented in the sunward direction.

Lassen (1972) has argued that there are several distinct components to the population of auroral forms which average together to produce the oval. Thus he distinguishes a system of forms oriented along a noon-evening-midnight spiral, a night system oriented along the oval near midnight and a system of Sun-oriented polar cap arcs.

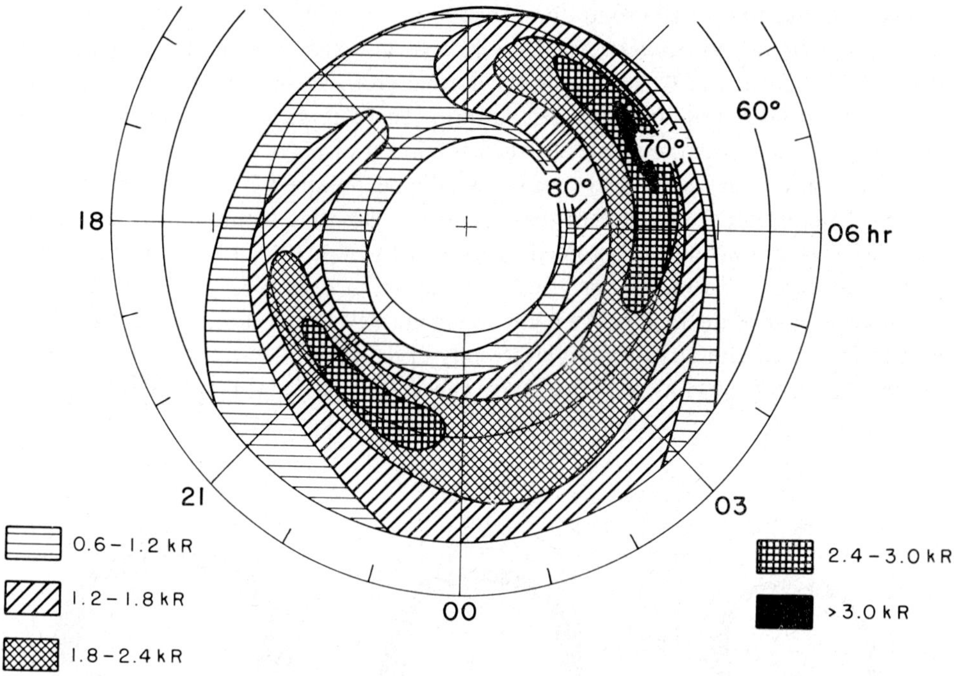

Fig. 3.3. Average intensity distribution of [O I] 5577 Å emission after Sandford (1964). (Courtesy *Canadian Journal of Physics*; reproduced by permission of the National Research Council of Canada.) (From Wiens and Vallance Jones, 1969.)

3.1.3. Conjugacy of Aurora

Although it was generally believed that displays of aurora borealis and australis were correlated it was not until the I.G.Y. that the effect could be properly demonstrated. Nagata and Kokobun (1961) and Wescott (1961) showed that magnetic variations were similar at conjugate points in the N and S auroral zones. Soon afterwards DeWitt (1962) showed from all sky camera observations that there was a high degree of correlation in brightness, forms and their motions, and time of break-up between stations near 60° geomagnetic latitude.

In recent years very detailed observations have been made from pairs of aircraft flown along magnetically conjugate flight paths in the two hemispheres. By means of all-sky cameras, photometers and television cameras very important results have been obtained. The main conclusions as given by Stenbaek-Nielsen *et al.* (1972, 1973) will be summarized.

(i) There exist two distinct auroral arc systems of which one is diffuse and lies equatorward in the latitude range below 65°. The other system lies poleward and consists of discrete arcs.

(ii) The equatorward forms are closely conjugate and this relationship is not strongly affected by substorm activity.

(iii) The poleward forms are also conjugate but with displacements of up to 300 km from the theoretical positions. Before the break-up phase of substorms the southern forms are displaced W of the conjugate point but at the onset of break-up the displacement reverses to the E. During high magnetic activity the antarctic poleward arc systems are displaced equatorward relative to the conjugate point

(iv) The demarcation between the two arc systems is clearly identifiable in the growth and break-up stages of substorms but not in the recovery phase.

(v) In the Alaskan sector the N-hemisphere equatorward system is ~30% brighter than the S-hemisphere system and the same difference is found (but statistically only) between the poleward systems.

(vi) Conjugate pulsing patches (see Section 3.3.7) have been identified although not all pulsing forms are conjugate. According to Davis (1969a) conjugate pulsations with a period of ~4 s were in phase within 0.1 s in the two hemispheres. Flickering was observed simultaneously in the two hemispheres.

Statistical studies from conjugate point stations such as that of Spitz and Watson (1972) appear to be consistent with the above conclusions. Other conjugate point effects are noted later with regard to such effects as auroral absorption, magnetic pulsations and substorm development.

3.1.4. Longitude effects

It has been widely assumed that auroral precipitation characteristics are statistically independent of longitude. If this weie so, then a plot such as Figure 3.9 for which the data were obtained from a meridian chain of stations would also give the average distribution as a function of geomagnetic latitude and longitude. If on the contrary the distribution is a function of the longitude of the observing station then the average pattern as a function of geomagnetic latitude and longitude becomes also a function of universal time.

The possibility of such effects has been considered by Loughnan (1961) and Davis (1967a). A universal time effect in soft electron fluxes at about 3900 km height was detected by Maehlum (1968) and ascribed to a dipole axis tilt variation.

However the conjugate point effects discussed above have been satisfactorily explained by Stenbaek-Nielsen *et al.* on the basis of the effect proposed by Loughnan, namely variations in the reflection height at conjugate points of trapped particles due to irregularities in the terrestrial magnetic field. This same effect should also give rise to significant longitudinal average intensity variations. Some evidence for such effects is given or cited in the above references. As shown by Stenbaek-Nielsen *et al.* (1973) the effect varies with the pitch angle distribution of the incident particles. As a detector of incident particles it is clear that the atmosphere acts as a 'scintillator' of which the sensitivity varies over the surface. Clearly many of the quantitative aspects of auroral morphology may need to be further studied with this effect in mind. Presumably dipole tilt may make an additional contribution to the longitudinal effect.

3.2. Individual Displays – Auroral Substorms

The statistical auroral oval gives only an average distribution for the appearance of aurora. While it was obvious to auroral observers from a single station that there was a considerable short time-scale movement and intensity variation in auroral displays, it was not until the work of Akasofu (1963) that a clear description was made of the overall nature of the developments of a typical display. The extensive all-sky camera observations of the I.G.Y. made it possible for Akasofu to show that the evolution of an outburst of bright aurora involved a time variation in the typical pattern of precipitation in geomagnetic latitude-time coordinates. He called this sequence of events (at the suggestion of Chapman) an *auroral substorm* by analogy with the term DP *substorm* which had been applied by Akasofu and Chapman (1961) to the analogous magnetic disturbance. (Because systematic magnetometer observations had been started earlier the analysis of the current patterns associated with magnetic disturbances was in advance of that of visual aurora.)

A recent version (Akasofu, 1968) of the typical substorm development is shown in Figure 3.4. In the initial condition there are one or more weak arcs aligned along the auroral oval. Then a rapid brightening of one of these arcs takes place (or the growth of a new rapidly brightening parallel arc in the same region). The bright arc then moves rapidly northward (or new bright arcs form successively to the north) so that the region between the original oval in the midnight region and the northward boundary is filled with active bright aurora. A bend in the bright northern arc system, called the *westward traveling surge* which appears in Figure 3.4C spreads west while in the post-midnight sector other folds and patches propagate eastward (Figure 3.4D). The midnight region activity gradually decreases while the disturbance propagates to the evening and morning sector. Considerable activity spreading well southwards, in the form of patches and rapidly changing but fainter active arcs, remains for some time in the post-midnight sector (Figures 3.4E and F) and is probably responsible for the bulge at this location in the statistical auroral oval (Figure 3.2b). Eventually the disturbance dies down and the situation returns to a state resembling the initial stage. In the midday sector, aircraft borne ASC observations reported by Akasofu (1972) indicate that there is an equatorward shift during the early phase of the substorm followed by a poleward shift during the recovery phase. Substorms tend to repeat themselves and several may occur in one night at intervals as short as 1 to 2h The time scale of the evolution of a typical substorm is indicated on Figure 3.4.

The recent success of the ISIS-2 satellite in obtaining nearly instantaneous pictures of aurora across most of the oval region is leading to a much improved description of the pattern of auroral precipitation. These results give 'snap shots' of the distribution of aurora at intervals of about 2 h. and consequently are not capable of revealing the time development of a substorm. They do reveal for the first time the complete large-scale spatial distribution. An example supplied by Dr C.D. Anger of a satellite picture is shown in Figure 3.5. Other examples may be found in recent publications by Shepherd *et al.* (1973a) and Anger and Lui (1973). Some of the main discoveries from the satellite results are the following:

(i) There is a continuous belt of diffuse aurora forming the equatorward boundary of auroral precipitation. This is probably related to if not identical with the equatorward arc system discussed in Section 3.1.3. In substorms this brightens and expands filling in the region up to the most poleward arc.

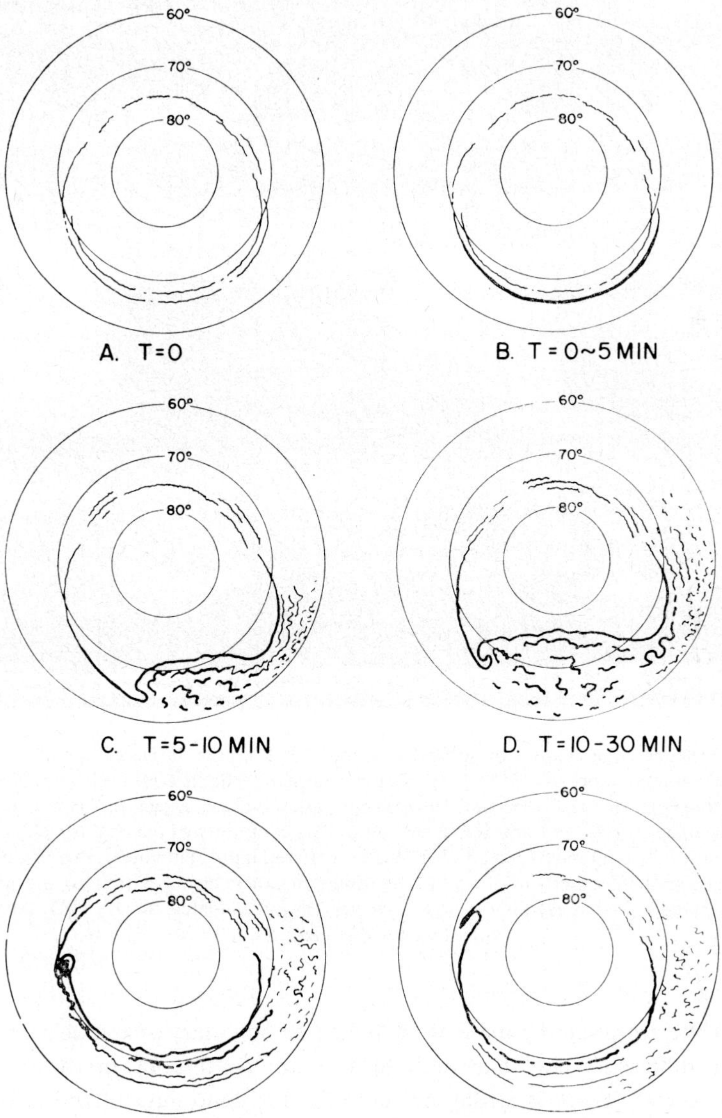

Fig. 3.4. Development of an auroral substorm according to Akasofu (1968). (Courtesy D. Reidel Publishing Co.)

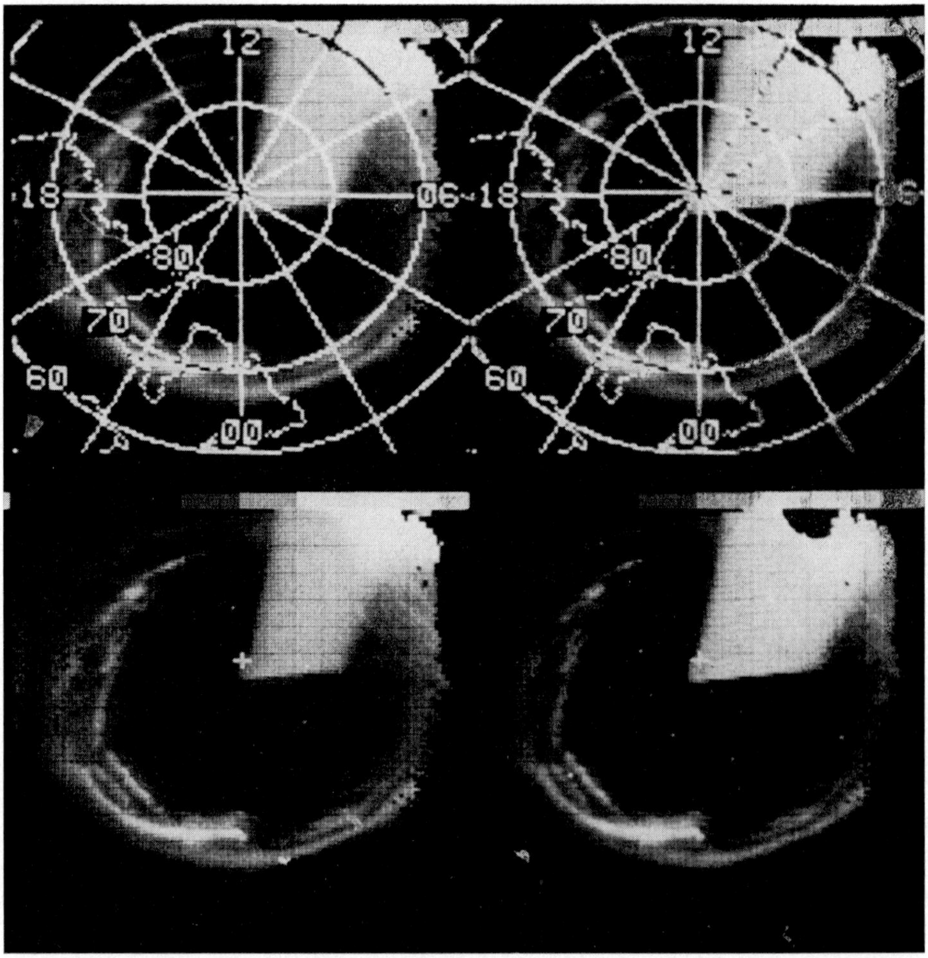

Fig. 3.5. ISIS-2 satellite pictures of auroral oval for 0402 UT pass on December 15, 1971. The left hand side shows the picture in λ 5577 (O I] and the right hand side in λ 3914 N_2^+ emission. The data have been transformed to a corrected geomagnetic latitude-time projection for which the grid is shown in the upper frames and omitted in the lower. The continuity of the oval is well demonstrated (except for part of the forenoon sector where baffle scattered light obliterates some of the data). The diffuse aurora at the southern limit of the nighttime aurora is clearly visible as are a number of discrete arc systems in the evening sector. I am very much endebted to Drs C.D. Anger and W. Sawchuk for this illustration.

(ii) Discrete arc systems appear at the poleward boundary of and also equatorward within the diffuse aurora. The discrete arcs are continuous in the evening and afternoon sectors. There is strong evidence for the continuity of the instantaneous oval. Discrete arcs frequently appear at very high latitudes in the midnight sector.

(iii) At times of intense disturbance very complex arc systems occur and there is evidence for a sharp discontinuity at midnight. Well defined western traveling surges as in Figure 3.4C are rare. A good example is discussed by Anger *et al.* (1973a). Arcs can show abrupt changes in direction and intensity.

(iv) The Sun aligned arcs of the polar cap can be intense and may join with the midnight aurora.

(v) In one intense substorm a large spiral structure was seen in the dawn sector of the polar cap $\Lambda_c = 75°$ to 85°.

In general, as might be expected, the real substorm pattern is more complex than the idealized one. Clearly a full understanding of the dynamics implied by the ISIS-2 pictures will require some time. Recently further pictures* from USAF meteorological satellites have become available (Rogers *et al.*, 1973). These have higher spatial resolution than those from ISIS-2 and show much remarkable detail in the discrete aurora including fine examples of the spirals discussed in Section 3.3.2.

This account describes the main visual aspects of the auroral substorm. Important magnetic effects which indicate currents flowing in the ionosphere and magnetosphere are also observed. In recent years observations of changes of particle fluxes and spectra within the magnetosphere have been obtained. Since it is now clear that the visual aurora is but one aspect of a complicated process further discussion is left for Section 3.4 in which the relation of the auroral and magnetospheric substorms is described. Before describing this we will examine the finer details of auroral forms.

Finally is should be noted in anticipation of Section 3.7 that it is now widely (although not unanimously) believed that there is a growth phase before $T = 0$ in Figure 3.4 during which a build-up for the substorm occurs in the magnetosphere with little visible auroral activity.

3.3. Detailed Morphology of Auroral Forms

As seen from by a ground observer the many variations of auroral forms, their movements and intensity changes are especially striking. A detailed description of these and their classification is to be found in Störmer's (1955) monumental work on 'The Polar Aurora'. A more complicated classification scheme and an outstanding collection of illustrative photographs is to be found in *The International Auroral Atlas* (International Union of Geodesy and Geophysics, 1963). A further interesting collection is the atlas of auroral forms of Kaneda *et al.* (1968) who include many very beautiful color photographs.

3.3.1. Classification of Visual Auroral Forms

Since references to Störmer's classification of forms occur frequently the meanings of the more common terms and their abbreviations are outlined in Table 3.1. The International Auroral Atlas system is organized more logically but has not been widely adopted. Some terms from the new system are commonly used, particularly (i) patch (P) for a diffuse surface $\leq 10°$ in extent, and (ii) veil (V) for a more extensive

* See example in frontispiece.

diffuse surface. Since the role of visual ground observations in auroral research is declining it is doubful whether there is much to be gained by a radical change in the descriptive system.

TABLE 3.1

Common auroral forms and symbols

Term	Abbrev.	Notes
Homogeneous arc	HA	Regular form with constant height, intensity and direction
Homogeneous band	HB	Like HA but varying in direction along length
Rayed arc or band	RA, RB	Like HA or HB respectively but with vertical ray structure
Rays	R	Narrow, almost vertical columns of luminosity (field aligned)
Diffuse surface	DS	Uniform emitting surface well above horizon
Pulsating surface	PS	Same as DS but varying in intensity with a period of several seconds
Pulsating arc	PA	Self-explanatory
Corona	C	R system viewed parallel to magnetic field and so appearing to converge in magnetic zenith
Flaming	F	Forms with intensity fluctuations moving up towards zenith

3.3.2. Detailed structure of auroral forms

Examination of auroral forms with image orthicon (or analogous) techniques has established the existence of fine structure not obvious to the eye. This structure is normally unrecognizable in direct photographs because of motions during the exposure. The image orthicon as described by Davis (1966) can record images of even weak aurora in 1/60 s. Maggs and Davis (1968) showed that, when viewed in the magnetic zenith, the thickness distribution of the elements of auroral structures had a median value of $\sim$200 m varying from $\leq$70 m to over 4 km. Thus the classical auroral arc with a thickness of perhaps 10 km has a fine structure which Davis has called ‘arc elements’ with a typical width of 200 m. Davis has also drawn attention to the fact that arcs themselves tend to occur in ‘arc systems’ that is sets of parallel arcs extending over perhaps 500 km.

The TV technique results led Hallinan and Davis (1970) to propose two new forms to describe structural features of auroral forms. As seen from below arc elements often form *curls.* These are *counter clockwise* (from below) folds spaced at from 2 to 10 km and wound up tightly into cylindrical forms 200 to 300 m across. As seen from the side curls undoubtedly appear as rays. Thus the rays of rayed arcs and bands are curls. The time constant for the formation of a curl is about 1/4 s. Curls have velocities up to 20 km s^{-1} parallel to the arc center and the motion may be in opposite directions on either side of the center line.

The other element recognized by Hallinan and Davis is the *spiral,* a usually *clockwise* (in the northern hemisphere) folding on a larger scale. These have an estimated diameter of 30 to 1400 km. They have a growth time constant of about 60 s and velocities of perhaps 10 km s^{-1} parallel to the arc in which they are formed. The western traveling surge may in fact be a spiral.

In general these observations show that the classical arc or band is really a much more complex structure with many narrow elements carrying out rapid and complicated motions.

3.3.3. Intensity Indices

The intensity of aurora is now referred to the intensity of the 5577 Å [OI] line according to the International Brightness Coefficient. The scale ranges logarithmically from IBC1 (1 kR) up to IBC4 (1000 kR). The brightness of a particular form would depend on the direction of observations although no attempt is usually made to correct for this effect since the value of the scale is mainly qualitative. For low-level aurora (< 100 km) where the 5577 Å line is weakened relative to other important emissions the brightness coefficient would not be a true index of the auroral intensity*.

3.3.4. Height Distribution of Normal Aurora

In this section the height measurements of 'normal' or type-c forms (see Section 3.3.5) are discussed. The heights associated with other auroral types will be discussed in Section 3.3.5.

The work of Störmer and others using the parallactic photographic method of height determination has provided comprehensive information about the height distribution of measurable points in different auroral forms. The results are detailed by Störmer (1955) and further analyses of Störmer's data have been carried out by Egeland and Omholt (1967).

The lower borders of arcs and bands should be accurately measurable by this technique. Chamberlain (1961) reproduces one of Störmer's figures which indicates a decrease in the mean height of the lower border of arcs from 114 km for weak aurora to 95 km for very strong aurora. Egeland and Omholt (1967) showed that there was some tendency for a diurnal variation in the average height of the lower limit of arcs and bands from 105 to about 95 km. The average value for all forms seems to fall in a narrow range from 105 to 108 km according to a survey of published results by Kinsey (1965).

The vertical extent of the luminosity has been studied by single and double station photography by Currie and Weaver (1955) who found the most frequent vertical extents to be 20 to 40 km. Vertical intensity profiles determined using 2-station photometry by Belon *et al.* (1966), although limited in number have a median value of about 110 km for the lower half maximum intensity level and a median value of

* In a recent (Kyoto, 1973) IAGA resolution it is recommended that IBC coefficients be reserved for visual estimates and that if possible auroral intensities should be reported in terms of measured emission intensities (preferably at least of the 4278 Å N_2^+ band). In Chapter 4 we shall use IBC3 as a shorthand for $I(5577) = 100$ kR in the presentation of some observational results.

about 40 km for the half intensity vertical extent. Many arcs however had an upper 10% maximum intensity level above 200 km.

Rays have a considerably greater vertical extent. In the dark atmosphere this varies from 100 to 600 km according to Störmer's (1955) results, the longer ones being observed in lower latitude aurora near sunspot maximum. Long rays appear to be associated with great aurorae and atmospheric heating.

The height statistics based on photographic triangulation may be representative of the brighter forms only. It is possible that lower intensity arcs could appear at different heights. Thus Vallance Jones and Gattinger (1972) and Akasofu and Yasuhara (1973) have presented evidence for the fairly frequent occurrence in the oval of low intensity auroral red arcs (LAR arcs) which they deduce must lie above 150 km, while Boyd *et al.* (1971) have given a series of 2-station photometer height measurements which shows that there is a systematic latitude and diurnal effect in the height of the stable HA's so detected. The results of this are shown in Figure 3.6. One must assume that the weaker higher arcs seen in the auroral oval were not included in the photographic studies. Egeland and Omholt (1967) as well as Störmer (1955) did find a type of high HA at heights between 160 and 230 km at latitudes below 62°.

3.3.5. Visual types of aurora

Visually several different varieties of coloured aurora have long been recognized. The types as classified by the *International Auroral Atlas* (International Union of Geodesy and Geophysics, 1963) are set out in Table 3.2; in brackets are given alternative terms which are more commonly used although less precise. The same lettering system was preserved where possible in the IAA system.

TABLE 3.2
Visual types of aurora

IAA color class	Description	Typical height (lower border) km
Type-a (type-A red)	Red upper region	
Type-b (type-B red)	Red lower border	80–100
Type-c (normal)	Green or whitish (below color threshold)	110
Type-d (type-A red)	Red overall	250
Type-e (type-B red?)	Red and green alternating along horizontal extent commonly with rapid motions of rays along arc.	
Type-f	Blue or purple dominant	

As discussed in Section 4.2.5 these color variations are largely height effects arising from variations in the energy spectra of the incoming particles.

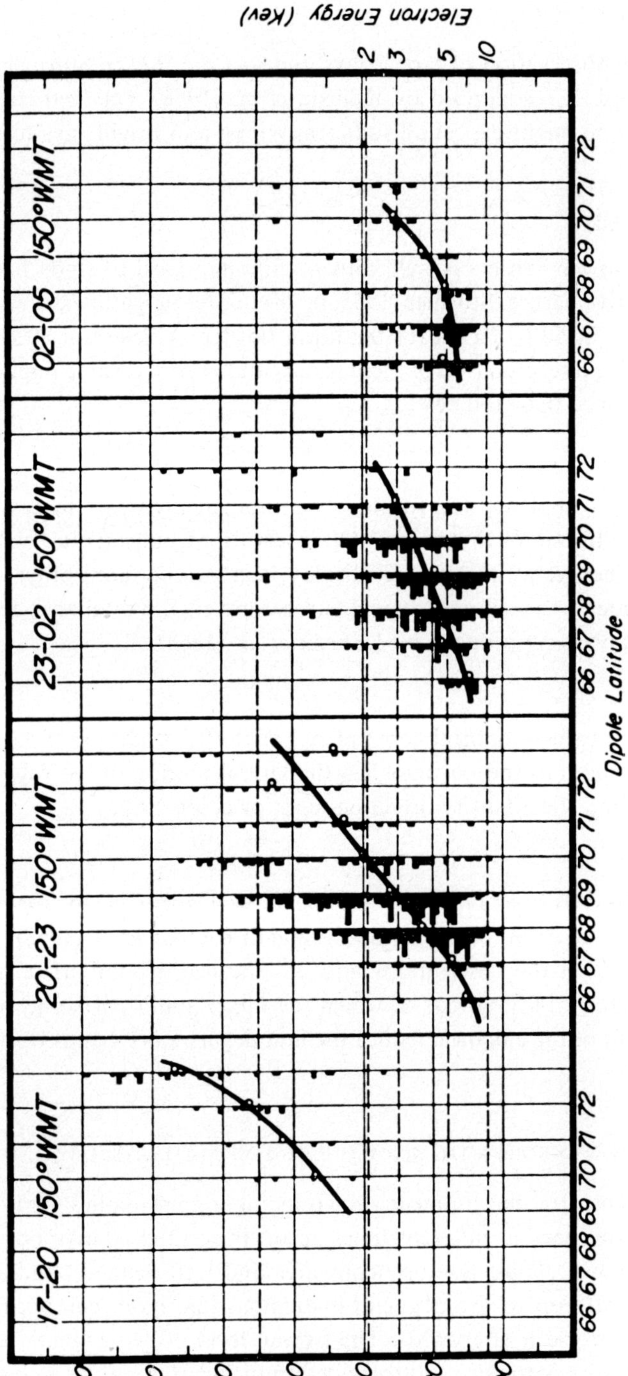

Fig. 3.6. Heights of homogeneous arcs as function of local standard (i.e. 150°W meridian) time and dipole latitude after Boyd *et al.* (1971) (Courtesy *Journal of Geophysical Research*; copyright by American Geophysical Union).

Type-a (Type-A red)

References to the observation of type-a rays appear on p. 142 in Störmer (1955) and a color photograph is reproduced by Kaneda *et al.* (1968). On theoretical grounds (Section 4.2.5.2) one might expect all sufficiently long and bright rays to show a red upper region.

Type-b (Type-B red)

Aurora with red lower borders is well known, although hard to study because of its transitory and active nature. Störmer (1955, p. 78) quotes height measurements which gave values as low as 60 to 70 km for the lower border. A height of 80 km might be a more conservative interpretation of these results and even this value implies electrons with energies of more than 100 keV.

Type-d (Type-A red)

The occurrence of completely red forms is well documented e.g., in Störmer (1955, p. 143 and p. 97). In one of these cases the observations indicated a height range of 250 to 450 km. The overall red type of display seems to be more often observed at low latitudes. Some further cases of type-d auroras are cited by Vallance Jones (1971). A color photograph is reproduced by Kaneda *et al.* (1968).

Type-e

This form appears to be a natural consequence of the 0.7 s lifetime of the upper state for the 5577 Å green line and occurs when the lateral motion of the rays in a RA is sufficiently large and the intensity bright enough for color vision.

Type-f

The occurrence of purple or blue aurora appears well documented, for example, in Störmer (1955, p. 143) or in the color photograph in Kaneda *et al.* (1968) (Figure 41). It appears likely from the descriptions and on physical grounds that such aurora usually takes the form of long rays of which the upper sunlit parts may appear blue or violet as a result of the enhancement of the first negative N_2^+ bands by fluorescence in sunlight.

3.3.6. Variations in Aurora with Geomagnetic Time and Latitude

The pattern of substorm development shown in Figure 3.4 implies some differences in the auroral forms seen at different times according to the relative position of the observer and the oval. Thus, homogeneous arcs and well defined band systems are seen more frequently up to geomagnetic midnight while more complicated, active, short-lived forms are seen afterwards. The overall level of geomagnetic activity and the time history of the particular substorm determine what is seen at a given time and place. Certain notable kinds of aurora are mostly associated with particular time-latitude regions.

(a) *Mid-Latitude Aurora*

This denotes the kind of display seen typically from geomagnetic latitudes below 50°. Such displays are seen only during very large magnetic disturbances and their characteristic features are probably due to the tendency for type-a, -d and -f aurora to occur in such storms. As discussed later in Section 4.2.5.5, unusually high temperatures are observed in the often red or blue forms of these displays. Spectral features due to neutral and ionized atoms are enhanced. These features suggest extensive atmospheric heating coupled with softer than normal particle influxes as discussed in Section 4.2.5.4. Some types of mid- or low-latitude displays may be due to the observation of the upper parts of forms beyond the poleward horizon; however, red type-d forms are often seen in the zenith.

(b) *Aurora in Midday Region of the Oval*

At the winter solstice the auroral oval is dark at midday especially for the region between NE Greenland and in the Arctic Ocean N. of Siberia where this portion of the oval lies furthest within the region of midday darkness. This locality is not good for ground observations but in recent years a number of aircraft borne expeditions have made possible more detailed studies.

Whalen *et al.* (1971c), Romick and Brown (1971), and Eather and Mende (1971a, 1972) have described the auroral phenomena seen in crossing the auroral oval near the midday geomagnetic meridian. Heikkila *et al.* (1972) have reproduced impressive color photographs of the midday aurora. According to Romick and Brown there is a region of diffuse emission which lies furthest equatorwards; polewards there is first a region of normal type-c arcs; still further polewards there are type-d rayed arcs and patches. Eather and Mende regard the poleward red arc region as the auroral oval and argue that the red forms there, are the only ones sufficiently bright to have been recorded with all-sky cameras in the original auroral oval studies. Romick and Brown suggest that the equatorward green forms lie along the oval and identify the poleward red arcs with the penetration of plasma sheath particles through the cusps, F, of the magnetosphere as shown in Figure 1.1. It seems generally agreed that the red arc region is produced in this manner and it is perhaps a matter of the definition of the 'oval' as to whether one identifies it with the red arc region or with the green arcs further south. This question is discussed further in Chapter 6.

(c) *Mantle Aurora*

As noted in Section 3.1.2 Sandford (1964, 1968) showed that a substantial fraction of the energy input from particle precipitation fell diffusely in a zone of radius about 20° around the geomagnetic pole. This appears to be produced by particles of higher than normal energy which drift eastwards at constant L after energization in the substorm processes of the midnight sector. (See Appendix 6A for discussion of such drifts.) It is discussed further in Section 4.2.5.3.

(d) *Polar Cap Glow*

This is a diffuse emission which occurs across the polar cap at the time of polar cap absorption events. It was discovered by Sandford (1961) from an analysis of I.G.Y. patrol spectrograph observations and is produced by 1 to 100 MeV protons.

(e) *Discrete Polar Cap Aurora*

Within the polar cap at geomagnetic latitudes north of 80° weak arcs have been observed with an orientation in the direction of the Sun. Lassen (1969) has reviewed this topic and has reproduced an ASC photograph of such an arc. These forms appear as short-lived diffuse bands or quiet long-rayed arcs. They appear to be most intense immediately north of the forms discussed in (b) but can occur even at the center of the polar cap. The zenith forms at Thule (86° N geomagnetic) are most often patches or fragments of bands with a lifetime of a few minutes only. Romick and Brown (1971) noted the Sun aligned forms in the polar cap and suggest that they are excited by electrons of energy about 2 keV superposed on a background of lower energy electrons. Eather and Akasofu (1969) have also described these forms and their spectroscopic observations also suggest excitation by electrons of energy ≤ 1 keV

(f) *Stable Auroral Red Arcs (SAR- or M-Arcs)*

These are stable, globe encircling arcs, consisting of almost entirely pure [O I] ($^3P - {}^1D$) ($\lambda\lambda$6300–64) emission which may appear in 45 to 55° geomagnetic latitude region during major geomagnetic disturbances. Their intensity varies from several hundred to several thousand rayleighs with a height of maximum emission around 400 km. There is always simultaneously bright but separate auroral activity poleward of the SAR-arc region. The observations and theory of SAR-arcs have been treated very fully in a recent review by Rees and Roble (1973). Since it is debatable whether these arcs should be included in an account of aurora, we confine ourselves to these brief remarks followed by a summary of the theory of their excitation in Section 6.2.6.

3.3.7. Rapid time variations – pulsing aurora

The fine spatial structure of some aurora and the rapid motion associated with it implies the existence of rapid time variations of the exciting particle fluxes within limited volumes of the atmosphere. These variations may even be quasi-periodic as in a rayed arc with a horizontal motion of the rays. Time variations which involve some kind of time dependent modulation of a more slowly varying auroral structure have been classified as *pulsing* in the *International Auroral Atlas* (1963). There are several varieties of pulsing aurora. The first traditionally called a *pulsation*, involves a synchronous quasi-periodic time modulation of the intensity of a more extended form or even of the luminosity of a wide region of the sky. The term pulsation has tended to be restricted to such variations with maximum power in the frequency range from 0.05 to 0.15 Hz and usually weak intensities of the order of 1 kR.

In recent years the term *flickering* has been applied especially to pulsing at a frequency of the order of 10 Hz as observed sometimes in bright (>10 kR) homogeneous forms near the break-up phase.

The term *flaming* has long been applied to an intensity modulation which travels upwards (or very occasionally downwards) through auroral forms towards the magnetic zenith. This has been studied quantitately by Carleton (1960), and more recently by Scourfield and Parsons (1969) and Cresswell (1969).

Finally there are the horizontal traveling intensity modulations known as *streaming* and *fast auroral waves*. The former runs along arcs or bands commonly near break-up while the latter involves a fast latitudinal motion of a succession of arc-like forms. Fast auroral waves have been the subject of work by Cresswell and Belon (1966), Cresswell (1968), Scourfield and Parsons (1971), and Boyd *et al.* (1972). In the study of Scourfield and Parsons a sharp transition was observed from the equatorward auroral wave motion (15 to 150 km s^{-1}) to an east–west streaming motion (20 to 90 km s^{-1}).

Pulsing aurora is of great interest for two main reasons.

(i) The time variation of the incident particle stream provides the opportunity to study the time constants of the excitation processes and to observe the effects of quenching processes on metastable excited states. Applications of this are discussed in Section 4.2.4.9. It is of particular importance in connection with the quenching of the $O(^1S)$ and $N_2(A^3\Sigma)$ states and in indirect excitation processes of $O(^1S)$ and $N_2(B^3\Pi)$.

(ii) The time variations must be closely related to the acceleration and precipitation mechanisms for auroral particles. The extent to which particles of different energies and mass are modulated and the possibility of variable phase shifts with particle energy could be critical in determining the nature and site of precipitation process. This aspect is discussed in Chapter 6 (see Section 6.3.3.).

Some points concerning the temporal and spatial morphology of pulsing aurora will be made in this section. Recent general reviews are given by Shepherd and Pemberton (1968), Omholt *et al.* (1969), and Omholt (1971).

Methods of observation. Visual and even photographic observations have been made of some of the slower pulsations. Recently the TV camera techniques discussed in Section 2.1.3 have made possible important advances. Filter photometer systems have also proved very valuable in making quantitative studies of the amplitude, phase relations and frequency spectra of the pulsing. On the other hand photometers usually do not reveal clearly which of the types of pulsing is involved. Clearly a combination of methods is required.

3.3.7.1. *Pulsating Aurora*

Geomagnetic time-latitude distribution of pulsating aurora. At the statistical level a study by Kvifte and Pettersen (1969) determined the probability of observing pulsations in the latitude range 65 to 68° and near 75°. In summary these results indicate a maximum probability near the equatorward edge of the oval particularly towards 0600 h. geomagnetic time. At 60° geomagnetic latitude Paulson and Shepherd (1966a) observed pulsations (modulation ~5%) in almost all quiet forms.

Relation of pulsations to substorms. As with aurora generally, the dependence upon substorm time is likely to be very important. Cresswell (1972) has drawn conclusions from a comprehensive study of Alaskan data. He distinguishes 3 species of pulsating forms:

(1) Those which occur within the poleward bulge near midnight (poleward of the original oval) after the scattered remnants of arcs and bands have been transformed into largely diffuse pulsating forms.

(2) Pulsating forms appearing equatorward of the original oval before 0300 geomagnetic time within a diffuse envelope which spreads equatorward.

(3) In the late morning oval, diffuse eastward drifting patches.

All these kinds of pulsating forms are characteristic of the later stages of substorms such as Figure 3.4D.

Frequency spectra and types of pulsations. Johansen and Omholt (1966) distinguished 3 types of pulsations on the basis of the general appearance of photoelectric records. Type-A (rare) consisted of a rapidly decaying train of pulses spaced at 1 to 3 s. Type-B consisted of a smaller higher frequency modulation atop a 20 to 40 s long, almost rectangular, pulse while type-C appeared as a longish train of either positive (C_1) or negative (C_2) pulses of quasi-periodic nature. Examples of pulsations are given by Johansen and Omholt (1966), Paulson and Shepherd (1966a), and Omholt (1971).

The frequency spectrum of pulsation trains has been extensively analyzed in papers just cited and also by Shepherd and Pemberton (1968). In general most ordinary pulsations have a principal peak in the power spectrum in the range 0.05 to 0.15 Hz and usually have little power above 0.4 Hz. Some show subsidiary peaks out to 0.5 Hz.

Rather higher frequency pulsations sometimes occur with appreciable power (~1% of that at the maximum) out to 10 Hz. These are usually subvisual. (The type-B pulsations noted above appear to be of this kind.) Another type of high frequency pulsation is observed with high magnetic activity in the morning after 0600 h; this resembles X-ray microbursts. Examples and spectra of these higher frequency pulsations are given by Omholt (1971),

The pulsations are often coherent for fields separated by 5 to 10 km and sometimes more in the cases of pulsating arcs. More detailed information of spatial coherency of pulsating patches has been obtained by Scourfield *et al.* (1972) who found different patches pulsate independently. At the other extreme Gowell and Akasofu (1969) and Murcray (1959) have observed pulsations coherent over a large part of the sky, the former with a long period of the order of 1 min and the latter somewhat faster. The important earlier study of Campbell and Rees (1961) was done with a 90° field centered on the zenith.

Relation of pulsations to other phenomena. There is sometimes a close relation between auroral pulsations and fluctuating X-ray emission. This is discussed in Section 5.6.3B. Further examples are discussed by Omholt (1971). It thus appears that sometimes the same mechanism can modulate or precipitate a wide range of particle energies.

There also appears to be a close relation between auroral pulsations and magnetic and possibly Earth current pulsations. Similarities ranging from a close ressemblance of the frequency spectra to a detailed 'peak to peak' correspondence have been reported. Because of the variability in spatial extent of the coherence of the auroral pulsations, different degrees of overall correspondence might be expected. The recent work of Campbell (1970) in which the outputs from a group $\lambda 4278$ N_2^+ photometers at Byrd were compared with local and conjugate point magnetometers is of great interest. In cases where the photometer outputs were in phase (corresponding to a large scale coherent auroral pulsation), the magnetic field pulsation was found to correspond closely with the optical records with a time lag of 1.3 s.

It has also been generally believed that pulsating aurora and auroral absorption (see Section 5.5.3) were related and even that the average frequency of the pulsation correlated with increasing particle energy. Brekke (1971, 1972) has argued that there is really no direct connection between these factors. As pointed out by Shepherd and Pemberton there is an inconsistency between evidence which suggests that pulsations are associated with a hardening of the incident electron energy spectrum and other evidence which suggests that the faint diffuse forms which show pulsations should lie at greater than normal heights. It may be that there is no direct connection between average particle energy and the occurrence of pulsations.

Direct measurements of particles producing pulsating aurora. An informative rocket flight, which was made through a pulsating aurora, was carried out by Bryant *et al.* (1971). In the rather strong aurora ($I(3914) \sim 10$ kR) modulation was observed for 4, 6 and >22 keV electrons as well as by a $\lambda 3914$ ground photometer. Modulation of the >22 keV electrons was somewhat stronger. Whalen *et al.* (1971b) also made direct measurements in morning pulsations and observed a 4 to $5\times$ flux increase for electrons in 25 to 80 keV range and a change to an isotropic pitch angle distribution at the intensity peaks. They did not measure the <25 keV electrons.

3.3.7.2. *Flickering Aurora*

As defined above in accordance with the suggestion of Beach *et al.* (1968) flickering appears to involve a somewhat ill-defined fluctuation or even motion of the fine structure in near-break-up aurora in the magnetic zenith. The 7 to 13 Hz intensity fluctuation was observed in taking the light from a small area of the TV playback. Essentially the same phenomenon was observed directly by Paulson and Shepherd (1966b) using a photometer with a field of about 320 m at auroral heights in the zenith. It appears to be closely related to a ~ 10 Hz fluctuation found in the incident particle flux by Evans (1967) and a similar fluctuation observed in HF radio reflections by Sofko and Kavadas (1969).

3.4. Proton Aurora

An optical emission with many distinctive characteristics is excited by proton bombardment. Usually this emission is faint or undetectable visually and consequently the preceding sections of this chapter have been concerned almost entirely with the morphology of aurora produced by electrons.

Two fundamental differences between protons and electrons ensure that proton excited emissions must be rather diffuse. First the radius of gyration of a proton in the Earth's magnetic field is about 200 m for 5 keV protons. Secondly, incident protons spiralling about the field can undergo charge capture processes in colliding with atmospheric particles as discussed in Section 4.3.1. The neutral H atom so formed escapes the control of the field and travels in a straight line in a direction determined by the pitch angle before ionization. This transverse motion is terminated by an electron loss collision. The resulting proton path is shown schematically in Figure 4.36. This tranverse diffusion process has been analyzed by Davidson (1965) who showed that a linear sheet of incident 10 keV isotropic protons could be broadened to a distribution with a half width of the order of 100 km. Higher energy protons would be less affected as would protons having an angular distribution more strongly peaked in the field direction. As a consequence, particularly of the diffusion effect, one would expect proton auroral forms to be broad and diffuse compared to those of electron aurora.

Figure 3.7 shows some all-sky camera (ASC) and meridian spectrograph observations of the initial stage of an intense substorm on April 28 and 29, 1960, as recorded from a geomagnetic latitude of 61.5° N by Montbriand and Vallance Jones (1962). The intensity variation along the spectrometer slit corresponds to that along a geomagnetic meridian which lies in the '11 o'clock' direction on the ASC photographs. The Hα line is clearly visible. In the 2108 to 2116 h spectrogram the Hα emission extends from about 45° south of the zenith to the N horizon while by the 2129 to 2138 spectrogram it reaches from near the S horizon to a little N of the zenith. In each case the Hα emission appears to correspond to a diffuse arc-like structure in the ASC photograph. In this display the Hα emission is rather strong so that the proton auroral form was registered by the camera. The spectrograph results suggest that the Hα emission is distributed in a broad horizontal sheet with a NS extent of 6° or more. The arc-like appearance of the Hα-emitting region (particularly in the second and third spectrograms) is probably due mostly to the van Rhijn effect, i.e., the apparent increase in brightness towards the horizon of a uniformly emitting sheet because of the longer path in the line of sight. The lower border of the 'arc' in the south probably corresponds to the equatorial edge of the proton precipitation zone. Discrete H-emitting arcs are never seen near the zenith. Direct meridian scans with photometers always show a slow angular intensity variation of the type shown in Figure 3.8 which shows some new data kindly provided by Dr R. H. Eather. These are similar to earlier results of Eather and Sandford (1966).

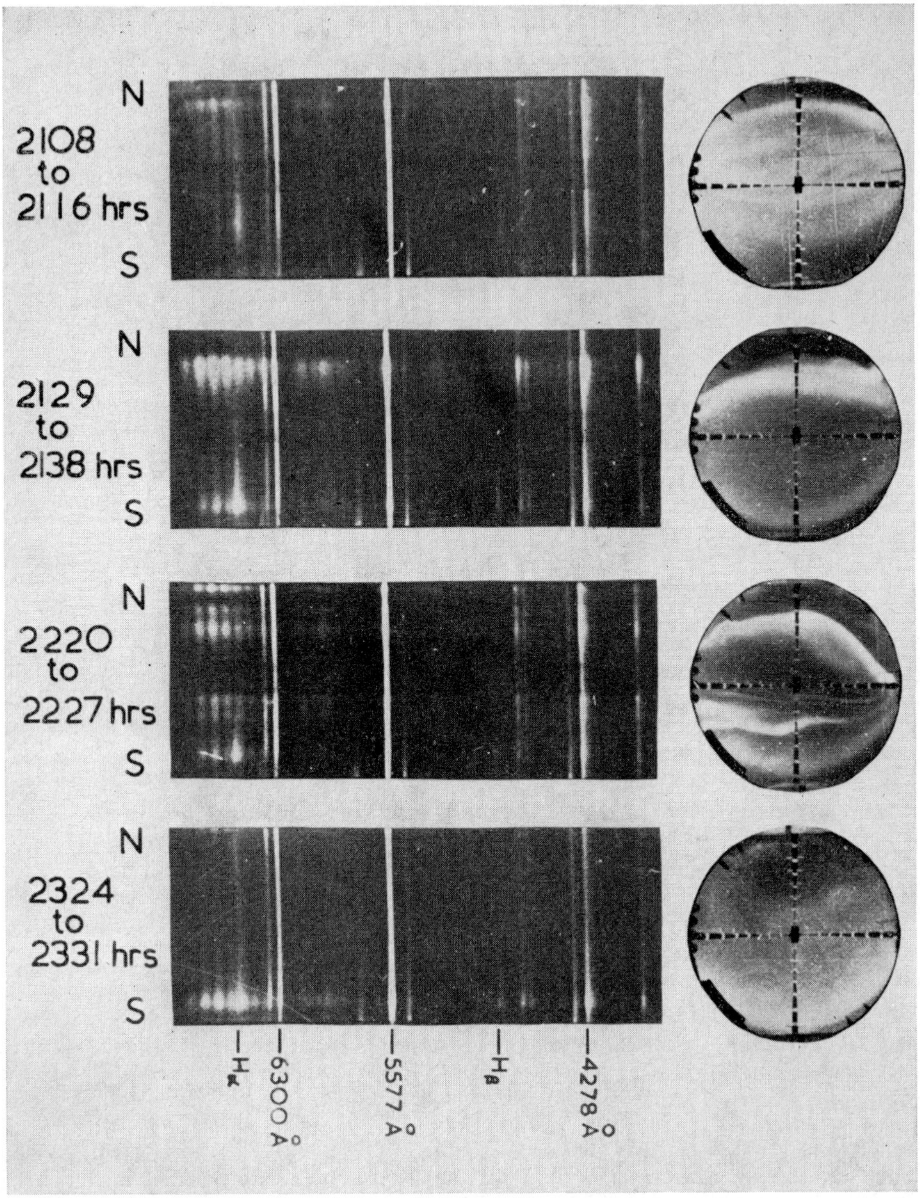

Fig. 3.7. All-sky photographs and meridian spectrograms of aurora of April 28 and 29, 1960 after Montbriand and Vallance Jones (1962). (Courtesy *Canadian Journal of Physics*; reproduced by permission of the National Research Council of Canada.)

Height of Proton Aurora

The height distribution is conveniently measured by rocketborne photometers which are most effective with diffuse extended forms. Values obtained by Miller and

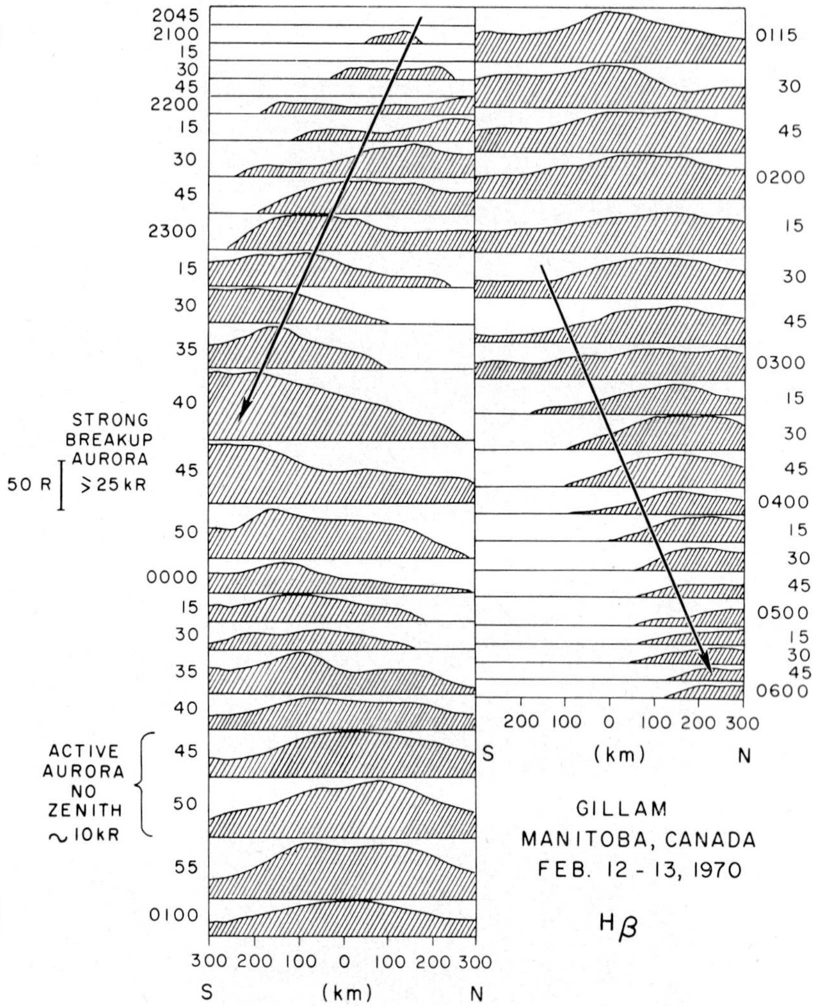

Fig. 3.8. Meridian scans of Hβ emission from Gillam (corrected geomagnetic latitude ∼68°) at local times shown. I am very much indebted to Dr R. H. Eather for providing these data in advance of publication.

Shepherd (1968), Murcray (1966), Wax and Bernstein (1970), Miller and Shepherd (1970), and Egeland *et al.* (1973) are, respectively, 116, 105, 130, 145 and 111 km. Wiens (1968) obtained 106 km as a mean value from ground based triangulation on the edges of bright proton fields. It appears that as with electron aurora, the height is variable and may be lower as the intensity increases.

Intensity and Intensity Fluctuations

The intensities of the Hα and Hβ lines have been extensively measured. Omholt (1971) has assembled a comprehensive summary of reported intensities. It appears that intensities up to 500 R or even 1000 R* may occur very rarely under exceptional conditions associated with great magnetic storms but that maximum zenith values in the auroral zone are rarely greater than 100 R and often no more than 25 R. Intensity variations are slow and no rapid pulsations ($\tau \sim 10$ s) of Hβ were detected by Eather (1968b) or Miller and Zeitz (1971) in a study of pulsating aurora. Slower $\tau \sim 1$ min pulsations were observed simultaneously in λ5577 and Hβ. Another type of irregular intensity variation has been reported by Beach *et al.* (1968) who, during a flickering aurora, observed a rise in I(Hβ) to about 650 R over a few minutes when the auroral intensity must have approached IBC3.

Geomagnetic Time-Latitude Variation – Proton Auroral Oval

Romick and Elvey (1958) and Galperin (1959) first recognized that H emission often precedes the appearance of visual forms in the evening. Rees *et al.* (1961) used one of the first high speed patrol spectrographs to show that the Hα emissions shift equatowards in the evening ahead of the visible electron aurora as shown in Figure 3.7 and then retreat polewards. A similar shift is to be seen in the photometer results of Figure 3.8. On the basis of such observations, Montalbetti and McEwen (1962) and Eather and Sandford (1966) made proposals for the location of a proton auroral oval. Both proposals call for the proton oval to lie slightly equatorwards of the electron and before midnight but Montalbetti and McEwen suggested the central lines of the two crossed over after midnight while Eather and Sandford proposed that the proton zone remained equatorwards at all times. Subsequent work by Derblom (1968) and Yevlashin (1970) has supported the former view while studies by Rees and Benedict (1970) and Eather and Mende (1971b) do not show the crossover. Some examples of the crossover have been found in satellite observations by Chubb and Hicks (1970). The most comprehensive statistical ground-based study is probably that of Wiens and Vallance Jones (1969) who examined the distribution of proton and electron aurora from 2 yr data from patrol spectrograms like those of Figure 3.7. The results are summarized in Figure 3.9 which show distribution contours for the highest intensity levels. On the average there appears to be after midnight a poleward displacement of the center of the proton oval with respect to the electron oval especially at high K_p. Both distributions are fairly wide and overlap considerably. The poleward boundary of the proton oval appears to change little with K_p.

The apparent disagreements between various observers as to the relative locations of proton and electron precipitation regions after midnight are probably due in part to the limited latitude range of some of the observations and partly to possible differences between average distributions and instantaneous ones. Thus the evening equatorward displacement of proton aurora appears to be maintained consistently whereas after midnight it is clear there is no such simple relation. It may be that the average intensity distribution is strongly influenced by intense but transient conditions. Such an interpretation is strengthened by the substorm behavior discussed below.

* At least some of the high values observed by photometers could be spurious and due to filter leaks or underlying electron excited bands.

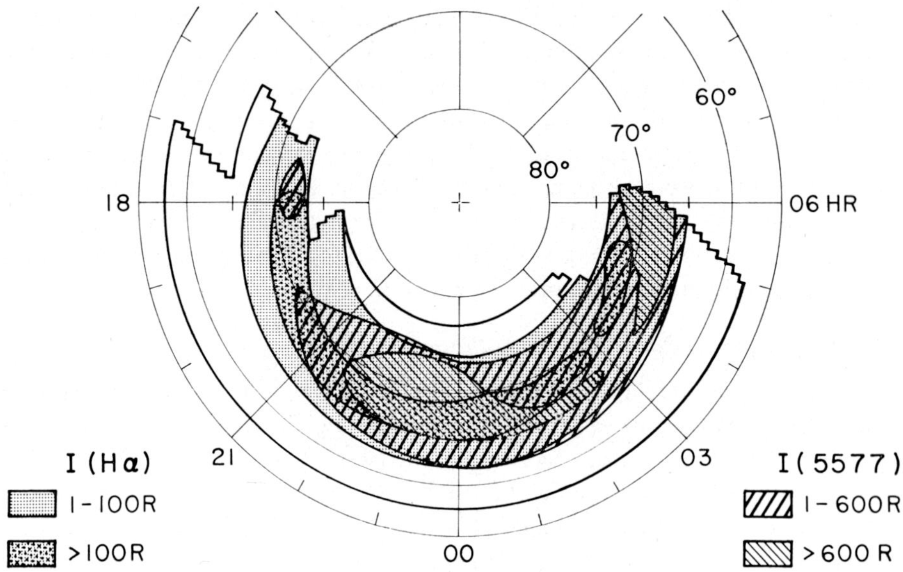

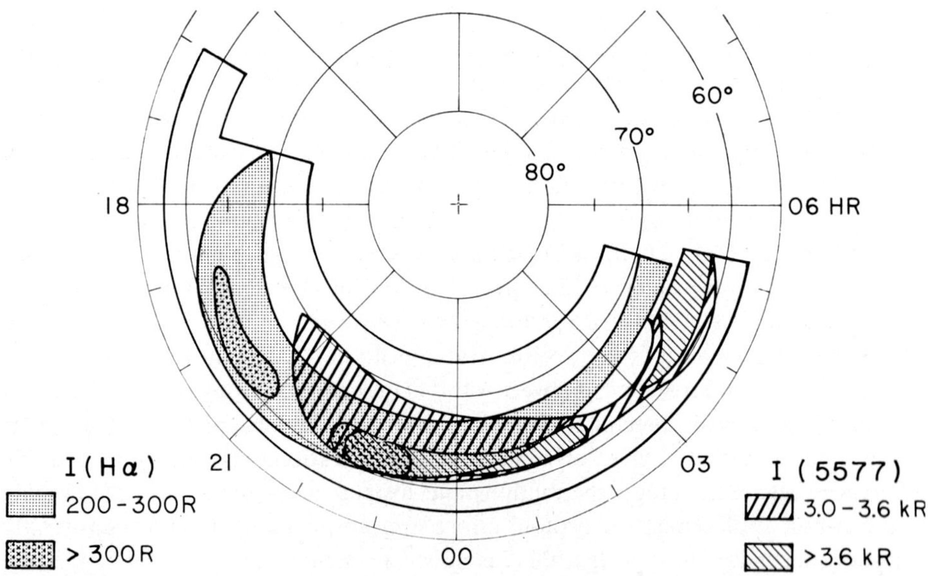

Fig. 3.9. Average intensity contours on plot of corrected geomagnetic latitude vs corrected geomagnetic time for [O I] 5577 Å and Hα for (a) $K_p < 2_o$ and (b) $K_p \leq 4_o$ after Wiens and Vallance Jones (1969). (Courtesy *Canadian Journal of Physics*; reproduced by permission of the National Research Council of Canada.)

In the daytime sector of the oval, H emission is observed coincident with the region of red rayed arcs. This emission is quite faint and does not exceed 10 R according to Eather and Mende (1972).

Proton Precipitation in Auroral Substorms

The behavior of the H emissions in auroral substorms is of great interest but has been difficult to study because of the problems of making good measurements from a sufficiently large network of ground stations. The most systematic attempt has been that of Montbriand (1969, 1971) who studied a very large number of patrol spectrograms and ASC photographs from the Canadian line of stations covering the geomagnetic latitude range 57° to 76°. Since it is not possible to observe the full latitude range of any single substorm in this way, the overall pattern must be

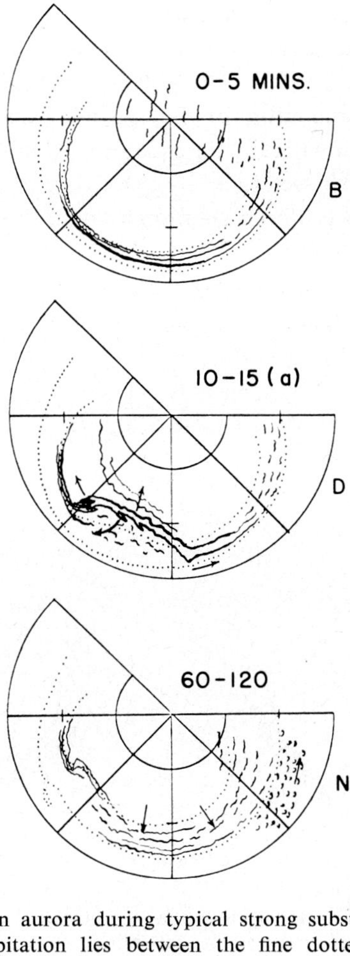

Fig. 3.10. Changes in proton aurora during typical strong substorm after Montbriand (1971). The region of proton precipitation lies between the fine dotted lines. (Courtesy D. Reidel Publishing Co.)

synthesized by combining the data from many substorms which occurred when the observing stations were at various geomagnetic times. This process is not easy because few substorms have the idealized characteristics of the scheme of Figure 3.4 and consequently the final results rely heavily on the judgement of the person who seeks to analyze the pattern.

The results of Montbriand's study of over 75 000 spectrograms and 2×10^6 ASC photographs are shown on Figure 3.10. The H emission lies at all times between the fine dotted lines. The darker lines depict electron excited auroral forms as described in Section 3.2 in connection with Akasofu's substorm scheme represented in Figure 3.4. The main points are as follows:

(i) At the initial stage of the substorm in the evening sector quiet arcs lie along the poleward edge of proton zone while at and after midnight they coincide.

(ii) In the expansive phase the H emission spreads into the polewards bulge behind the advancing electron auroral front and behind the westward traveling surge.

(iii) At and after the expansive phase in the midnight and early morning sectors patchy aurora appears equatorward of the H emission region.

If this scheme is correct, the post midnight 'crossing' of the statistical proton and electron ovals may be characteristic of the expansive and recovery phases only of substorms and may be dominated by an equatorward movement of patchy electron aurora particularly in 0300 h sector.

Recently this description has been confirmed and elaborated by Fukunishi (1974) who summarized his conclusions in Figure 3.11. Fukunishi distinguishes between diffuse-type and break-up type proton aurora. It is the latter which occurs in the poleward region after midnight.

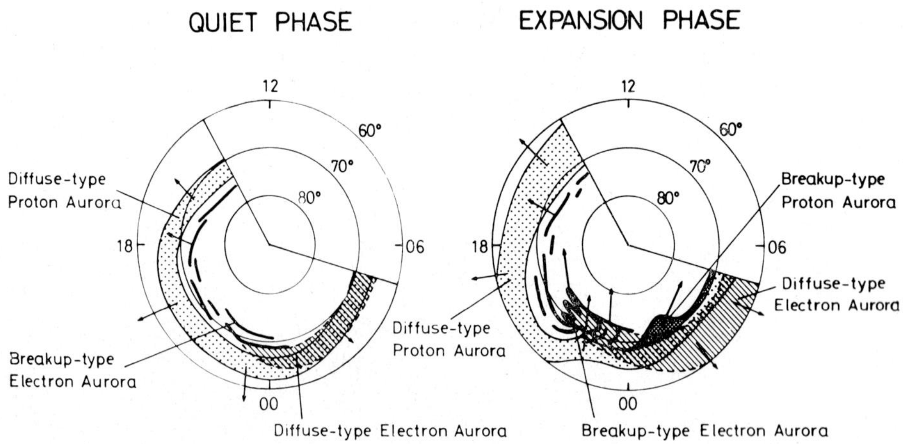

Fig. 3.11. Distribution of diffuse and break-up type proton and electron aurora for quiet (pre-breakup) and expansion phases of substorm after Fukunishi (1974). (Courtesy *Journal of Geophysical Research*; copyright by American Geophysical Union.)

3.5. Relation of Aurora to Solar Events

Statistical studies of the occurrence of auroral displays revealed many years ago that there is a good correlation with overall solar activity although correlations with particular solar events are not always so clear. The study of the statistical periodicities of auroral and magnetic disturbances has been pursued very actively in the past and the results are to be found in detail in such books as Störmer (1955) and Chamberlain (1961).

3.5.1. Periodicities and Recurrences

The main periodicities are as follows:

(i) *11 yr Cycle*

There is no doubt of the existence of an 11 yr cycle in the frequency of the overall occurrence of aurora. There is some evidence for a 1 yr lag on the sunspot number curve (Stetson, 1939).

(ii) *Semi-Annual Variation*

At least for stations on or equatorwards of the auroral zone, there is no doubt of the seasonal variation with maxima at the equinoxes.

(iii) *27–30 Day Periodicities*

The results of Meinel *et al.* (1954) indicated that there was a small $\sim$27 day recurrence tendency associated with the apparent rotation period of source regions on the Sun. It was necessary first to subtract a 29.5 day periodicity ascribed to the effect of the Moon on visual auroral observations. Stoffregen (1968) has claimed from I.G.Y. all sky camera data that the 29.5 day periodicity is real with a maximum near new Moon. However the reality of the latter effect is not yet generally accepted. The 27 day recurrence tendency could not be shown to persist beyond a simple solar rotation in the auroral data of Meinel *et al.* Lassen *et al.* (1964) did not find the 27 day recurrence in Danish visual data from 1957–60 a period of very high solar activity. It seems possible that recurrence is a property of a certain class of magnetic disturbance which is associated with the long-lived centers of magnetic fields on the Sun which give rise to the sector structure of the interplanetary magnetic field as discussed in Section 3.5.3. As demonstrated by Wilcox and Colburn (1972) the sector boundaries are associated with enhanced geomagnetic activity and consequently it appears likely that there must be a 27 day recurrence relation. Superposed on this however are the strong disturbances originating in a random manner from solar flares (see Section 3.5.2). It is plausible that the latter could swamp the recurrent disturbances especially at a great sunspot maximum such as that of 1957–58.

3.5.2. Correlations with Solar Events

The direct relation between auroral displays and solar events such as flares and sunspots is important as an indication of the mechanism of aurora. An association between various ionospheric disturbances and solar flares is generally accepted. Statistically as shown, for example, by Warwick and Hansen (1959) there is an excellent correlation between high values of magnetic disturbance indices and flares of importance* 3 or more at least for the maxima of the sunspot cycle. The peak magnetic disturbance is on the average 2–3 days after the flare. As shown by Hartz and McAlpine (1962) and Maeda *et al.* (1962) flares of importance $\geq 2+$ which are accompanied by noise bursts of frequency below 200 MHz are followed by ionospheric disturbances with a delay most frequently of 2 to 3 days.

Intense low latitude displays of aurora can usually be associated with solar flares occurring up to 1 to 4 days earlier although the converse is not always true. Hartz and McAlpine showed that the ionospheric disturbance magnitude was highest for flares occurring about 20° W of the central meridian of the Sun.

The relation between sunspots and aurora is complex. Pecker and Roberts (1955) showed there was a negative correlation between the central meridian passage of radio quiet sunspot groups and geomagnetic disturbances. They suggested that such groups interfered with the ejection of plasma from the solar surface. On this interpretation some weaker recurrent disturbances could be correlated with plasma ejected from the normal solar surface between sunspot groups. Mariska and Oster (1972) have recently shown that most of the correlations with sunspots and other *photospheric* features are dubious.

3.5.3. Relation of Aurora to Solar Wind and Interplanetary Magnetic Field

The measurement by satellite of plasma streaming out from the Sun in interplanetary space has provided direct evidence of the link between solar events and magnetospheric disturbances. The solar wind (Axford, 1968) has the characteristics shown in Table 3.3. The higher values are generally associated with solar disturbances.

TABLE 3.3

Properties of solar wind

Particle flux	10^8–10^{10} cm^{-2} s^{-1}
Density	2–120 cm^{-3}
Velocity	200–1000 km s^{-1}
Ion temperature, T_i	10^4–10^6 K
Electron temperature	1.5–5 T_i
Magnetic field	2–40 γ

* See Ellison (1957), Ratcliffe (1960, p. 194), or Öhman (1958) for the classification of flares.

The interplanetary field (Wilcox, 1968) is directed predominantly away or towards the Sun at the spiral hose angle, i.e. the angle which would be assumed by a *streamer* of plasma ejected from a fixed point on the rotating Sun and moving *radially* outward at the solar wind velocity. The field direction is usually fairly close to the plane of the ecliptic but can deviate 10 to 20° above or below. The interplanetary field is divided into two or more sectors alternating between directions away and towards the Sun. These sectors originate in areas of predominately N or S polarity on the Sun. Wilcox and Colburn (1972) give references to a series of reports on the evolution of the sector structure from the minimum to the maximum of solar activity. They find a strong tendency for magnetic (and presumably auroral) activity to be highest within a few days of the Earth crossing a sector boundary.

Many attempts have been made to relate the occurrence of auroral substorms to observable variables of the solar wind such as plasma density, velocity or magnetic field strength, direction and variability. References to these studies may be found in a recent paper by Arnoldy (1971) who has succeeded in obtaining a high correlation between the auroral zone magnetic activity index and a particular function of the interplanetary field. This function, $\Sigma B_s\tau$ the time integral of the southward component

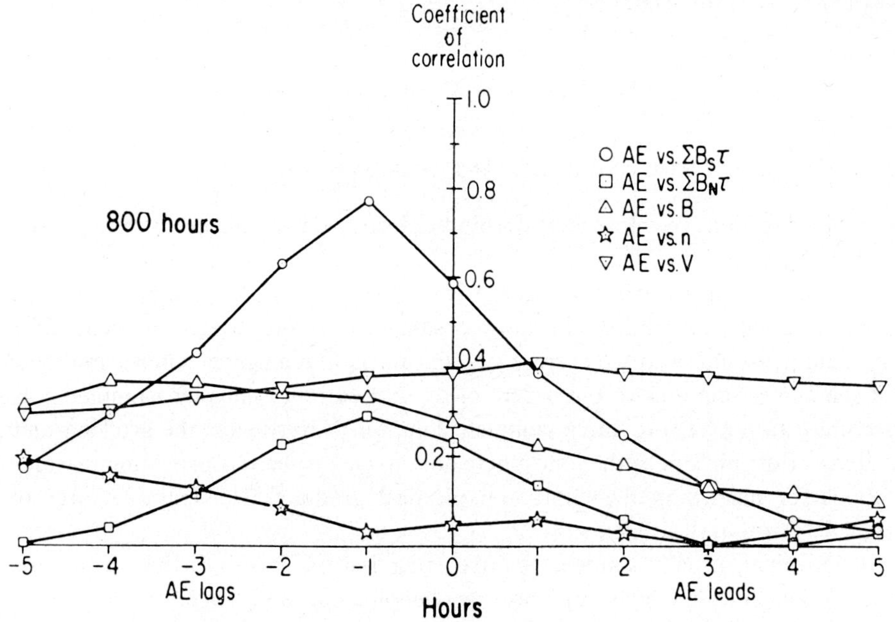

Fig. 3.12. Cross correlation functions of various solar wind parameters against auroral electrojet index AE from Arnoldy (1971). $\Sigma B_s\tau$ is the sum over 1 h of the southward component of the interplanetary magnetic field B_s, where τ is the time interval at which the sum is evaluated. $\Sigma B_N\tau$ is the corresponding quantity for the northward component. B, N and V are the values of total interplanetary field, solar wind ion concentration and velocity, respectively. (Courtesy *Journal of Geophysical Research*; copyright by American Geophysical Union.)

of the interplanetary field over the preceding hour is shown plotted against the magnetic index AE* in Figure 3.12. The interplanetary field is referred to solar magnetospheric coordinates, i.e., a system with its z axis along the dipole axis of the Earth's field with the Sun in the xz plane. The southward component is then the negative value of B_z. This index appears to provide the best correlation with AE; there is a time lag of about 60 min between the variations of $\Sigma B_s\tau$ and AE. As discussed in Section 6.1.4 $\Sigma B_s\tau$ is highly significant in the theoretical treatment of substorms. World wide effects such as magnetic storm sudden commencements seem to be associated with discontinuities in the solar wind but individual substorms are generally observed only after periods of southward interplanetary field. There is at present some debate as to whether substorms can be triggered by storm sudden commencements and sudden impulses without an initial period of southward interplanetary field. In fact $\Sigma B_s\tau$ does not appear to be so unique in its correlation with substorm activity as is suggested by Figure 3.12 according to a recent analysis by Intriligator (1973).

Recently it has been found that the E–W component of the interplanetary field is involved in a related reconnection process which is responsible for certain polar cap magnetic effects. It has proved possible to determine the polarity of the sector structure from high latitude magnetograms. (See Jørgensen *et al.* (1972) and Svalgaard (1973) for references.)

3.6. Magnetic Disturbances and Aurora

3.6.1. Magnetic storms and equivalent current systems

In previous sections the close relationship between magnetic disturbances and aurora has been frequently mentioned. Such disturbances indicate the presence of enhanced electric currents which can flow in the ionosphere and in the magnetosphere and which are usually associated with auroral substorms. Ground measurements do not give a unique solution to the current system causing the magnetic effects and there is still some uncertainty as to the nature of auroral currents since measurements from spacecraft do not yet provide a complete solution. Nevertheless the growing understanding of the physics of the ionosphere and magnetosphere is providing reasonable grounds for narrowing the choice on physical grounds. The main features of a magnetic storm are as follows:

(i) An initial, usually sudden, positive excursion of H; this is the storm sudden commencement (SSC) and is a worldwide phenomenon simultaneous within a minute or less. It is attributed to a compression of the Earth's magnetic field by an increase in the pressure of the solar wind plasma. The electric currents involved flow at the magnetopause and they are sometimes called collectively the DCF system after

* See Appendix 3B for definition.

Chapman and Ferraro. (Similar currents must flow continuously to maintain the magnetopause.)

(ii) The main phase, a worldwide negative excursion of H due to the trapping of solar plasma within the closed part of the Earth's field. The trapped energetic protons drift in longitude and constitute a ring current which may persist for one to several days. The ring current may build up during each magnetic substorm. The ring current system is sometimes called the DR system.

(iii) During the various stages of the one or more substorms which can occur during a single magnetic storm, complex current systems appear in the auroral oval region and have associated with them very strong but localized current systems flowing at least partly in the ionosphere. These have been called the DP systems. Much analysis of surface magnetograms has been done with the assumption that the currents responsible for the auroral oval disturbances are *closed* in the ionosphere. Such a solution may be regarded as an *equivalent current system* and is a convenient representation of the surface magnetic observations. Two rather different average current systems have been derived for the magnetic substorm. These are shown in Figure 3.13 taken from a review by Obayashi and Nishida (1968). (a) is the 'classical'

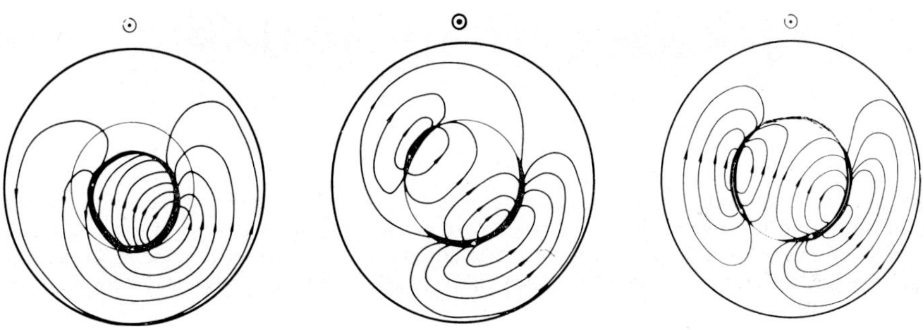

Fig. 3.13. Different possible idealized *equivalent* current systems for magnetic substorms after Obayashi and Nishida (1968). Cases (a), (b) and (c) as discussed in the text are displayed from *right to left*; the Sun symbol indicates local noon. (Courtesy *Space Science Reviews*.)

two vortex scheme which has strong E and W currents along the auroral oval called electrojets. (c) is the scheme recently advocated by Akasofu *et al.* (1965) who find only a western electrojet. (b) is an intermediate case. Feldstein (1966) has obtained another version of the average current system, somewhat like (b). Boström (1971) has shown that similar equivalent current systems may be plotted during individual substorms as is shown in Figure 3.14. Grafe (1969) has demonstrated that the two-cell pattern like Figures 3.13a or b can be quite strong in intense substorms.

Formally, the average magnetic variations during storms were originally resolved (e.g. Vestine, 1960) into a solar time dependent component S_D and a solar time independent component D_{st}. Equivalent currents systems can be constructed in the

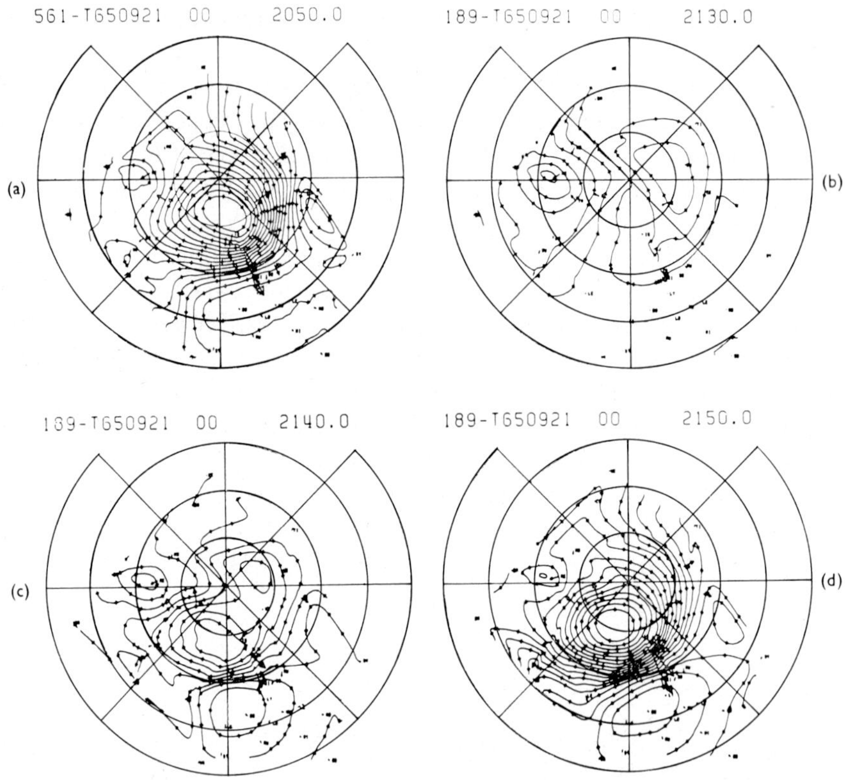

Fig. 3.14. Instantaneous equivalent current systems derived by Boström (1971) at various epochs of actual substorms. Geographic latitude circles are drawn at 10° intervals. 0° geographic longitude is vertically downward. UT is given at top right hand of each frame. (a) peak phase of substorm, (b) quiet phase, (c) growth phase, and (d) new peak phase. (Courtesy D. Reidel Publishing Co.)

ionosphere for each. S_D is similar to DP and was thought to be largely due to ionospheric currents whereas D_{st} was recognized to be largely due to a ring current.

The problem with the average current systems is that the instantaneous distributions are smeared out; on the other hand there are not enough observing stations to resolve all the details necessary to obtain the instantaneous distribution from a single event. It is certain, however, that the *equivalent* current system for a substorm is typified by Figure 3.13b or c with at least a strong western electrojet flowing along the oval. Heppner (1967) has argued strongly for the reality of the eastern electrojet. This current pattern has been recently called the DP1 system.

Another current system has been derived by Nishida (1968) and is shown in Figure 3.15. This has been called the DP2 system and is associated with the growth phase of substorms (Kokubun, 1971), with certain worldwide magnetic fluctuations and with quiet polar cap conditions. DP2 is a twin vortex system without auroral electrojets. This may be the system appearing between substorms in Figure 3.14b.

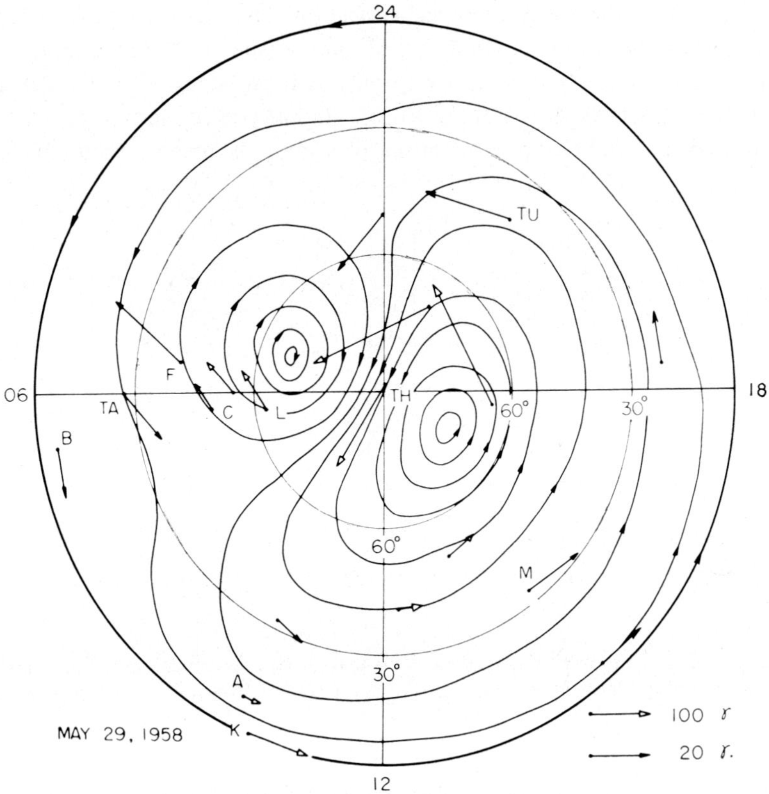

Fig. 3.15. DP2 current system from Obayashi and Nishida (1968). (Courtesy *Space Science Reviews*.)

3.6.2. Real current systems in aurora

The DCF and DR systems probably exist somewhat as discussed above. The DP1 system is almost certainly a mathematical fiction as has been emphasized by Boström and his collaborators (see Boström (1971) for discussion and references). As pointed out first by Fukushima (1969) the vertical loop current, A of Figure 3.16 can be shown to produce the same magnetic field below as a 'dipolar' current system C having the same horizontal 'electrojet' segment. Thus the vertical (Birkeland) currents in A produce the same field on the ground as does that of the extensive system of horizontal return currents of C. Physically it is easier to suppose that auroral particles can produce the conducting path for pattern (a) than for (b). Heppner *et al.* (1971) have shown from electric and magnetic field measurements in the polar cap that the transpolar currents of Figure 3.13 cannot be driven by the observed electric fields. Boström (1971) and Kisabeth and Rostoker (1971) have shown that the ground magnetometer observations can be produced by horizontal current segments of limited extent in longitude connected to the magnetosphere at their ends by Birkeland currents (parallel to the geomagnetic field lines). In some cases the horizontal currents are single linear segments, in others multiple or sheets of appreciable extent in latitude. Kisabeth and Rostoker have given a detailed scheme for the development of the current system in a typical substorm. Heppner *et al.* have suggested that the overall pattern in a substorm may consist of a large number of such current loops.

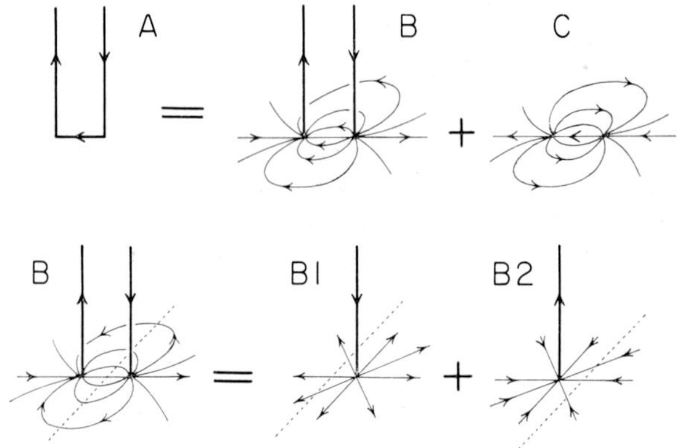

Fig. 3.16. Outline of proof by Fukushima (1969) that current systems A and C produce identical effects below at ground level. B may be regarded as the sum of B1 and B2 which produce zero field below. (Courtesy *Reports Ionosphere and Space Research*, Japan.)

Potter (1970) has inferred from rocket measurements that the westward electrojet current often consisted of multiple filaments parallel or coincident with visual forms. Burrows *et al.* (1971) found likewise that the electrojet current was concentrated in auroral forms when these were present while Pendrel *et al.* (1973), from rocket magnetometer measurements, have questioned the existence of localized currents sufficient to account for the ground magnetograms.

In principle a current segment such as that of Figure 3.16a could have associated with it a sheet current loop (down in front and up behind). Such a sheet current loop would produce little magnetic effect below if the spacing between the sheets is small compared to their width, but could be detected by rockets or satellites crossing the sheets.

Recent measurements by Park and Cloutier (1971) have detected just such current sheets associated with an auroral arc. They found the current distribution shown in Figure 3.17. The electron flux actually measured between 2 and 18 keV corresponded to about half of the upward current. Armstrong and Zmuda (1970) from satellite borne magnetometers have observed somewhat similar sheet currents on the dayside although more widely spaced and with the downward current usually to the north. Armstrong (1974) has reviewed the observations of field aligned currents including some more recent polar-orbit satellite results. These satellite measurements show that in the premidnight sector the major flow is always upwards at the poleward boundary of the oval. Almost always this is balanced by downward flows at lower latitudes. The satellite measurements do not show double sheets more closely spaced than ~400 km so that the system shown in Figure 3.17 must be either extremely transient or else correspond to a small portion of the overall current distribution. Measurements of field aligned currents further back in the tail have been reported by Aubry *et al.* (1972) and Fairfield (1973); the latter study indicated a predominant inward current flow at the dawn side and an outward flow at the evening side and directed towards the poleward boundary of the oval.

There is no direct observation of the magnetospheric source of the Birkeland currents. Meng and Akasofu (1969) have postulated that the westward electrojet is a partial short circuit of the ring current in the midnight sector. Kawasaki and Akasofu (1971) have calculated the surface magnetic field from such a current system

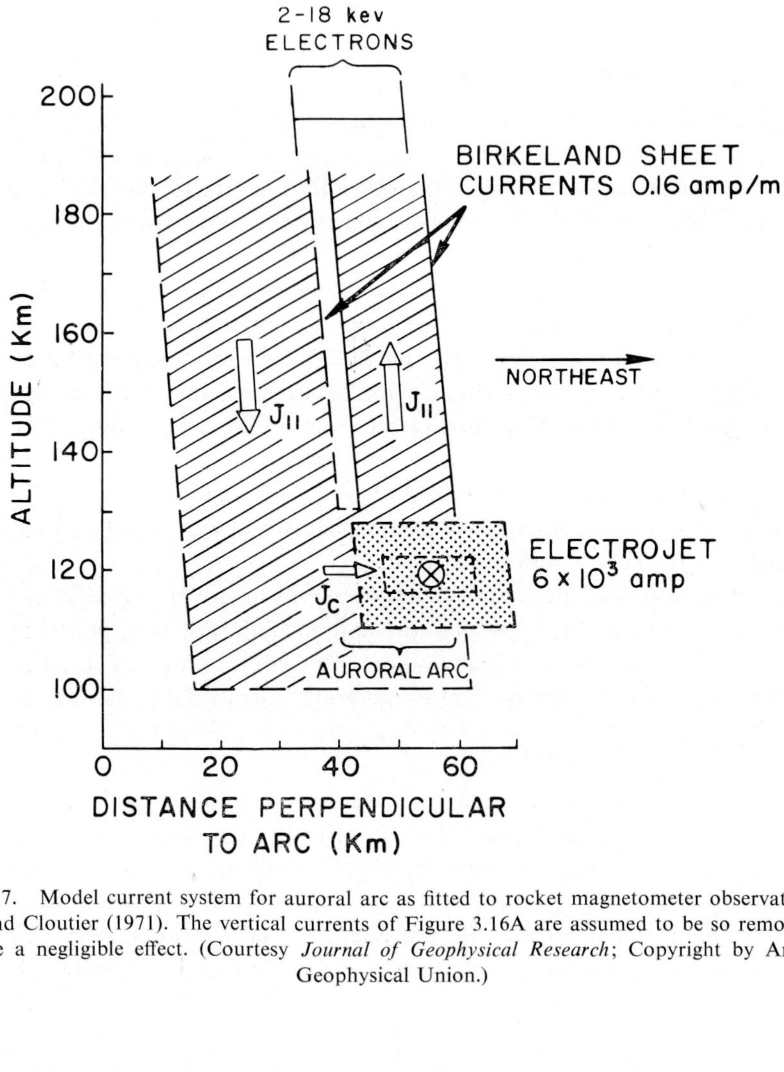

Fig. 3.17. Model current system for auroral arc as fitted to rocket magnetometer observations by Park and Cloutier (1971). The vertical currents of Figure 3.16A are assumed to be so remote as to produce a negligible effect. (Courtesy *Journal of Geophysical Research*; Copyright by American Geophysical Union.)

and found fairly good agreement with observations. Another possibility is that the earthward edge of the current sheet, *G*, in the tail of the magnetosphere is directed into the ionosphere. These questions will be further discussed in Chapter 6.

3.6.3. Electric fields in aurora

Many measurements have been reported of electric fields in or near auroral forms. These include direct rocket probe measurements by Potter and Cahill (1969), Potter (1970), Aggson (1969), Mozer and Fahleson (1970), Kelley *et al.* (1971a), and Mozer and Bruston (1967), Ba ion cloud measurements described by Haerendel and Lüst (1970) and Wescott *et al.* (1969), balloon measurements by Kelley *et al.* (1971b) and Mozer and Manka (1971) and satellite probe measurements by Cauffman and Gurnett (1971). Very recently Whalen *et al.* (1974) have reported results by the ion flow method. The observations may be divided into those of electric field components parallel and perpendicular to the magnetic field.

(a) *Electric Field Perpendicular to* **B**

The highest spatial resolution is given by rocket probe measurements. Aggson (1969) found fields up to 50 to 100 mV m^{-1} near auroral forms but <10 mV m^{-1} within. This was confirmed by the results of Wescott *et al.* (1969) and Potter (1970), although Mozer and Fahleson (1970) did not find any such decrease within forms. Whalen *et al.* (1974) find an anticorrelation between particle flux and electric field.

The balloon measurements referred to give the most comprehensible indications of the variations of **E** in substorms although with a resolution of the order of 100 km. One may question the reliability of this method but the results do appear consistent with those of the direct probes. Mozer and Manka found that the W component of the field turned positive with a value of the order of 30 mV m^{-1} about 1 h before a substorm and that the S component changed from perhaps -50 to $+100$ mV m^{-1} at the onset of a negative bay (western electrojet) in the midnight sector. The rocket experiments have mostly reported southward fields or at least significant southward components to the field in negative bays.

In a positive bay (which normally occurs well before midnight) Westcott *et al.* (1969) observed a northward field equatorwards and a southward field poleward of the break-up region. Similar transitions were observed regularly by Cauffman and Gurnett (1971) along the auroral oval in traversals near to the dawn dusk meridian.

(b) *Electric Fields Parallel to* **B**

A certain amount of experimental evidence has been obtained in favor of the occurrence of parallel fields although the conductivity in this direction would normally be expected to relax such fields very rapidly. Two different kinds of observations suggest their existence.

First, Mozer and Fahleson (1970) and Kelley *et al.* (1971a) have found a spin modulation in their perpendicular probe data when the spin axis was inclined to **B** and which corresponded to a parallel component of about 20 mV m^{-1} downward. Mozer and Bruston (1967) earlier reached the same conclusion. Recently Bering *et al.* (1973) have presented further data.

Secondly, the particle flux measurements of Rème and Bosqued (1971), Hultqvist *et al.* (1971) as well as some earlier ones of Mozer and Bruston (1966) suggest the presence of a parallel field, from anti-correlated time variations in proton and electron fluxes.

It is difficult to decide whether the direct observations of parallel fields should be accepted at face value. There are some theoretical reasons discussed in Section 5.2.1 for believing that anomalous resistance along the magnetic field lines may occur in connection with auroral break-up, but the height range involved is higher than that in which the direct measurements were made. O'Brien (1970) has criticized the conclusion that auroral particles may be accelerated by parallel fields.

The recent ion drift measurements of Whalen *et al.* (1974) who find parallel downward ion velocities of $\sim$2 km s^{-1} inside an auroral form suggest a parallel field of the order of 0.1 mV m^{-1}.

3.7*. Relation Between Auroral Substorms and the Magnetosphere

In the preceding sections of this chapter many aspects of auroral displays have been considered separately. It is obvious however that the auroral substorm is only part of a process which affects the whole of the magnetosphere and which is initiated by variations in the solar wind and the interplanetary field outside. Thanks to ground and satellite studies the main outlines of the overall phenomenon have been observed and it is now possible to give a general description of the whole process.

The configuration of the magnetosphere was shown in Figure 1.1. As determined by satellite observations the regions shown may be distinguished.

Many observations of magnetospheric phenomena in relation to ground observations have been reported. One interesting study is that of Mende *et al.*[5] who compared detailed optical observations with those from several satellites. The following is an attempt to summarize the most important changes observed in this and other work (cf. also the very recent and valuable study of McPherron *et al.*[30]). This presentation is partly derived from the scheme of Davis[6]. A detailed summary of the observed phenomena has been given by Rostoker[16].

It is convenient to classify the state of the magnetosphere, over the period of a substorm, into several phases.

I. *Quiet Phase*

(1) Interplanetary field has N component[4,5].
(2) Sun aligned arcs in polar cap $\Lambda > 80°$[6].

* Since many references are cited several times numerical superscripts will be used to refer to the papers listed on p. 75 at the end of this section.

(3) Weak arcs in auroral oval $\Lambda = 68–70°$; these are conjugate with opposite hemisphere[6].

(4) Auroral structures in oval drift W before midnight and E afterwards; there is also an equatorward drift[6].

(5) Stable trapping boundary[7] and inner edge of plasma sheet[8] at about 10 R_E. Plasma sheet[9] electrons peaked at few 100 eV with fluxes 10^8 cm^{-2} s^{-1} sr^{-1} keV^{-1}; protons peaked ≤ 1 keV with fluxes 10^6 cm^{-2} s^{-1} sr^{-1} keV^{-1}.

(6) Magnetic field at 6 R_E in tail ressembles dipole[10], i.e., current sheet well back in tail.

(7) Bimodal DP-2 type ionospheric current system (?)[6].

II. *Growth Phase*

The existence of a growth phase was first suggested by McPherron (1970) from a study of magnetograms over the period before the occurrence of the typical negative bays of substorms. It is not certain that all substorms have growth phases although Burch[24] has argued that a number of apparent exceptions studied by Kawasaki *et al.* (1971) did in fact possess them. There does appear to be good evidence that some kind of 'charging' process takes place before the spectacular surface manifestations of substorms; however, Akasofu and Snyder (1972) have expressed a different view. Coroniti and Kennel (1973) have given reasons why the growth phase could be very variable in duration. Rostoker (1974) has argued that ground based observations indicate that frequently one longitude sector of the magnetosphere can be experiencing a growth phase while an adjacent sector is experencing expansion. McPherron (1974) has further considered the concept of the growth phase and acknowledged that real substorms may be more complicated than originally supposed. The following changes have been proposed as typical of the 'growth phase'.

(1) Interplanetary field turns south[4,5,11] to initiate the growth phase which takes 1/2 to 2 h. (This change may be accompanied by an increase in the solar wind pressure which can produce a world-wide sudden impulse[12] in the H component of the surface field.)

(2) Polar cap arcs vanish and dayside aurora shifts equatorward[6].

(3a) Auroral structures in oval remain weak at first but eventually brighten a little. N and S hemisphere forms are conjugate but the projections of the poleward S-hemisphere forms fall poleward and westward of their N-hemisphere counterparts[6]. There is a tendency for the instantaneous oval to broaden and for the weak HA and RA forms observed to move equatorwards[6].

(3b) Proton aurora is found equatorward of the electron oval particularly in the premidnight sector[13].

(4) The same drift motions noted under I.(4) continue[6]. The equatorward drift accelerates[14].

(5) The inner edge of the plasma sheet moves[15] in by several Earth radii. The plasma pressure at 18 R_E increases steadily for a time to about $10 \times$ previous value. Toward the end of this phase the plasma pressure at 18 R_E usually decreases by a large

factor[18]; this has been interpreted as a thinning[18,19,30] of the plasma sheet so that the satellite is left 'high and dry'. (It is usually not exactly on the theoretical axis of the sheet.) At 6.6 R_E a gradual increase in particle pressure is also observed[5] corresponding to an increase in particle density.

(6) The magnetic field in the tail lobes increases by 30 to 50% during this phase[19,30] while the field at 6.6 R_E decreases[5,30]. These changes correspond to an inward movement[15] of the current sheet which causes the near field to become more 'tail-like'[10] as shown in Figure 3.18. The growth phase appears to be associated with a stripping of flux from the nose of the magnetosphere (with a decrease in the nose distance) and its transfer into the tail. This will not involve any compression of the nose field unless the solar wind pressure also increases. The surface field[5] has a worldwide small decrease corresponding to an increase in ring current.

(7) It appears that there is a somewhat intensified bimodal DP2 type current system[6,20,22].

(8a) There is an intensified westward electric field across the tail of the magnetosphere[21,17,18].

(8b) In the auroral oval ionosphere there is also a westward component to the electric field with a north- or southward component correlating with the west- or eastwards drift[23].

III. *Break-Up or Expansive Phase*

This marks the beginning ($T = 0$) of the strong auroral and magnetic effects observable from the ground as depicted in Figure 3.4. It appears to begin as the near-Earth plasma sheet reaches nearly zero thickness[30]. In the same order as before we list the salient developments.

(1) The interplanetary field does not directly appear to determine the initiation of the break-up although a sudden increase in solar wind pressure may trigger this phase[24].

(2) No special phenomena occur within the polar cap.

(3) A sudden brightening of the aurora occurs following the pattern of Figure 3.4.

(4) The auroral forms undergo rapid movement with the appearance of the phenomena described in Section 3.3.2. Type-e aurora (Section 3.3.5) with rapidly moving vertical striations usually appears and even type-b red forms. Narrow field photometers may record flickering, a 10 ± 3 Hz intensity fluctuation[6]. The westward traveling surge appears at the W end of the bright aurora[5]. The brightening of the aurora spreads as new forms appear successively poleward of the old and intensify rapidly.

(5) At 18 R_E the particle pressure remains low initially indicating either that the plasma sheet is still thin or has collapsed in toward the Earth[5,26]. Later the plasma usually reappears suggesting the appearance of plasma on more poleward field lines as shown in Figure 3.18[25]. At 6.6 R_E spectacular ($\sim 10\times$) increases in electron energy density occur, mainly as a consequence of increased particle densities[5,27]. In the case described by Mende *et al.*[5] there was a rapid increase in electron flux ($E < 6$ keV) and

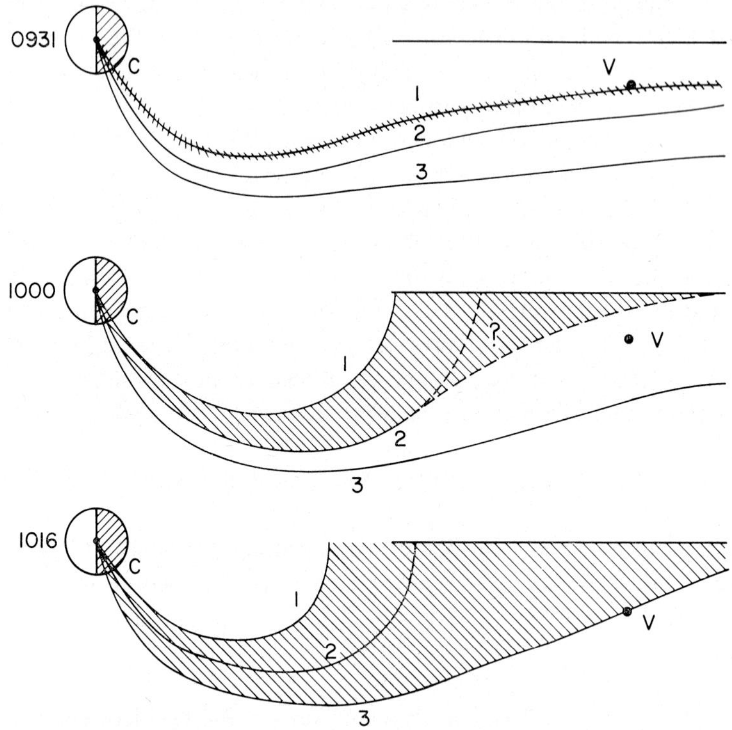

Fig. 3.18. Postulated variations in tail field and plasma sheet configurations during substorm according to Hones *et al.* (1971b). The upper diagram shows the situation at the end of the growth phase; the middle and lower diagrams probably correspond to early and late times in the expansive phase. (Courtesy *Journal of Geophysical Research*; copyright by American Geophysical Union.)

proton flux ($E > 15$ keV) at local break-up followed by a large increase in higher energy electron flux ($E > 10$ keV) in association with the passage of the western traveling surge across the satellite field line.

(6) The magnetic field in the tail lobes decreases and the field returns to a more dipolar configuration[10]. The tail current sheet moves further from the Earth and perhaps the near parts are diverted down across the midnight sector of the auroral oval.

(7) The ionospheric current system takes on the DP1 pattern i.e., there is an intense westward electrojet or jets correlated with aurora across the midnight region[6,22]. There is an upward current associated with the westward end at the westward traveling surge and probably a downward current at the eastward end[5]. There appear also to be sheet currents to the N and S of the arcs as discussed in Section 3.6.2.

(8) The electric field in the oval develops a strong southward component just at the time of break-up[23]. This is also the time at which parallel electric fields may occur[28].

IV. *Recovery Phase*

This begins when the aurora reaches its most poleward point. In the same order the principle points are:

(1) The interplanetary field is not involved.

(2) The polar cap is not directly affected.

(3) Auroral forms follow the pattern of Figure 3.4; in the midnight and early morning sectors short lived rayed forms come and go. Rather strong H emissions may be observed within or towards the poleward part of the emission region after midnight[29].

(4) The drift motions of I.(4) are resumed. Pulsating aurora may occur in the morning sector particularly on the equatorward edge of the oval[6].

(5) At 18 R_E the flux recovers with electron fluxes typically around 10^7 cm^{-2} s^{-1} sr^{-1} keV^{-1} at 1 to a few keV and proton fluxes about 10^5 cm^{-2} s^{-1} sr^{-1} keV^{-1} at 10 keV[9]. At 6.6 R_E, in the case described by Mende *et al.* the particle energy densities decayed in the 90 min following the break-up phase. The harder electrons (E>20 keV) dropped by a factor of 10 within 10 to 15 min[5].

(6) The tail field probably remains in the more dipolar configuration[10].

(7) There may be a widespread westward electrojet in the morning sector and an eastern electrojet growing in the afternoon and evening sector[6].

---

1. Hones (1969)
2. Heikkila and Winningham (1972)
3. Unti and Atkinson (1968)
4. Arnoldy (1971)
5. Mende *et al.* (1972)
6. Davis (1972)
7. Rothwell and Lynam (1969)
8. Vasyliunas (1968)
9. Hones *et al.* (1971a)
10. Fairfield and Ness (1970)
11. Aubry *et al.* (1970)
12. Burlaga and Ogilvie (1969)
13. Montbriand (1971)
14. Snyder and Akasofu (1972)
15. Coroniti and Kennel (1972b)
16. Rostoker (1972)
17. Mozer (1971)
18. Hones *et al.* (1971c)
19. Aubry and McPherron (1971)
20. Kokobun (1972
21. Kelley *et al.* (1971b)
22. Iijima and Nagata (1972)
23. Mozer and Manka (1971)
24. Burch (1972)
25. Hones *et al.* (1971b)
26. Aubry *et al.* (1972)
27. DeForest and McIlwain (1971)
28. Kelley *et al.* (1971a)
29. Fukunishi (1974)
30. McPherron *et al.* (1973)

3.8. Particle Fluxes in the Auroral Oval

Since bombardment of the atmosphere by 1 to 20 keV electrons and protons is probably responsible for the major features of the aurora, observations of such particles immediately above the emitting region are particularly important. Satellite measurements are beginning to give extensive information about the instantaneous particle energy spectra during passes across the day- and night side ovals. However, since satellites travel with a speed of the order of 8 km s^{-1} there is difficulty in obtaining information about individual elements of auroral forms which may have a thickness of no more than 200 m. In this respect rocket measurements are superior. Unfortunately, it is difficult to fire rockets over the most active and transient auroral forms and consequently progress toward a complete knowledge of the particles really responsible for different types of auroral forms has been slow.

A detailed review of observations of auroral particles has recently been published by Whalen and McDiarmid (1974). Some typical auroral particle energy spectra are shown in Figures 3.19 and 3.20. Hones *et al.* (1971a) have given figures summarizing

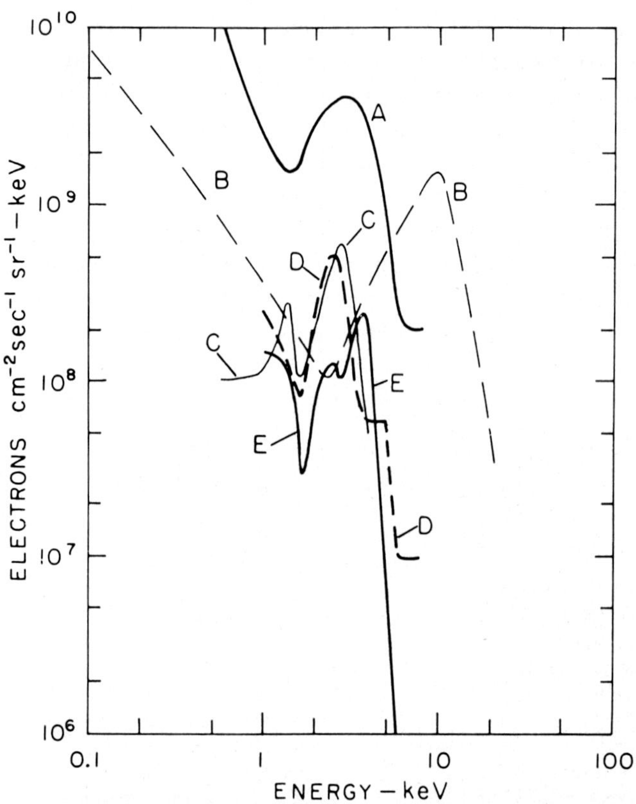

Fig. 3.19. Electron energy spectra observed in rocket flights over aurora. A, poleward boundary of folded band (field aligned) after Whalen and McDiarmid (1972); B, IBC 2–3 arc with isotropic angular distribution (> 1 keV) after Westerlund (1969); C, IBC 2 arc after Evans *et al.* (1972); D, active break-up aurora (isotropic, peak variable 1–4 keV) after Evans (1969); E, quiet pre-break-up aurora (isotropic, stable spectrum) after Evans (1968).

almost all the results to date. Many of the electron spectra obtained over auroral forms show a peak between 3 and 15 keV indicating that most of the energy of the incident primaries responsible for structured auroral forms was carried by particles in the peak. Sometimes the peak fluxes have been observed to be associated with an

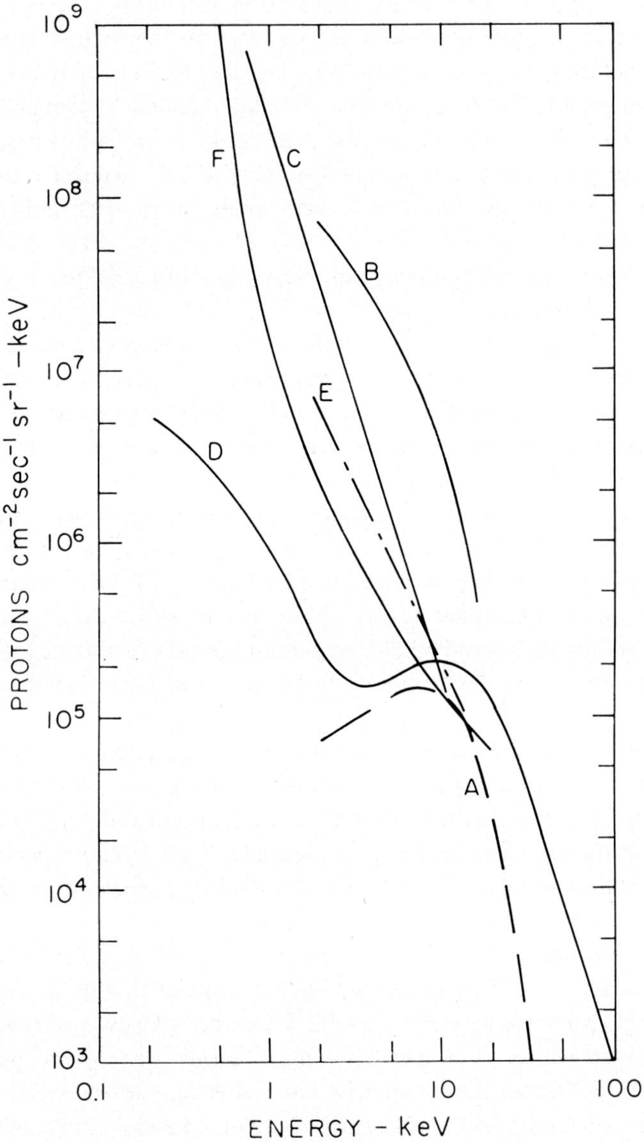

Fig. 3.20. Proton energy spectra observed in rocket flights over aurora. A, early evening proton aurora observed by Whalen *et al.* (1971a); B, IBC 1–2 quiet arc by Chase (1968); C, post break-up aurora by Bernstein *et al.* (1969); D, recovery phase aurora after Reasoner *et al.* (1968); E, recovery phase after Johnstone (1971); F, break-up aurora after Chase (1970).

isotropic pitch angle distribution and sometimes with a distribution largely parallel to the field. In some instances the higher energy (>25 keV) electrons have been found to have a 'trapped' pitch angle distribution, i.e., peaked at pitch angles near 90°.

The study of the variation in electron energy spectrum with substorm time has been the object of many rocket experiments. Whalen and McDiarmid (1969) concluded that there is a significant hardening of electrons of energy greater than 10 keV at break-up. They also concluded that the electron angular distributions at break-up and after varied between peaking at 90° to isotropic. In a later study (Whalen and McDiarmid, 1972), they found strong field aligned fluxes towards the poleward edge of an active folded band for electrons with energies less than 10 keV. Rocket measurements of proton fluxes have usually indicated a pitch angle distribution either isotropic or peaked at 90° (Whalen and McDiarmid, 1974). The proton energy spectra, as shown in Figure 3.20, are generally rather featureless; they often exhibit a broad peak in the 5 to 10 keV region.

In summary, the electron fluxes exciting aurora show considerable variations in energy and angular distributions. The typical spectrum of particles exciting discrete forms appears to show a peak between 2 and 15 keV superposed on a continuous spectrum rising towards lower energies. The latter may be responsible for the diffuse emission usually found between discrete auroral forms.

Satellite particle observations are yielding a great deal of coarser scale information on particle fluxes and spectra. Some measured electron spectra which appear to be connected with auroral emission are plotted in Figure 3.21. In comparison with the rocket spectra from individual forms, these spectra show only broad secondary maximum or shoulders beyond 5 keV. Apparently the sharp maxima associated with auroral forms are traversed too rapidly to be recorded by the satellite instruments. Ackerson and Frank (1972) have associated the emergence of the shoulder as in Figure 3.21 curve C with a visual auroral form. This spectrum is associated with an occurrence of a feature which they have called an 'inverted V'. This term describes the visual appearance of features which are seen when differential particle flux is shown as a color contour on a plot of energy vs latitude. Such a feature would indicate a significant rise to a maximum in the flux of 1 to 10 keV particles at a particular point followed by a drop as latitude varied. Frank and Ackerson (1972) have found in a survey of passes from quiet periods that the 'inverted V' fluxes lie poleward of the trapping boundary for 45 keV electrons. Equatorward of this boundary particularly around midnight they found electrons with a broader energy spectrum and a more diffuse spatial structure. A somewhat similar result was found by Deehr and Egeland (1972). This does not necessarily mean that the auroral electrons poleward of the trapped population of 45 keV electrons are on open field lines; indeed from the pitch angle distribution of electrons with $E > 25$ keV, McDiarmid and Hruska (1972) have shown that the tail field lines are closed to 10° higher latitudes than normal during the expansive phase of substorms.

The particle spectra above the midday sector of the oval are distinctively different from the night sector. The region of red type-d arcs and patches (discussed in Section 3.3.6b) is associated with soft proton ($0.3 < E < 3$ keV) fluxes and soft electron fluxes with $E < 1$ keV (Heikkila *et al.*, 1972). These are believed to be magnetosheath plasma penetrating into the magnetosphere in the vicinity of the dayside clefts. Equatorwards there is a wide band of electron fluxes with $E > 1$ keV especially on the

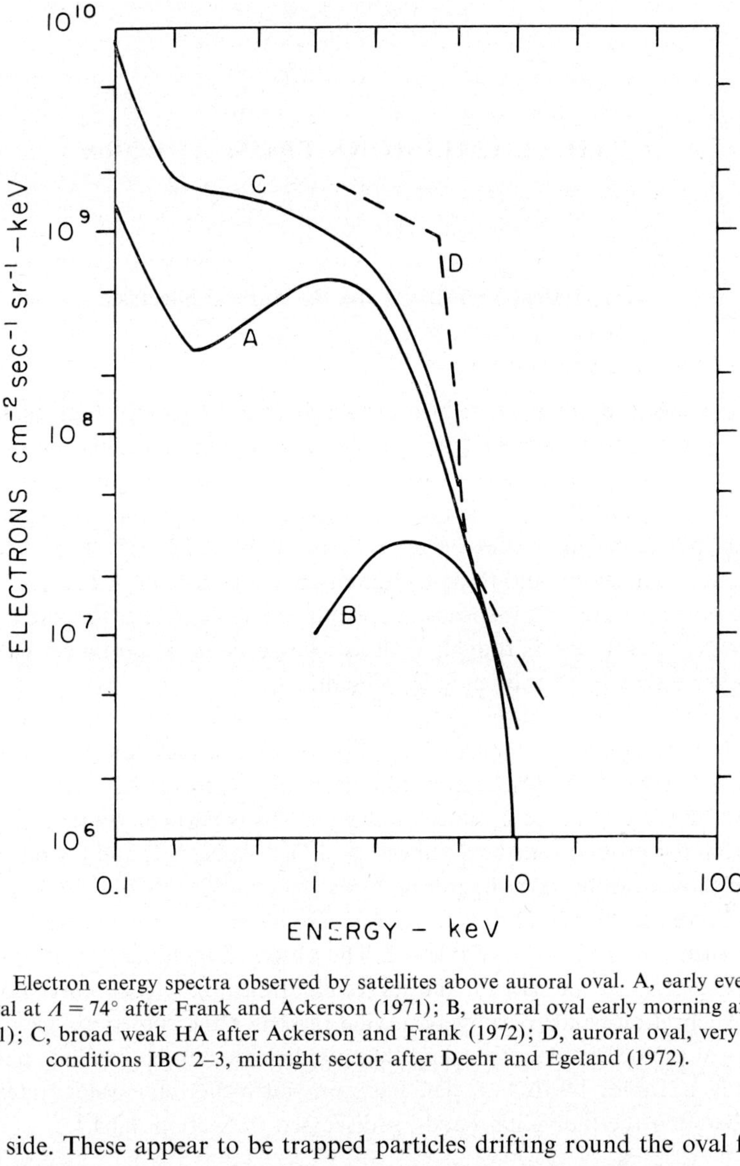

Fig. 3.21. Electron energy spectra observed by satellites above auroral oval. A, early evening over auroral oval at $\Lambda = 74°$ after Frank and Ackerson (1971); B, auroral oval early morning after Sharp *et al.* (1971); C, broad weak HA after Ackerson and Frank (1972); D, auroral oval, very disturbed conditions IBC 2–3, midnight sector after Deehr and Egeland (1972).

morning side. These appear to be trapped particles drifting round the oval from the nightside. McDiarmid *et al.* (1972) have studied the angular and energy spectra of these particles. Although the more energetic (equatorward) band is clearly on closed field lines, at this time it is not clear whether the magnetosheath-like particles are on open or closed lines.

CHAPTER 4

OPTICAL EMISSIONS FROM AURORA

4.1. Optical Transitions and the Auroral Spectrum

4.1.1. The observed spectrum

The auroral spectrum obtained at medium resolution is shown in Figures 4.1 to 4.7. The atomic lines and molecular bands of the spectrum are indicated on the figures. The relative intensities in the spectra are typical of medium-to-bright aurora (IBC 2–3) at heights near 110 to 120 km, the normal intensity maximum for such aurora. Some variations in relative intensities occur as a result of height effects or changes in atmospheric composition and these will be discussed in Section 4.2.5.

The observed features of the spectrum are almost all due to lines or bands of neutral or ionized N_2, O, O_2, and N roughly in descending order of importance. There is also some evidence that NO bands may be present.

Energy level diagrams* of the species involved are given in Figures 4.8 to 4.14 and the observed transitions are indicated. Because of the low collision rates and wall-less conditions at auroral heights, certain low probability ('forbidden') transitions from metastable states are able to produce radiation. This is particularly true of the states arising from the ground state configurations of O, O^+, N, N^+, and O_2 for which the levels are shown on the same figures.

The observed forbidden atomic transitions are listed in Table 4.1, while permitted atomic transitions are listed in Table 4.2. The observed molecular transitions are set out in Table 4.3. In the case of the molecular transitions, each corresponds to a system of vibration-bands occupying a wide region of the spectrum. The origins of the bands of the observed systems and their predicted or observed relative intensities are set out in Tables 4.9 to 4.18. The basis on which the total system intensity was divided between the individual bands is discussed in Section 4.2.4.1.

4.1.2. Excitation and ionization cross sections

Excitation Cross Sections

Most auroral emission is due to the radiation of excited states produced directly or indirectly as a result of collisions between atmospheric particles and primary auroral electrons, protons or secondary electrons. Consideration of proton and H impact cross sections will be left to Section 4.3. In recent years laboratory measurements have

* The molecular energy levels are fully designated in these diagrams but in subsequent references +, − and g, u symmetry indices are often omitted.

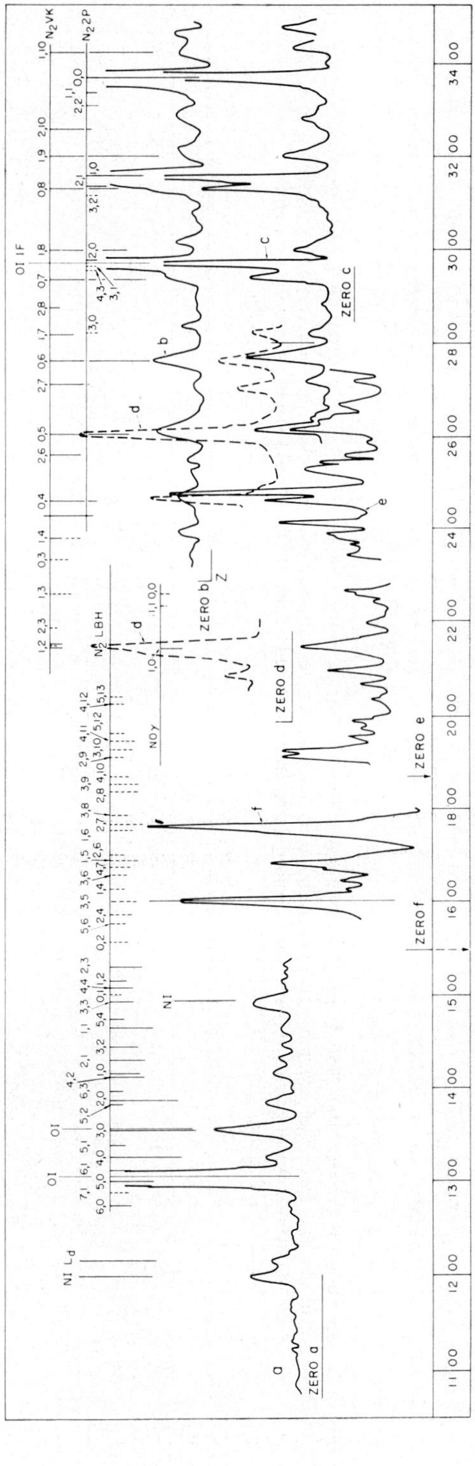

Fig. 4.1. Spectrum of aurora 1100–3400 Å: (a) Miller *et al.* (1968); (b) Fastie *et al.* (1961); (c) Sharp (1971); (d) Duysinx and Monfils (1972); and (e, f) Sharp and Rees (1972).

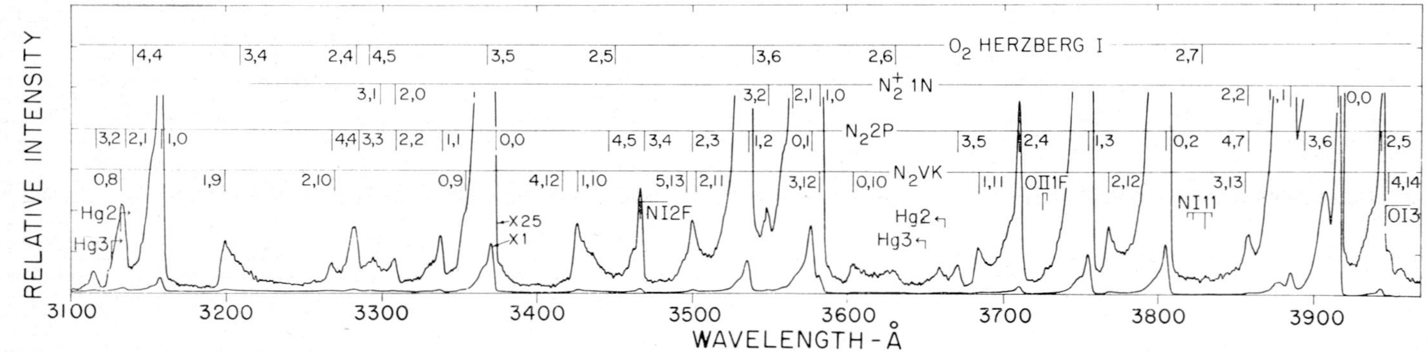

Fig. 4.2. Spectrum of aurora 3100–3950 Å and 3 Å resolution. The two curves are at the gains indicated. Bands and atomic features present or possibly present are indicated. The ordinates are relative intensity uncorrected for atmospheric transmission. I am very much indebted to Dr R. L. Gattinger for preparing this spectrum from data recently obtained at the NRC Auroral Observatory, Ft. Churchill.

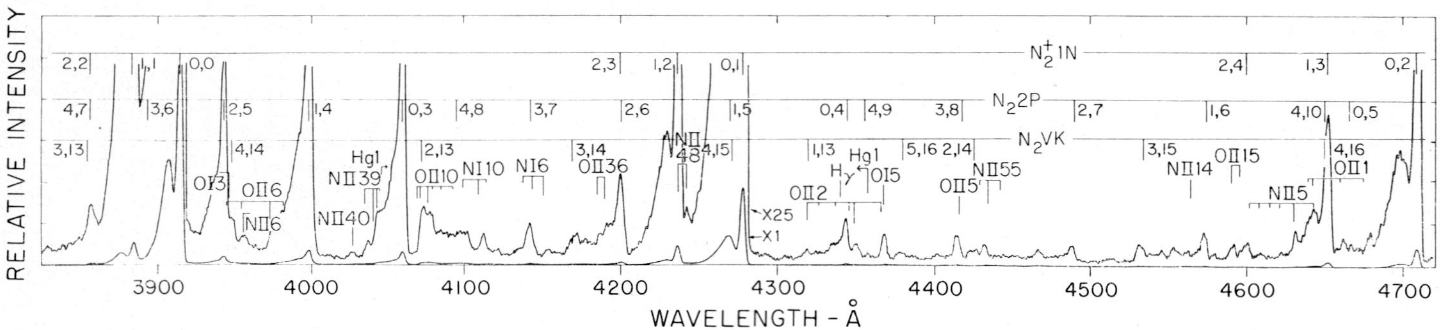

Fig. 4.3. Spectrum of aurora 3850–4700 Å as for Figure 4.2. I am very much indebted to Dr R. L. Gattinger for preparing this spectrum from data recently obtained at the NRC Auroral Observatory, Ft. Churchill.

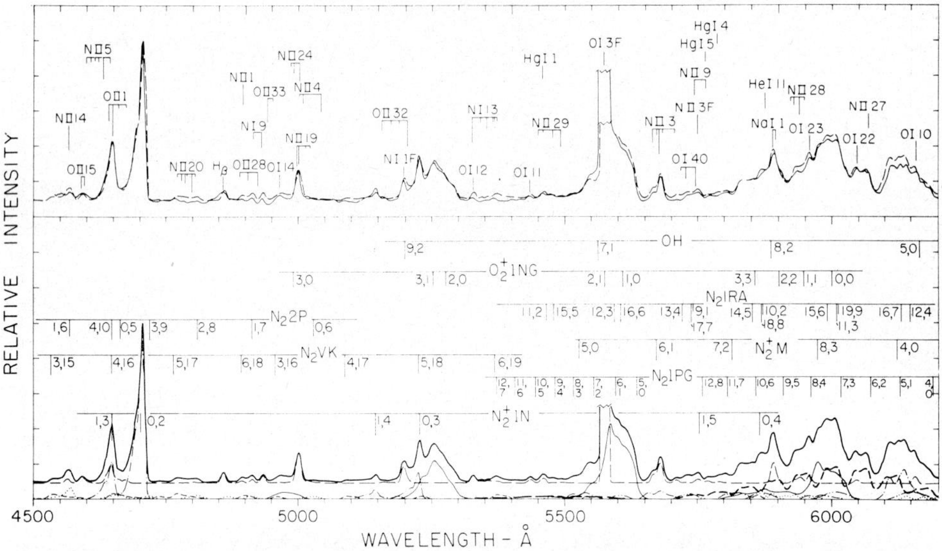

Fig. 4.4. Spectrum of aurora 4500 Å to 6200 Å at 10 Å resolution. Upper panel: solid line is the observed averaged spectrum of aurora with *I*(5577) in the range 20–60 kR; dashed line is total synthetic spectrum. Lower panel: Heavy solid line is the total synthetic spectrum; N_2 1P – heavy long dashes; N_2^+ M and 1N – alternating dot and dash; OH-dots; O_2^+ 1N – light solid line; atomic lines plus continuum-light long dashes; O_2 At-short dashes; N_2 IRA – double dot and single dash; N_2 2P – dots; N_2 VK short dashes. From Gattinger and Vallance Jones (1974). (Courtesy *Canadian Journal of Physics*; reproduced by permission of the National Research Council of Canada.)

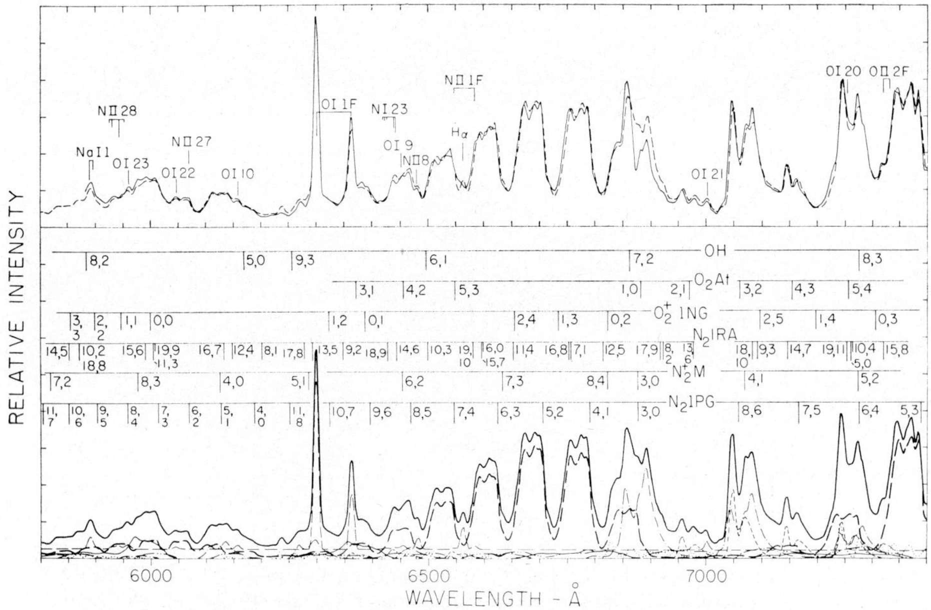

Fig. 4.5. Spectrum of aurora 5800–7400 Å; (continuation of Figure 4.4).

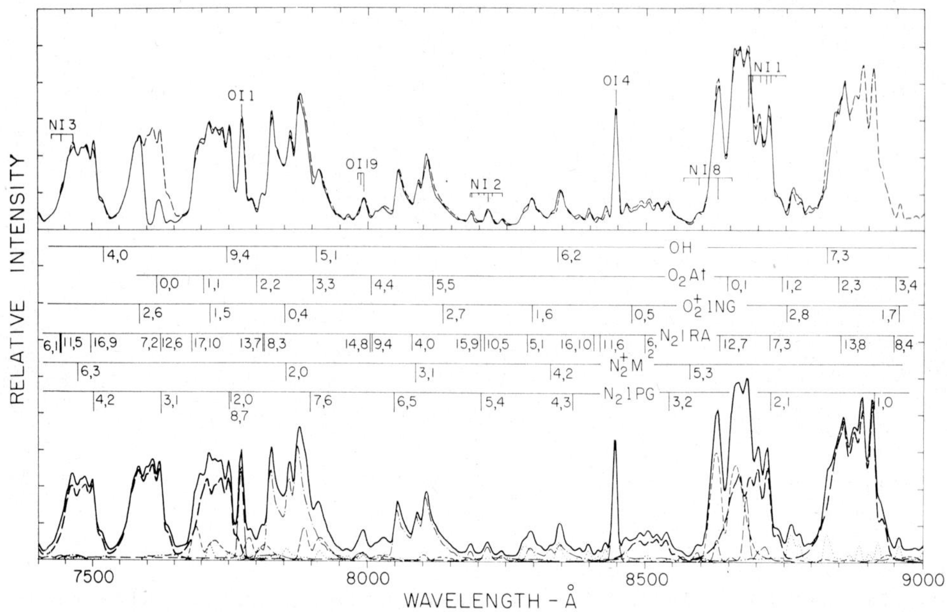

Fig. 4.6. Spectrum of aurora 7400–9000 Å; (continuation of Figure 4.4).

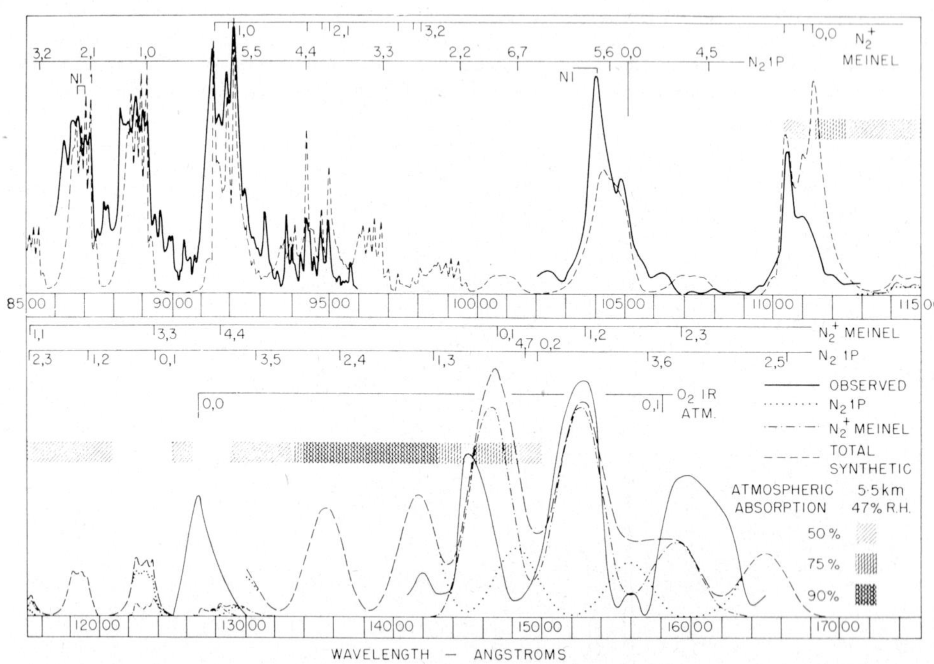

Fig. 4.7. Spectrum of aurora 8500–16300 Å. Results to 9500 Å from Fedorova (1963), 10300–11300 Å from Harrison (1969) and 14300 to 16300 from data of Harrison and Vallance Jones (1959). The synthetic spectrum was calculated by Vallance Jones and Gattinger (1971) from whose paper the figure is reproduced. The relative intensity of the Meinel system should be decreased by about 35%. (Courtesy D. Reidel Publishing Co.)

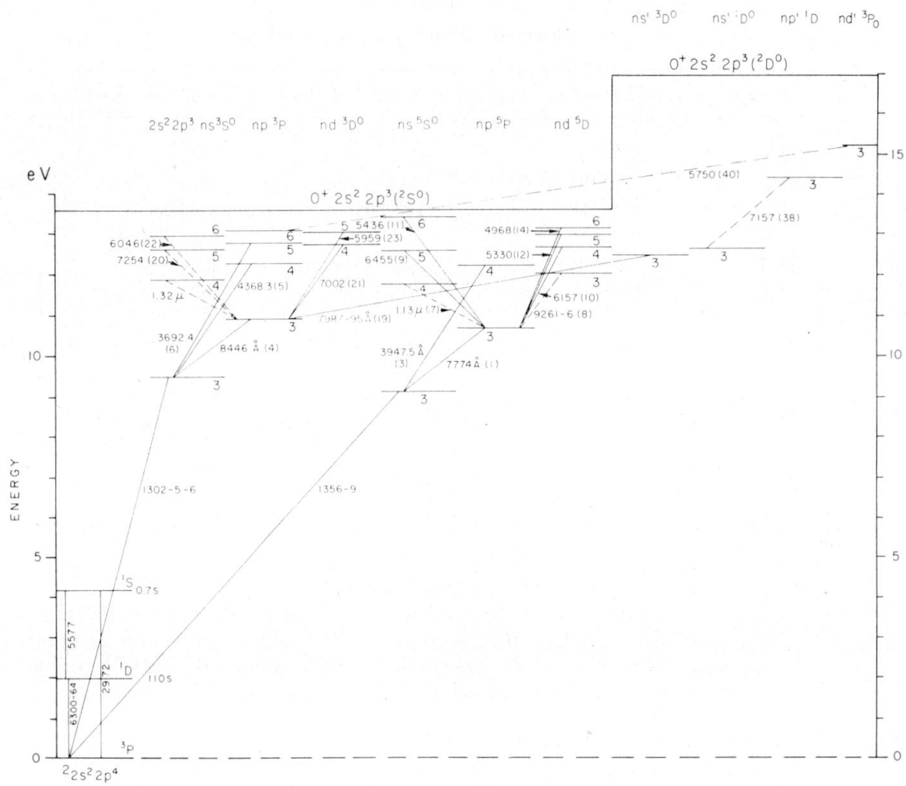

Fig. 4.8. Energy levels of O. Solid transition lines correspond to lines observed in aurora; the dashed lines indicate doubtful or predicted emissions. The numbers in brackets are multiplet numbers from Moore (1945).

provided direct data for many of the important cross sections. Some of the results are summarized in Figure 4.15 and in Table 4.4 in which the references to the curves for Figure 4.15 are given. The excitation cross sections for electron impact are to some extent related to optical transition probabilities and it may be possible to estimate missing collision cross sections from optical data or from semi-empirical family relationships. Quantum theoretical data alone are available for the metastable levels of O.

According to the Born approximation, electron impact excitation cross sections behave asymptotically at high energies as follows (Henry *et al.*, 1969).

(i) Optically allowed transitions vary as $E^{-1} \ln E$ (maximum at 4 to $5\times$ threshold energy).

(ii) Optically forbidden transitions, $\Delta S \neq 0$, vary as E^{-3} (maximum close to threshold).

(iii) Optically forbidden transitions $\Delta S = 0$, fall off as E^{-1}. (ΔS is the change in spin quantum number).

TABLE 4.1

Observed forbidden atomic multiplets

Emitter	Transition	Wavelength (Å)	Intensity (kR)
[O I]	$^1D-^1S$	5577	(100)[e]
[O I]	$^3P-^1S$	2972	(6)[f]
[O I]	$^3P-^1D$	6300–6364	2–100[g]
[O II]	$^2D-^2P$	7319–7330	0.4[h]–100[d]
[O II]	$^4S-^2D$	3727–3729	~1[i]
[N I]	$^2D-^2P$	10395–10404	6[a]
[N I]	$^4S-^2P$	3466	1[b]
[N I]	$^4S-^2D$	5199–5201	0.1–2[c]
[N II]	$^1D-^1S$	5755	≤0.04[g,h]
[N II]	$^3P-^1D$	6584	≤0.2[g,h]

[a] Harrison (1969, 1972).
[b] Broadfoot and Hunten (1964) observed $I(3466) \approx 1/24 \times I(2P; 0, 0)$.
[c] Covaries with [O I] 6300 Å $I(5200) \sim 1/30\ I(6300)$ (Gérard and Harang, 1973).
[d] Normally undetectable; the 100 kR estimate is from a comparison of a spectrum of the red aurora of February 10–11, 1957 (Vallance Jones, 1960) with the 1P N_2 5,3 band. In this spectrum the [O II] lines are at least 3 times the intensity of the 5,3 band.
[e] Reference value.
[f] Theoretical from transition probability.
[g] Very height dependent.
[h] Estimate of Gattinger and Vallance Jones (1974) for normal aurora.
[i] Normally undetectable. Rough estimate from Wallace (1959), type-d red aurora spectrum.

TABLE 4.2

Allowed multiplet transitions[b] with intensities greater than 0.1%[c] of λ5577

Emitter	Multiplet number	Wavelength	Observal intensity for $I(5577) = 100$
O I	1	7774	9.6
	4	8446	11.5
	5	4368	0.5[g]
	9	6454	0.14
	10	6158	0.14
	19	7995	1.2
	21	7002	0.3
	23	5959	0.13
	$2^3P-3^3S^0$	1302–6	$\sim 30 \pm 20$[d,e]
	$2^3P-3^5S^0$	1356–9	~2[d,e]

Table 4.2 (Continued)

Emitter	Multiplet number	Wavelength	Observed intensity for $I(5577) = 100$
O I	1	4649	0.6
	3	3749	$\sim 1^{f}$
	5	4415	0.5^{f}
	6	3973	0.3^{f}
	15	4591	0.13
N I	1	8680	10.5
	2	8216	3.3
	3	7468	0.55
	8	8629	3.84
	9	4935	0.19
	13	5329	0.22
	23	6442	0.11
	2^4S-3^4P	1200–1	$\sim 2^{e}$
	2^2D-3^2P	1493–5	$\sim 1^{e}$
N II	3	5680	0.65
	8	6482	0.4
	12	3995	$\sim 1^{f}$
N II	19	5001	0.6
	28	5942	0.2
	39	4044	0.5^{f}
	47–48	4242	0.5^{f}

[a] The observed intensities above 4500 Å are from Gattinger and Vallance Jones (1974).
[b] Excitation potentials and transitions can be found from Figures 4.8 to 4.11 which are labelled with wavelengths and transition numbers.
[c] The intensity criterion is only qualitative for wavelengths below 4500 Å; only transitions prominent on photographic spectra are included.
[d] Donahue and Strickland (1970).
[e] Miller *et al.* (1968).
[f] Order of magnitude from comparison with nearby bands on photographic spectra.
[g] Dick (1970b).

Ionization Cross Sections

Ionization processes are similar to those for excitation except that a continuum of levels is possible above the ionization energy, I, since the secondary electron may have a variable energy E_S. Thus the cross section for production of secondaries of energy E_S to $E_S + dE_S$ may be expressed as $\sigma_i(E, E_S)\, dE_S$ so that the total ionization cross section could be written

$$\sigma_i(E) = \int_0^\infty \sigma_i(E, E_S)\, dE_S . \tag{4.1a}$$

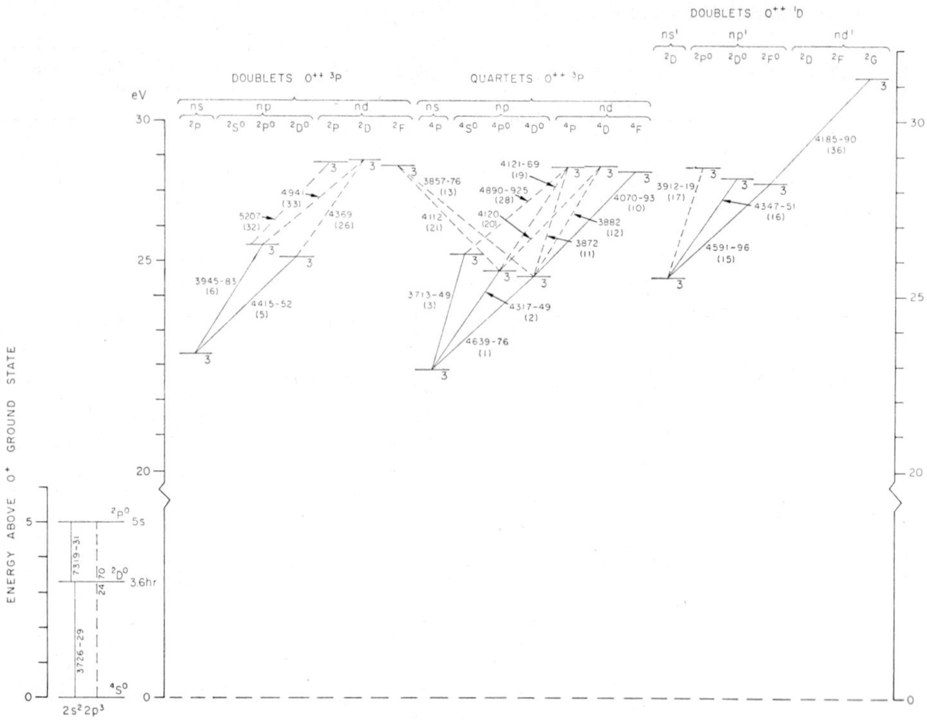

Fig. 4.9. Energy levels and transitions of O^+ indicated as for Figure 4.8.

TABLE 4.3
Molecular transitions in aurora and observed system intensities for $I(5577) = 100$ kR

System	Intensity kR[b]
N_2 1P	880
N_2 2P	110
N_2 LBH	200–400
N_2 VK	55
N_2^+ 1N	150
N_2^+ M	630
O_2^+ 1N	26
O_2 At	1300
O_2 IR At	2500[c]
NOγ	20–60
N_2 BH, WK	~2000[a]

[a] This is a rough value deduced with some assumptions from the photometer data of Paresce *et al.* (1972).

[b] In general the intensities are rounded values from Tables 4.9 to 4.18.

[c] Steady state value derived from Gattinger and Vallance Jones (1973c) analysis of observed time variation.

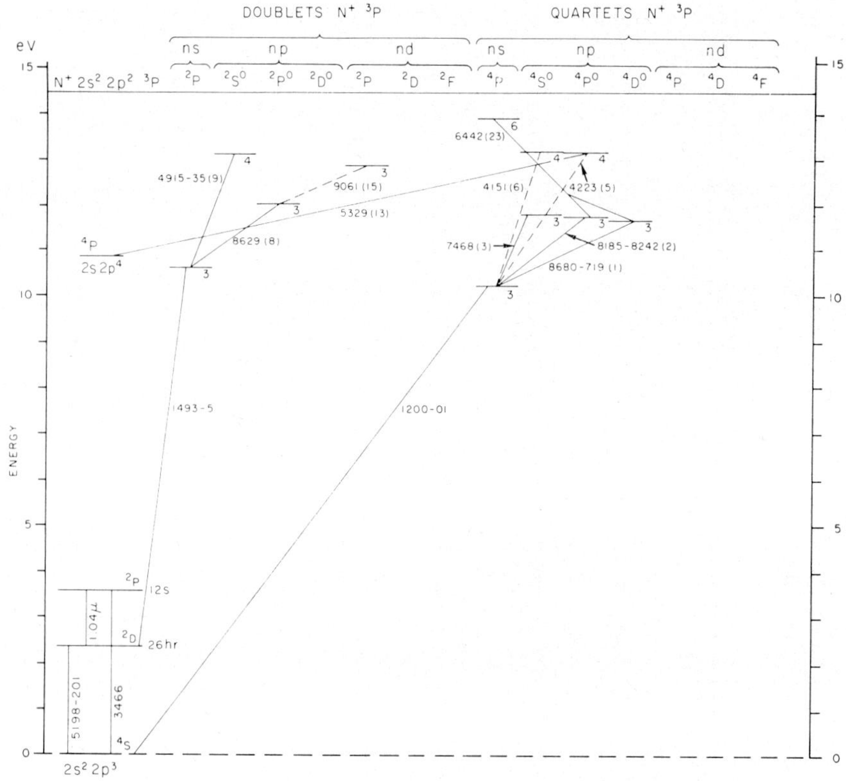

Fig. 4.10. Energy levels and transitions of N indicated as for Figure 4.8.

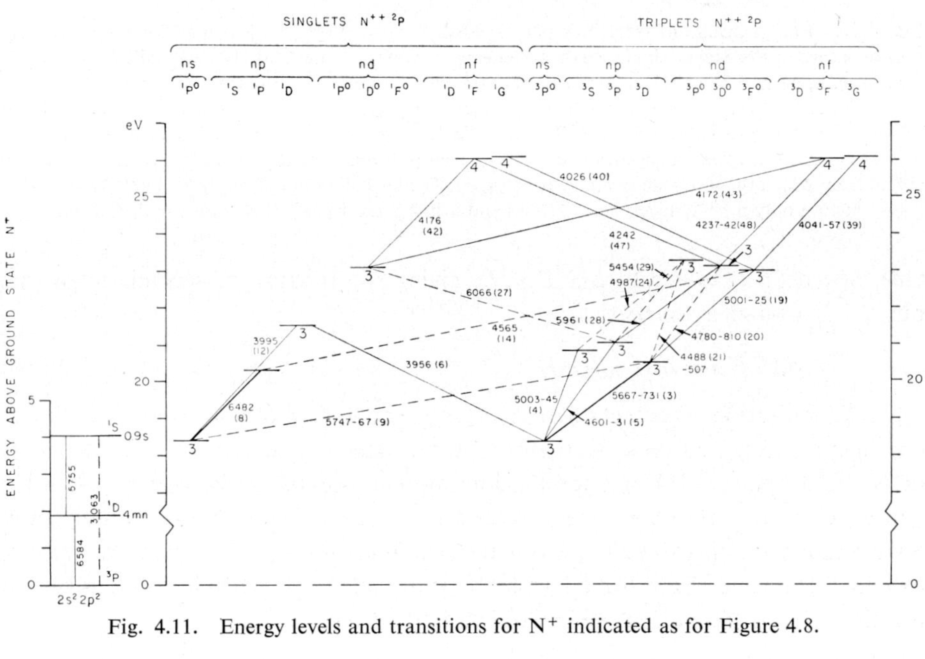

Fig. 4.11. Energy levels and transitions for N^+ indicated as for Figure 4.8.

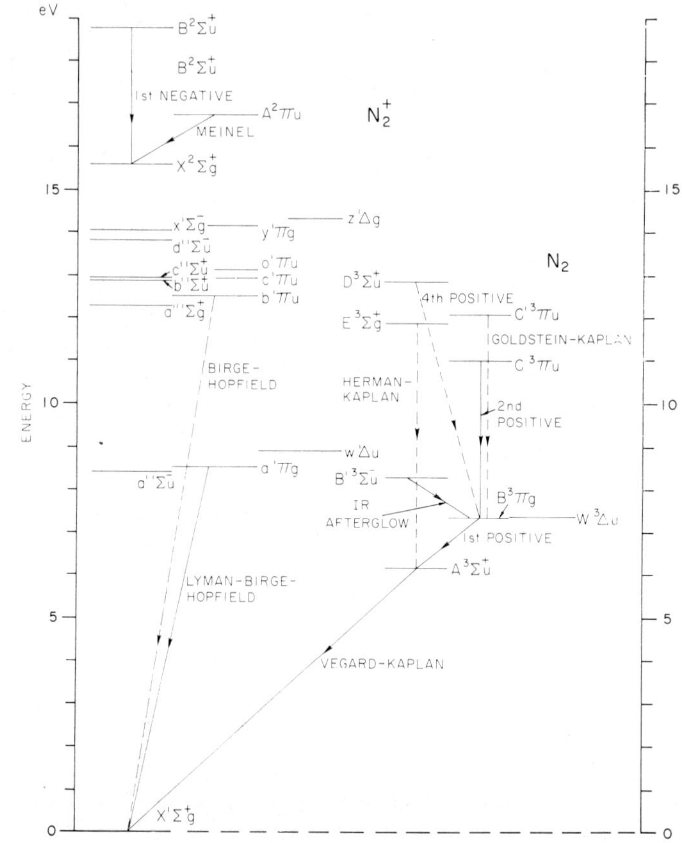

Fig. 4.12. Energy levels and electronic band systems for N_2 and N_2^+ indicated as for Figure 4.8. The nomenclature of the higher singlet band systems is confusing. Some authors have called the systems from o, c, b, c′, and b′ levels to X the Birge-Hopfield system, and from these levels to a, the Janın system. Transitions to the groundstate from Rydberg series electronic levels lying in the neighborhood of 14 eV have been called the Watson-Koontz, Worley, Carroll-Yashino, Worley-Jenkins and Hopfield systems. The latest classification of the N_2 energy levels is the compilation edited by Rosen (1970) from which the levels shown have been taken. The level diagram given by Stolarski *et al.* (1967) having only a $b^1\Pi_u$ level near 12.7 eV and a $h^1\Sigma_u^+$ level near 14 eV is a simplification.

One may define a quantity $S(E,E_S)\,dE_S$ giving the fraction of secondaries in the range dE_S around E_S so that

$$\sigma_i(E,E_S) = \sigma_i(E)\,S(E,E_S). \tag{4.1b}$$

The total ionization cross section $\sigma_i(E)$, for the production of ions from N_2 is shown on Figure 4.15b. The cross sections for the production of particular excited ions of N_2^+ (see Figure 4.12) and for the ionization processes of O_2 (see Figure 4.13) appear to be proportional to one another except near their thresholds. Consequently it is a reasonable approximation, within the experimental error of the measured values, to adopt the same shape for all with the peak values as given in Table 4.4. The functions $S(E,E_S)$ are discussed in Section 4.2.1 below.

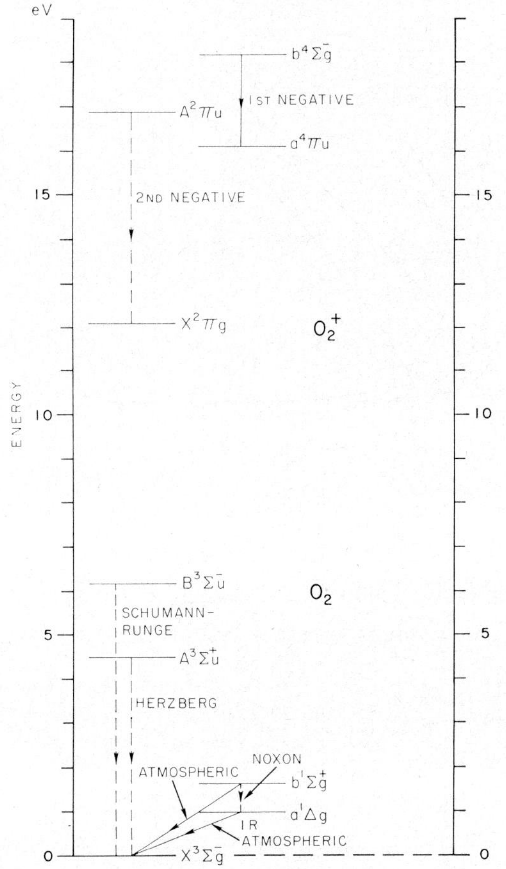

Fig. 4.13. Energy levels and electronic band systems detected in aurora for O_2 and O_2^+ depicted as for Figure 4.8.

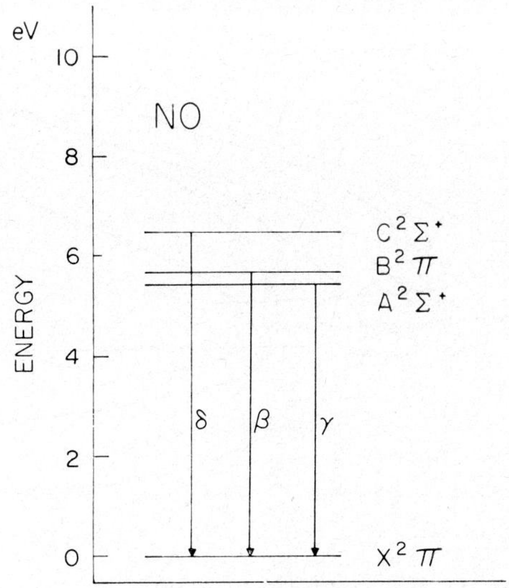

Fig. 4.14. Energy levels and electronic band systems of NO. The γ system may be present in aurora.

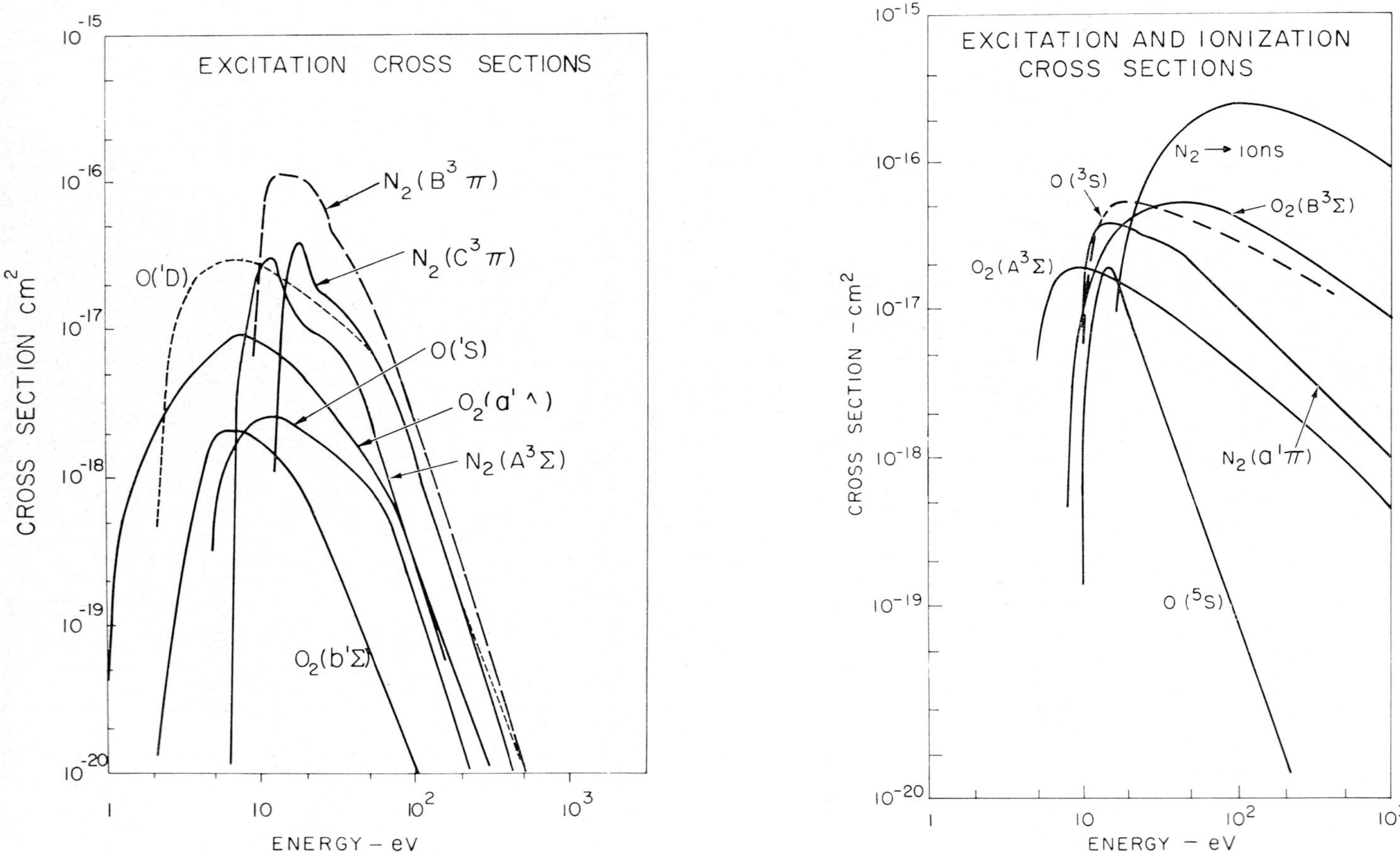

Fig. 4.15. Excitation and ionization cross sections for electron impact for some states of importance in aurora. References are given in Table 4.4.

TABLE 4.4
Data on cross sections important in aurora

Excited state or emission	Initial[a] state	σ_{max}	E_{max} (eV)	Reference
$O(^1S)$		0.25(−17)	10	Henry *et al.* (1969)
$O(^1D)$		0.28(−16)	5.6	Henry *et al.* (1969)
$O(3^3S)$	$O(^3P)$	0.54(−16)	19	Stone and Zipf (1974)
	$O_2(X^3\Sigma)$	0.34(−17)	83	Mumma and Zipf (1971a)
$O(3^5S)$	$O(^3P)$	0.2 (−16)	15	Stone and Zipf (1973a)
	$O_2(X^3\Sigma)$			Wells *et al.* (1971)
$N_2(a^1\Pi)$		0.4 (−16)	15	Ajello (1970) and Brinkmann and Trajmar (1970)[d]
$N_2(A^3\Sigma)$		0.28(−16)	10	Rees and Jones (1973)[b]
$N_2(B^3\Pi)$		0.11(−15)	12[c]	Shemansky and Broadfoot (1971a)
$N_2(C^3\Pi)$		0.38(−16)	16	Shemansky and Broadfoot (1971a)
$N_2(D^3\Sigma)$		0.4 (−17)	14.5	Cartwright *et al.* (1972a)
$N_2(E^3\Sigma)$		0.4 (−17)	14	Cartwright *et al.* (1972a)
$N_2(W^3\Delta_u)$		0.20(−16)	[e]	Chutjian *et al.* (1973)
$N_2(w^1\Delta_u)$		0.25(−17)	[e]	Chutjian *et al.* (1973)
$N_2(B'^3\Sigma_u^-)$		0.33(−17)	[e]	Chutjian *et al.* (1973)
$N_2(a'^1\Sigma_u)$		0.34(−17)	[e]	Chutjian *et al.* (1973)
$O_2(a^1\Delta)$		0.85(−17)	6.5	Trajmar *et al.* (1971)
$O_2(b^1\Sigma)$		0.2 (−17)	6	Trajmar *et al.* (1971)
$O_2(A^3\Sigma)$		0.19(−16)	10	Watson *et al.* (1967)
$O_2(B^3\Sigma)$		0.55(−16)	40	Watson *et al.* (1967)
$N(3^4P)$	$N(^4S)$	0.25(−15)	30	Stone and Zipf (1973)
	$N_2(X^1\Sigma)$	0.7 (−17)	90	Mumma and Zipf (1971b)
N_2^+ 1N(0, 0)	$N_2(X^1\Sigma)$	0.17(−16)	100	Borst and Zipf (1970a)
ions	$N_2(X^1\Sigma)$	0.24(−15)	100	Borst and Zipf (1970a)
O_2^+ 1N(1, 0)	O_2	0.43(−17)	100	Borst and Zipf (1970b)
ions	O_2	0.27(−15)	100	Borst and Zipf (1970b)
$N_2^+(A^2\Pi)$	$N_2(X^1\Sigma)$	0.94(−16)	100	Cartwright *et al.* (1974)
		≥1.2 (−16)	100	Shemansky and Broadfoot (1971a)

[a] If not ground state of same species.
[b] This value was obtained privately from a paper to be published by Cartwright, Chutjian, Trajmar and Williams. There are also recent data by Borst (1972).
[c] There are also data quoted by Rees and Jones (1973)[b].
[d] See also Borst (1972).
[e] These cross sections have so far been determined only at 20 eV.

Partial Cross Sections for Vibrational Levels

The excitation cross sections make it possible to calculate the total production rate of diatomic molecules in a particular excited electronic state if the exciting electron flux is known. Thus $\sigma(M,S,E)$ denotes the cross section for exciting molecule M to state S by electrons of energy E. However the state S, if stable, has vibrational levels indexed by the quantum number v' while the ground state, usually denoted X, has vibration levels with quantum number v'''. Consequently the cross section from level v''' of the ground state to v' in the upper state could be denoted $\sigma(M,S,v',v''',E)$.

The basic wave mechanical equations for collisional excitation (Nicholls, 1962) indicate that the probability of excitation by electron impact between v''' and v' should to a good approximation be proportional to the Franck-Condon factor,

$$q(v',v''') = \left| \int \psi_v^*(v',S)\psi_v(v''',\mathrm{X})\,\mathrm{d}r \right|^2, \tag{4.1c}$$

where $\psi_v(v',S)$ is the vibrational wavefunction of state S for quantum number v' and r is the internuclear distance. As shown by Nicholls a better approximation is that the excitation probability is proportional to $[G(\bar{r}_{v',v'''})]^2 q(v',v''')$ where G is a perturbation integral analogous to the electronic transition moment $R_e(\bar{r}_{v'v''})$ defined in Section 4.1.3 for optical transitions. In principle, G could be obtained by comparing laboratory values of $\sigma(v',v''')$ with theoretical values of $q(v',v''')$. As yet no extensive measurements of G's appear to be available and as noted by Shemansky and Broadfoot (1971a) there is evidence that the transition probabilities are indeed closely proportional to $q(v',v'')$. Since the overlap integrals have the property that

$$\sum_{v'} q(v',v''') = 1 \tag{4.1d}$$

we may write that

$$\sigma(\mathrm{M},S,v',v''',E) = \sigma(\mathrm{M},S,E)\, q(v',v''') \tag{4.1e}$$

and find that the sum of the partial cross sections from a given ground state level v''' is equal to the total excitation cross section.

4.1.3. Optical transition probabilities

The atoms and molecules which enter the various excited states are eventually de-excited by emission of radiation or by losing the energy of excitation in collisions. The probability of radiation is determined by the Einstein transition probability A_{nm}. In general this probability is given by the expression* (e.g., Eyring *et al.*, 1944)

$$A_{\mathrm{nm}} = \frac{32\pi^3\bar{v}^3_{\mathrm{mn}}}{3\hbar}\Big\{|\langle m|\,\mathbf{M}_e\,|n\rangle|^2 + \\ + \Big|\langle m|\sum_{i}^{N}\frac{e}{2mc}\,\mathbf{r}_i\times\mathbf{p}_i\,|n\rangle\Big|^2 + \tfrac{3}{10}\pi^3\bar{v}^2\Big|\langle m|\sum_{i}^{N}e\mathbf{r}_i\mathbf{r}_i\,|n\rangle\Big|^2\Big\}, \tag{4.1f}$$

where the summations are over all N electrons of the atom or molecule. $\mathbf{M}_e = \sum_i^N e\mathbf{r}_i$ is the dipole moment of the electrons. $\mathbf{p}_i$ is the momentum operator of the ith electron. $\langle m|\mathbf{M}_e|n\rangle$ is called the electronic transition moment and is often denoted by $\mathbf{R}_e(n,m)$. Sometimes $|\mathbf{R}_e(n,m)|^2$ is denoted by S_{nm}, the line strength. The first term of Equation (4.1f) represents the electric dipole contribution to the transition probability and is normally dominant with a value of the order of $10^8\ \mathrm{s}^{-1}$. The second term

* The wavenumber for the transition from state n to m is here represented by $\bar{v}_{\mathrm{nm}}$ since σ (cf. Chapter 2) here denotes cross section.

represents magnetic dipole radiation and the third electric quadrupole; these terms are of the order of 10^{-5} and 10^{-8} the value of the electric dipole term. The two latter terms can become important if the electric dipole term is zero as a consequence of the symmetry properties of the wave functions. Such low probability transitions are examples of 'forbidden' transitions and are particularly important in the aurora as noted in Section 4.1.1 above. States connected to lower energy levels only by forbidden transitions are *metastable* and may have radiative lifetimes of many seconds or even hours. One important class of metastable states are the terms of O, O^+, N and N^+ belonging to the ground-state electron configuration. These are strictly forbidden by symmetry for electric dipole radiation but can occur weakly through electric quadrupole or magnetic dipole radiation. Their lifetimes are shown on Figures 4.8 to 4.11. Another important class of 'forbidden' transition is forbidden to the approximation that electron spin and orbital angular momentum do not interact. In this approximation transitions involving a change in spin quantum number do not occur. Since some interaction does occur, the electric dipole transition moment for such transitions can depart from zero.

The principal energy levels, their radiative lifetimes and modes of decay may be found in Tables 4.1 to 4.3 and 4.7 and in Figures 4.8 to 4.14.

The intensities of emission features (in photon units) is given by the relation,

$$I_{\mathrm{nm}} = N(n)A_{\mathrm{nm}}\,, \tag{4.1g}$$

where $N(n)$ the population in the upper state depends on the balance between the various production and loss rates for the state. When the upper and lower states are degenerate, i.e. consist of g_{n} and g_{m} states of equal or approximately equal energy, the individual transitions between these levels are conveniently taken together to give an effective A-value for electric dipole radiation of the form

$$A_{\mathrm{nm}} = \frac{32\pi^3\bar{\nu}^3_{\mathrm{nm}}}{3\hbar} S_{\mathrm{nm}}/g_{\mathrm{n}}\,, \tag{4.1h}$$

where

$$S_{\mathrm{nm}} = \sum_{\mathrm{i,k}} |R_{\mathrm{e}}(n_{\mathrm{i}}, m_{\mathrm{k}})|^2 \tag{4.1i}$$

and the summation is over all the intercombinations between the sub-levels. Since a given atom (or molecule) can only be in one excited substate the sum must be divided by the statistical weight of the upper level g_{n}. S_{nm} is known as the line strength.

Atomic Lines and Multiplets

For atomic multiplets the A-values for the different possible components are related to one another through the expression

$$S_{\mathrm{nm}} = \text{const.}\ \mathscr{S}(\mathscr{M})\ \mathscr{S}(\mathscr{S}), \tag{4.1j}$$

where the constant depends on an integral involving the radial wavefunctions of the active electron. Values of $\mathscr{S}(\mathscr{M})$ which give the relative strength of the multiplets belonging to a particular configuration array have been tabulated by Goldberg (1935, 1936) while values of $\mathscr{S}(\mathscr{S})$ which give the relative strengths of the components of individual multiplets have been tabulated by Russell (1936) and also by Allen (1963). The values of the constant in Equation (4.1j) can be determined from data for key transitions of many arrays given in Allen. This treatment is valid only for L-S coupling which is however a good approximation for the multiplets of O and N encountered in auroral emissions.

Electronic Band Systems

The intensities of individual bands are determined by the equation

$$I(v',v'') = N(v')\,A(v',v''), \tag{4.1k}$$

where $N(v')$ is the number of molecules in the v'th level of the upper state and $A(v',v'')$ is the Einstein transition probability for the transition from this level to the v''th level of the lower state. The value of $N(v')$ is determined, in the steady state, by balance between the production and loss mechanisms for the excited level (e.g., the equations given in Section 4.2.3).

The transition probabilities may be shown to be proportional to $q(v',v'')$ (see above) according to the relation

$$A(v',v'') = \text{const.}\ \bar{\nu}^3(v',v'')R_e^2(v',v'')\,q(v',v''). \tag{4.1l}$$

R_e, the electronic transition moment is given (for electric dipole radiation) by

$$R_e(v',v'') = \frac{\int\psi_v(v')\left[\int\psi_e'\mathrm{M}_e\psi_e''\,\mathrm{d}\tau_e\right]\psi_v(v'')\,\mathrm{d}r}{\int\psi_v(v')\psi_v(v'')\,\mathrm{d}r}, \tag{4.1m}$$

where ψ_e, the electronic wavefunction, is a function of the position and spin coordinates of all the electrons within the molecule and is in addition, a slowly varying function of the internuclear distance, r. ψ_v, the vibrational wavefunction is a function of r alone. $\mathrm{d}\tau_e$ is the volume element for the position and spin coordinates of the electrons. Fraser (1954) showed that normally $R_e(v',v'')$ can be expressed as a function of $\bar{r}(v',v'')$, the r-centroid, the average value of r for a given v',v'' transition given by

$$\bar{r}(v',v'') = \frac{\int\psi_v(v')r\psi_v(v'')\,\mathrm{d}r}{\int\psi_v(v')\psi_v(v'')\,\mathrm{d}r}. \tag{4.1n}$$

Consequently the transition probability can be reduced to the form

$$A(v',v'') = \text{const.}\ \bar{\nu}^3(v',v'')\mathrm{R}_e^2(\bar{r}(v',v''))\,q(v',v''). \tag{4.1o}$$

Values of $q(v',v'')$ and $\bar{r}(v',v'')$ have been calculated for most states of molecules active in aurora (see e.g., Nicholls, 1964a; Benesch *et al.*, 1966a, b). By comparison with laboratory measurements of relative band progression intensities it is possible to determine relative values of the function $R_e(\bar{r})$ so that the relative values of $A(v',v''$

can be obtained. Finally measurements of the lifetime of the molecule in an excited state can be used to place the A-values on an absolute basis since

$$\tau(v') = 1/\sum_{v''} (A(v',v'')). \tag{4.1p}$$

(Lifetime measurements, as a function of v', can be interpreted to give values of $R_e(\bar{r})$.) Alternatively, when the lower state is the ground state absolute values of $A(v',0)$ may be derived from absorption measurements. Sometimes the data are quoted in terms of band strengths

$$p(v',v'') = R_e^2(\bar{r})\, q(v',v''). \tag{4.1q}$$

In a few cases for which $R_e(\bar{r})$ values have not been determined, a reasonable approximation is to set $R_e(\bar{r})$ constant.

Rotational Fine Structure of Electronic Bands

Each vibrational band has fine structure arising from the quantized rotational energy of the molecule. A full discussion of this rotational structure of bands can be found in Herzberg (1950). Some further discussion of this topic is to found in Section 4.2.5.5 where the effect of atmospheric temperature on rotational temperature is discussed.

4.1.4. Quenching and Energy Transfer Processes

Metastable atoms and molecules may also lose their energy in collision processes. The probability of such a transfer taking place in a given collision varies from close to unity to as small as 10^{-10}. In some cases the deactivating particle may be itself excited in the process. These processes will be discussed in Section 4.2.3.

4.2. Electron Aurora

4.2.1. Interaction of an Energetic Electron Beam with the Atmosphere

A satisfactory theory of auroral emission may be divided into two largely separate parts. First, there are the mechanisms responsible for producing a downward beam of energetic electrons at the 'top' of the atmosphere (that is at a height of 300 km or more). This is discussed in Chapter 6. Secondly there is the description of the processes occurring in the penetration of such a beam into the atmosphere.

A complete solution to the problem of calculating, as a function of height, the electron velocity distributions which arise from a given incident flux, is extremely difficult. This is due partly to the complexity of the interactions and partly to lack of exact knowledge of many of the cross sections. Several different procedures have been proposed; in general each has value in particular aspects of the problem. The main methods are:

(i) the analytical method of Spencer,

(ii) the method of Stolarski and Green,
(iii) the semi-empirical method of Rees,
(iv) the Monte Carlo method of Maeda,
(v) the Fokker-Planck method of Walt, and
(vi) the extended Fokker-Planck method of Banks.

These treatments have been labelled for convenience by the name of one of their principal exponents. Full references will be given below. The method of Spencer has been described by Chamberlain (1961) but has not been developed in recent years. It has been applied to the calculation of energy deposition curves for monodirectional, monoenergetic electrons. Method (ii) originally gave only height integrated fluxes. Methods (iii) and (iv) provide complementary information while (vi) is an extension of (v). The basic ideas of each will be discussed in turn.

Semi-Empirical Method of Rees

The original development proposed by Rees (1963b) appealed to laboratory data on the ionization produced by electron beams, to provide information on the distribution of ion production in the atmosphere caused by an incident electron beam of arbitrary energy and angular distribution. Because of the similarity in the shape of the cross sections for most ionization and many optically allowed excitation processes (noted in Section 4.1.2) this information provided the production rates for the N_2^+ 1N bands, the O_2^+ 1N bands and probably the N_2^+ M bands. In a further step taken by Rees *et al.* (1967), the ionization rate was used to infer a production rate for secondary electrons which were assumed to be absorbed locally. Originally it was assumed that all secondaries were produced at a single energy but later theoretical or empirical data were used for their initial energy distribution. Secondary electron fluxes were derived from a consideration of the various energy loss processes so that the excitation rates of most of the remaining auroral emissions could be calculated.

This model by itself provides no information about the changes in energy distribution of the primary electrons. Moreover it relied originally on the laboratory measurement of the ion production vs fraction of particle range traversed. Ion production was related to energy deposition by the approximation, valid over a wide range of energies and media, that ~ 35 eV are required to produce each ion. These theoretical gaps were filled by the Monte Carlo calculations of Maeda who was able to follow electrons entering the top of the atmosphere and calculate numerically the energy deposition (and the ion production) distribution with height. The Monte Carlo calculations also gave the variation with height of the energy distribution of the primary electrons down to 5% of their initial energy. The original version of this latter calculation was not accurate but it has recently been improved (Berger *et al.*, 1974). References and details of this treatment will be presented below.

Extended Fokker-Planck Method

In its original form this approach was intended to provide a method of calculating the angular and energy distribution of an electron beam incident on the atmosphere.

In brief the effect of collisions with atmospheric particles is considered on the distribution function in phase space of the electron beam. In its original form (Walt, 1967) it was particularly useful in calculating the fluxes of primary electrons above 10 keV including the effects of magnetic field convergence and collisions, on the back-scattered flux. It was extended down to 1 keV by Donat *et al.* (1971) who used improved angular and energy loss cross sections.

Banks *et al.* (1974) have extended this treatment to 500 eV and combined it with a further computation of electron fluxes down to 3 eV. The lower energy range was covered by solving the flux continuity equations for upward and downward fluxes. In a particular energy range, the divergence of those fluxes may be equated to the resultant of production by ionizations caused by higher energy electrons and of losses and gains arising in inelastic and elastic collisions. This technique provides values for the electron fluxes (no distinction between primary and secondary) as a function of height and so permits auroral excitation profiles to be calculated as described in Section 4.2.2. To keep the computation tractable Banks *et al.* assumed a mean pitch angle of 68° for the upward and downward fluxes below 500 eV. This method is particularly valuable for auroral emissions above 200 km where the assumption of local absorption of secondary electrons is not valid. It is also more likely to give reliable results for electron fluxes in the 100 to 500 eV range where there is a transition between the two parts of the Rees-Maeda theory.

Model of Stolarski and Green

As originally formulated by Stolarski and Green (1967) this procedure made it possible to calculate the number of secondary (and tertiary) electrons produced in each energy range by each incident primary electron. Thence height integrated values of excitations produced could be derived. This procedure was extended by Stolarski (1968) to give height profiles by means of the approximation that the energy of each primary is degraded in accordance with the energy deposition function for the beam as a whole. This is a reasonable first approximation but should not be as accurate as the newer methods.

A more detailed account will now be given of some concepts useful in the auroral electron flux problem.

4.2.1.1. *Empirical Energy Deposition Function*

This theory was first applied by Rees (1963 b) in the calculation of ionization rates due to various assumed incident spectra of primary electrons. The starting point is the energy deposition function $\lambda(z/R_0)$ defined by the relation,

$$\lambda(z/R_0) = (\mathrm{d}E/E_0)/(\mathrm{d}z/R_0)$$

in which dE is the average energy lost by electrons of initial energy E_0 in traversing an increment of mass thickness (gm cm^{-2}) dz, and R_0 is the electron range (in mass thickness) for energy E_0. The definition of R_0 (see Berger *et al.*, 1973) is to some extent arbitrary but one must be careful to use the same definition for R_0 in

Equations (4.2.1a) and (4.2.1b) etc. λ is thus the fraction of the initial energy deposited per increment in fraction of the total range.

λ has been determined experimentally (Grün, 1957) in laboratory measurements of the luminosity produced by a narrow beam of electrons in air. The measurements give the luminosity distribution of the N_2^+ bands as illustrated in Figure 4.16a. The emission of the N_2^+ bands is proportional to total ionization which is itself (except at low energies <200 eV) proportional to total energy deposition. Consequently by integrating the luminosity at constant z across the beam λ can be determined directly

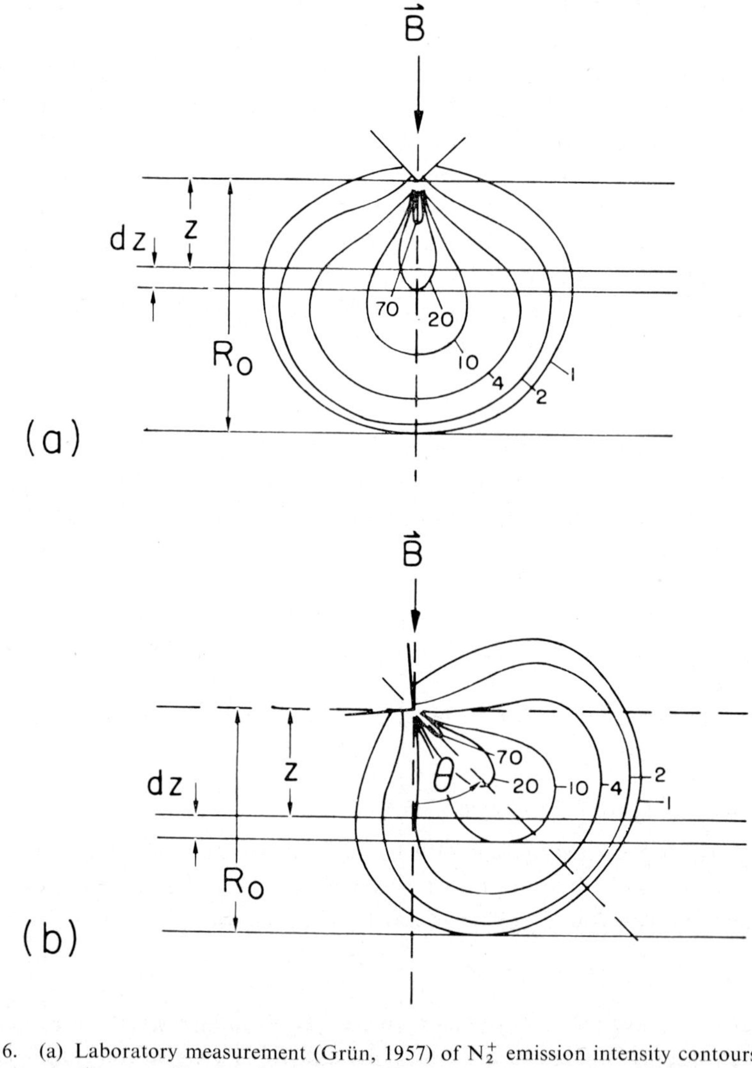

Fig. 4.16. (a) Laboratory measurement (Grün, 1957) of N_2^+ emission intensity contours produced by mono-energetic, monodirectional pencil of electrons in air. In a magnetic field, **B**, the electron paths are confined to helices around the dashed line. (b) Same as (a) but with direction of incidence at angle θ to **B**. The beam would still be confined close to the dashed line if a field is present.

for the case of the monoenergetic, initially parallel beam of electrons. In the presence of a magnetic field the motion of the electrons in the z direction (assumed parallel to the field) would not be altered but the motions in the xy plane would be wound up in a helix close to the beam axis (z-direction) in such a way that the integrated values of energy distribution $\lambda(z/R_0)$ would be unaltered. In the case of a narrow beam making an angle, θ, with the field direction $\lambda(z/R_0)$ would be obtained by integrating across the luminosity pattern in slices perpendicular to the field direction z as shown in Figure 4.16b. In the case of an arbitrary initial angular distribution about the field direction, the effective $\lambda(z/R_0)$ function can be found by integrating the contributions to λ. In this way energy deposition functions for a variety of initial angular distributions were derived by Rees (1963b) and are reproduced in Figure 4.17. The usefulness of the function λ is increased by the experimental result that it is only weakly dependent on E_0 and is largely independent of the changes of composition typical of the atmosphere in the range of interest, i.e., from 80 to 500 km.

4.2.1.2. *Calculated Energy Deposition Function*

The function λ as well as other characteristics of the electron diffusion problem have been investigated by the Monte Carlo method of Berger *et al.* (1970, 1974). The procedure is to simulate possible paths (according to the probabilities of the various collision processes) of a sufficiently large number of electrons penetrating into a gas. If the differential cross sections for elastic and inelastic scattering processes are known,

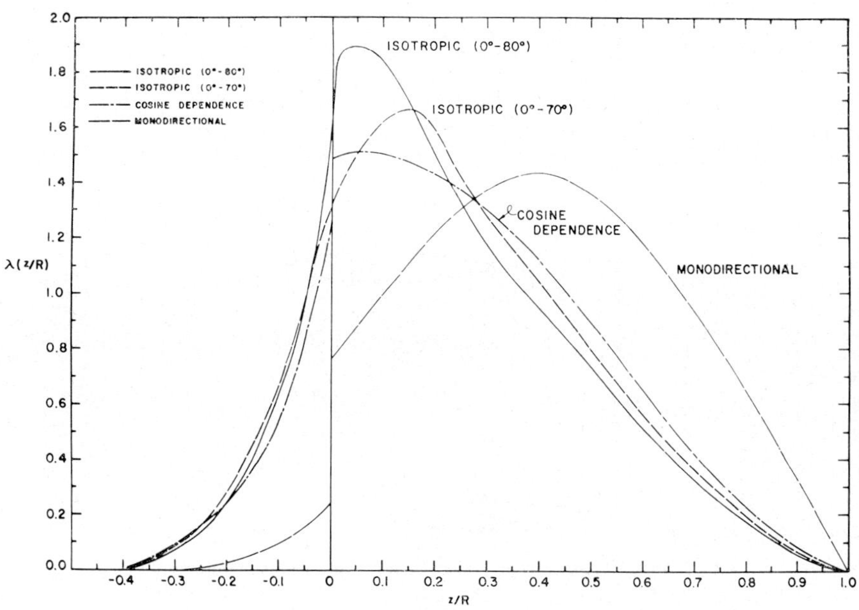

Fig. 4.17. Energy deposition function, $\lambda(z/R_0)$, for various initial angular distribution functions as derived by Rees (1963b) from Figure 4.16. The negative values of z/R_0 correspond to back scattering. (Courtesy *Planetary and Space Science*.)

possible paths and energy losses for each electron can be calculated and averaged so that λ can be determined for any initial angular distribution. In principle the angular distribution and energy distribution of the electrons of the primary beam can also be followed. In each inelastic ionizing collision the higher energy electron after the collision would be considered the primary. In practice the Monte Carlo calculations performed have involved approximations arising from lack of exact knowledge of the relevant cross sections and because of limitations of computing time. For example Berger *et al.* (1970) used a method to determine λ in which the path of each electron was followed over segments involving many collisions but short enough to limit the angular deviation per segment to a small range. By means of the calculations based on angular differential cross section for elastic scattering, the Bethe stopping power formula and the Landau expression for distribution of energy loss, it was possible to define the probability of a given angular deviation and energy loss at the end of each segment. This method involves approximations, but the final results for λ were in satisfactory agreement with the empirical curves of Figure 4.17.

4.2.1.3. *Ionization Rate Height Profiles*

These may be calculated in two ways (i) by using the energy deposition function λ to calculate an energy deposition height profile in the atmosphere together with the approximation that 35 eV on the average is able to produce an ion-pair, and (ii) directly from the electron fluxes calculated by the method of Banks *et al.* (1974) by means of Equation (4.2.2a).

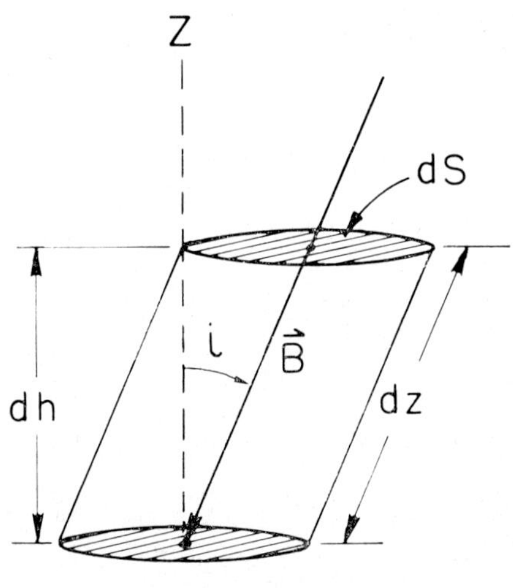

Fig. 4.18. Geometry of ion production in a volume element.

In the first approach the energy deposited per unit volume for an initial flux of $F(E_0)$ particles/unit area, if ρ is the atmospheric density, will be (referring to

Figure 4.18)

$$\frac{dE}{dV} = \lambda F \frac{dS\, E_0\, dz}{R_0 dV}. \tag{4.2.1a}$$

Now

$$dz = \rho\, dh \sec i$$

$$dV = dS\, dh$$

$$\therefore \quad \frac{dE}{dV} = \lambda \left(\frac{z}{R_0}\right) F\rho(h) \sec i E_0/R_0, \tag{4.2.1b}$$

where

$$z = \int_h^\infty dh\, \rho \sec i. \tag{4.2.1c}$$

This takes into account the inclination, i, of the magnetic field, B, about which we must assume the angular distribution of the incoming particles is symmetrical. For auroral zone geomagnetic latitudes $i < 10°$ and the sec i term has negligible effect.

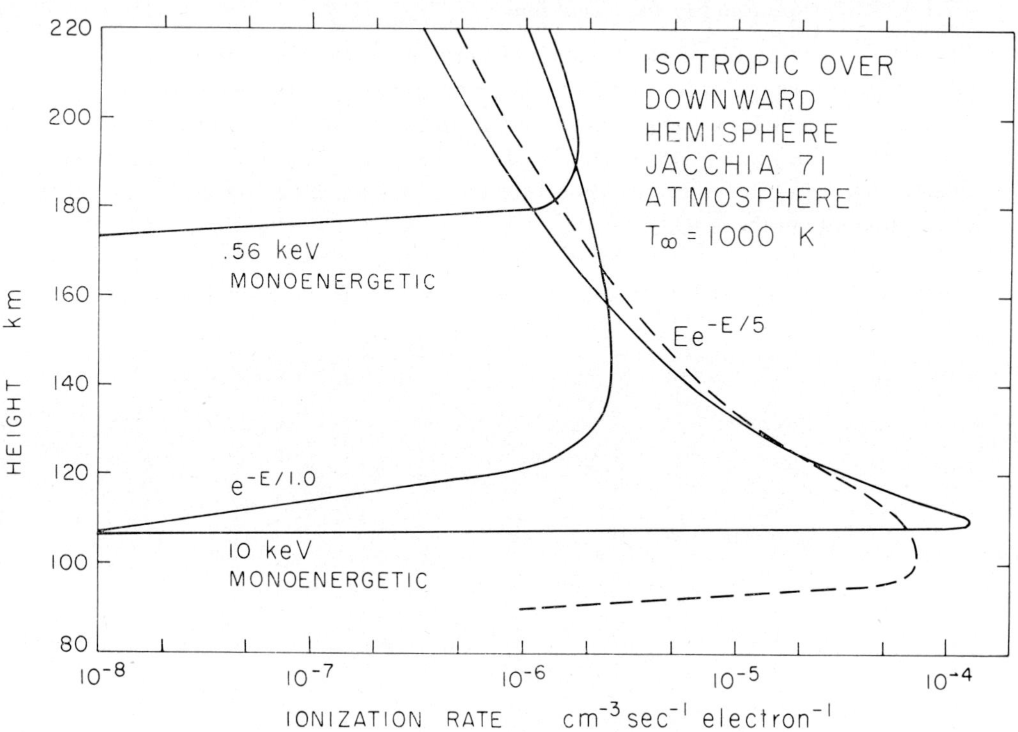

Fig. 4.19. Calculated ion production rates for electrons of various initial energy spectra. E is in keV.

Then the ionization rate becomes

$$\eta_i = \frac{FE_0}{\varepsilon_0} \lambda\, \rho(h)/R_0 \tag{4.2.1d}$$

$$= \frac{F\, E_0\, \lambda}{\varepsilon_0 R_0} m_{O_2} \left[n(O_2) + \frac{m_{N_2}}{m_{O_2}} n(N_2) + \frac{m_O}{m_{O_2}} n(O) \right] \tag{4.2.1e}$$

m_{O_2} is the mass of an oxygen molecule and $\varepsilon_0 = 35$ eV ion^{-1} pair.

Results for ionization rate profiles calculated from this relation are shown in Figure 4.19. Other distributions calculated by the same method for various atmospheres and electron fluxes may be found in the papers of Rees (1963b) and Berger *et al.* (1974).

The paper of Banks *et al.* (1974) gives several ionization profiles calculated directly from the derived electron fluxes. These compare well with those from the energy deposition method down to energies of 2 keV. For lower energies there are deviations, ascribed in part to failure of the assumption of local absorption of secondary electrons and of the constancy of ε_0 at 35 eV. Nonetheless even the 0.56 keV curve of Figure 4.19 is reasonably correct.

4.2.1.4. *Production Rates of Individual Ions*

The rates of ionization of the constituent particles of the atmosphere, i.e., of O, O_2 and N_2 are important in calculating the production rates and energy distributions of resulting secondary electrons. One must distinguish also the production rates of various ions which are needed in studies of ion chemistry. We can denote the former quantity by $\eta_i(X)$ i.e., the ionization rate of X and the latter by $\eta(X^+)$ i.e., the production rate of the ion X^+.

The peak values of the important total ionization cross sections are given in Figure 4.20 together with branching ratios, *f*, for various types of ions. Since the

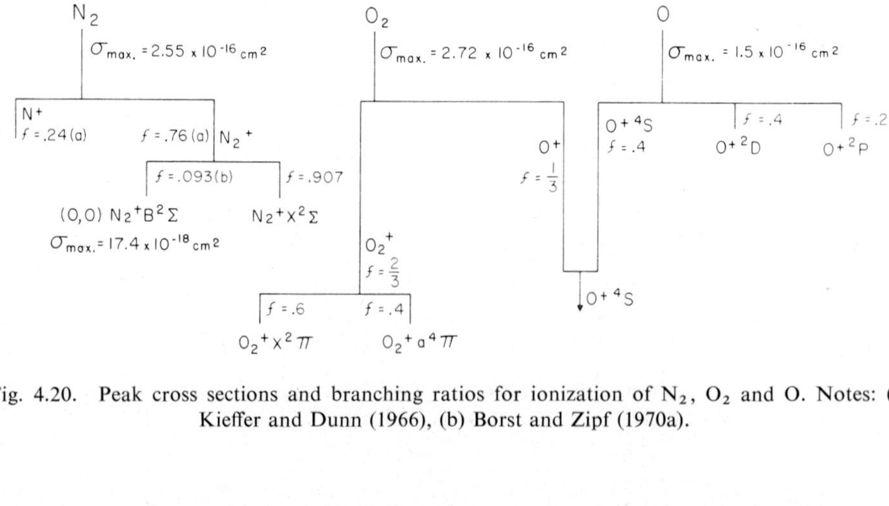

Fig. 4.20. Peak cross sections and branching ratios for ionization of N_2, O_2 and O. Notes: (a) Kieffer and Dunn (1966), (b) Borst and Zipf (1970a).

shapes of all the cross sections are similar (except in small region near threshold), one can assume that the relative frequency of the occurrence of any ionization process will be proportional to the product of the number of target particles and the peak cross section for the process. Thus we can write

$$\eta_i(N_2) = \eta_i \frac{n(N_2)\sigma_i(N_2,\text{total})}{n(N_2)\sigma_i(N_2,\text{total})+n(O_2)\,\sigma_i(O_2,\text{total})+n(O)\sigma_i(O)}$$

$$= \eta_i \left[\frac{0.94n(N_2)}{n(O_2)+0.94n(N_2)+0.55n(O)}\right] \tag{4.2.1f}$$

$$\eta(N_2^+) \simeq 0.76\eta_i(N_2) \tag{4.2.1g}$$

$$\eta(N^+) \simeq 0.24\eta_i(N_2) \tag{4.2.1h}$$

$$\eta_i(O_2) = \eta_i(N_2)\frac{n(O_2)\sigma_i(O_2\ \text{total})}{n(N_2)\sigma_i(N_2\ \text{total})} = 1.07\,\frac{n(O_2)}{n(N_2)}\,\eta_i(N_2) \tag{4.2.1i}$$

$$\eta(O_2^+) = \eta_i(O_2)\frac{\sigma_i(O_2 \rightarrow O_2^+)}{\sigma_i(O_2\ \text{total})} = 0.67\eta_i(O_2) \tag{4.2.1j}$$

$$\eta_i(O) = \frac{\eta_i(N_2)n(O)\sigma_i(O)}{n(N_2)\sigma_i(N_2\ \text{total})} = 0.59\eta_i(N_2)\,\frac{n(O)}{n(N_2)} \tag{4.2.1k}$$

$$\eta(O^+) = \eta_i(O) + 0.33\eta_i(O_2) \tag{4.2.1l}$$

$\eta(O_2^+)$ can be further divided into the production of $X^2\Pi$ ions and metastable $a^4\Pi$ ions. According to the cross sections of Watson *et al.* (1967) the branching ratios for these ions are 0.6 and 0.4, respectively. Likewise O^+ can be produced in the ^4S, ^2D and ^2P states (see Figure 4.9). According to the values shown in Figure 4.20 the branching ratios in the direct ionization of O are 0.4, 0.4 and 0.2 respectively, for these ions. Walker and Rees (1968) assumed that only this component of the O^+ production rate produced excited ions which leads to the relations

$$\left.\begin{aligned} \eta(O^+\ {}^4S) &= 0.4\eta_i(O) + 0.33\eta_i(O_2) \\ \eta(O^+\ {}^2D) &= 0.4\eta_i(O) \\ \eta(O^+\ {}^2P) &= 0.2\eta_i(O) \end{aligned}\right\} \tag{4.2.1m}$$

4.2.1.5. *Production Rate of Secondary Electrons as Function of Energy*

In this and the following section we explore the approach of Rees *et al.* (1969) in calculating a secondary electron flux distribution. This is the flux of all electrons which are produced in ionization processes. Formally, in any ionization one may regard the lower energy of the two post-collision electrons as being the secondary. The flux of primary electrons which have lost energy in successive collisions must be derived separately. Experimentally the primary and secondary fluxes cannot be distinguished.

The total rate of production of secondary electrons is equal to the total ionization rate. To calculate their effects in producing excitation it is first necessary to know their distribution in energy, $S(E_P, E_S)$ as defined by Equation (4.1b). In the original work of Rees *et al.* (1969) a Born approximation formula for S was used. Recently experimental values of the cross section have been measured by Opal *et al.* (1971) who find that the distribution can be well represented by an expression of the form

$$\begin{aligned} S(E_P, E_S) &= 1/\{\tilde{E} \arctan[(E_P - I)/2\tilde{E}]\,[1 + (E_S/\tilde{E})^2]\} \\ &\simeq 2/\{\pi\tilde{E}\,[1 + (E_S/\tilde{E})^2]\}. \end{aligned} \tag{4.2.1n}$$

$\tilde{E}$ is a parameter which depends somewhat on the primary electron energy E_P, but at higher energies can be taken as 14.1 for N_2 and 18.9 eV for O_2. For atomic O there is no experimental value although Banks *et al.* (1974) have adopted a value of 15.0 in this case. These semi-empirical values of S are generally 10 to 20% lower than those from the Born approximation formula out to 40 eV and considerably larger at higher energies.

Then the production rate, $\eta_S(E)$, of secondaries of energy E is

$$\eta_S(E) = \sum_j \eta_{ij}\, S_j(E_P, E), \tag{4.2.1o}$$

where j indexes the ionization rates η_{ij} and the distribution functions S_j of the atmospheric particles O, O_2 and N_2.

4.2.1.6. *Flux of Secondary Electrons*

The flux of secondary electrons $\phi_S(E)\,dE$ in the energy range E to $E + dE$ may be calculated by the expression

$$\Phi_S(E)\left[-\frac{dE}{dx}\right] = \int_E^{E_{S\max}} \eta_S(E)\,dE, \tag{4.2.1p}$$

where $-dE/dx$ is the energy loss per unit path. One may justify this expression by noting the flux may be written as $\eta_S(E)\,v$ so that the left hand side becomes

$$n_S(E)\,\frac{dx}{dt}\left(-\frac{dE}{dx}\right) = -\frac{n_S(E)dE}{dt} = -\frac{dn_S(E)}{dt},$$

i.e., the rate of loss of electrons of energy E per unit volume. $n_S(E)$ is the number density of secondary electrons in the energy range dE at E and v is the velocity corresponding to E. If one assumes that the secondaries lose energy continuously then this loss rate must be equal to the right hand side which is the rate at which electrons with energy greater than E are produced which in the steady state is the rate at which electrons of energy E appear. Of course the assumption of continuous slowing down is not strictly true but it has been supposed that it does not affect the results very seriously.

The quantity $-\mathrm{d}E/\mathrm{d}x$, the energy loss per unit path can be divided into two terms:
(i) losses in elastic collisions with ambient electrons; and
(ii) losses in inelastic collisions with atmospheric particles.

The former involve somewhat complicated interactions but may be calculated from formulae given by Schunk and Hays (1971). For the 1 to 100 eV range the theoretical formulae may be replaced by a more convenient expression suggested by Swartz *et al.* (1971),

$$\left(-\frac{\mathrm{d}E}{\mathrm{d}x}\right)_{\mathrm{ee}} = 3.37 \times 10^{-12}\,\frac{n(e)^{0.97}}{E^{0.94}}\,. \tag{4.2.1q}$$

The electron-neutral contribution to $\mathrm{d}E/\mathrm{d}x$ may be expressed directly in terms of the cross sections for the jth excitation process in which energy W_{kj} is absorbed,

$$\left.\begin{aligned}\left(-\frac{\mathrm{d}E}{\mathrm{d}x}\right)_{\mathrm{en}} &= \sum^{\mathrm{k}} n_{\mathrm{k}} \sum^{\mathrm{j}} \sigma_{\mathrm{kj}}(E)\, W_{\mathrm{kj}} \\ &= \sum^{\mathrm{k}} n_{\mathrm{k}} L(k \mid E) = nL(E)\end{aligned}\right\} \tag{4.2.1r}$$

where n_{k} is the concentration of the kth species of particle and σ_{kj} is the cross section of the jth process for the kth particle. n is the total particle concentration. $L(k \mid E)$ defined as in the second line of Equation (4.2.1r), is called the loss function or stopping cross section for the kth particle while $L(E)$ is the total loss function.

Figure 4.21 shows plots of the values of $L(\mathrm{N_2} \mid E)$, $L(\mathrm{O_2} \mid E)$ and $L(\mathrm{O} \mid E)$ as calculated from the cross sections of Figure 4.15 plus contributions in the 1 to 5 eV range from vibrational and rotational excitation of O_2 and N_2 as given by Takayanagi and Itikawa (1970).

For ionization processes the energy loss is $W = I + E_{\mathrm{S}}$ where I is the ionization energy. Since E_{S} is a variable the loss function contribution for the ith ionization process becomes

$$\left.\begin{aligned}L(i \mid E) &= \int_{\mathrm{I}}^{\mathrm{E_{S_{max}}}} \sigma_{\mathrm{i}}(E, E_{\mathrm{S}})\,(I + E_{\mathrm{S}})\,\mathrm{d}E_{\mathrm{S}} \\ &= \sigma_{\mathrm{i}}(E)\,\overline{W}\end{aligned}\right\} \tag{4.2.1s}$$

where $\overline{W}$ is the average value of $I + E_{\mathrm{S}}$ weighted by $\sigma_{\mathrm{i}}(E, E_{\mathrm{S}})$ (or $S(E, E_{\mathrm{S}})$). Thus a loss function of the same form as Equation (4.2.1s) may be written if the average value $\overline{W}$ is used. Ionization contributions thus have been included in the curves of Figure 4.21 (by means of the formula of Dalgarno *et al.*, 1963) and indeed make the largest contributions beyond 50 eV. Other sets of loss functions have been given by Dalgarno *et al.* (1969) and Rees and Jones (1973).

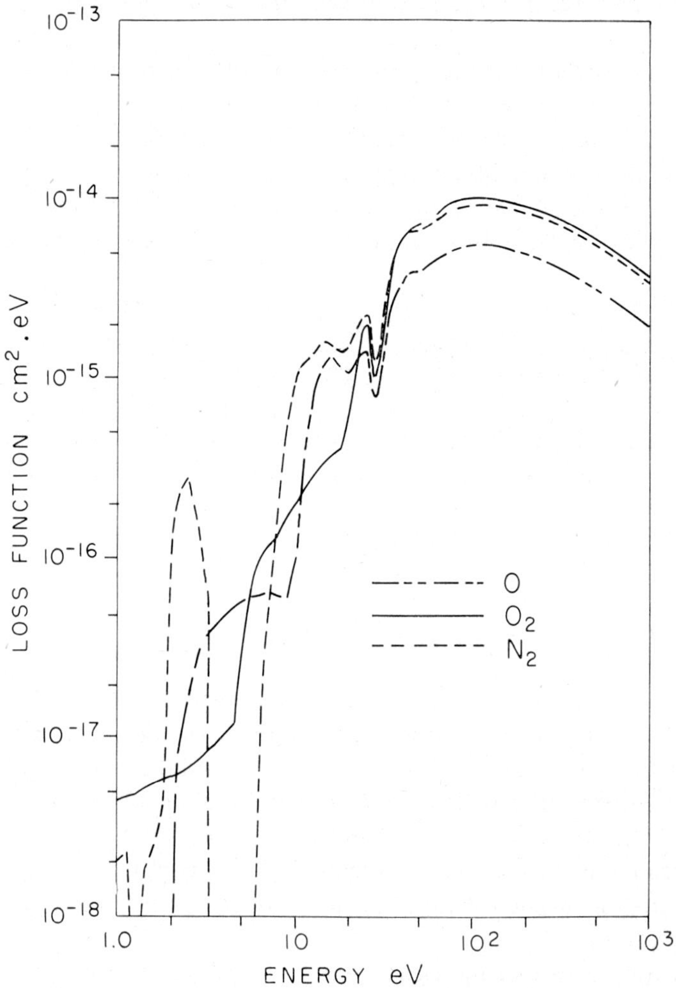

Fig. 4.21. Electron-neutral loss functions $L(N_2 \mid E)$, $L(O \mid E)$ and $L(O_2 \mid E)$ calculated from the cross sections of Figure 4.15.

Secondary electron fluxes may be calculated by means of Equation (4.2.1p) after the secondary production rate $\eta_S(E)$ has been obtained from Equation (4.2.1o). (The electron concentration is needed to evaluate Equation (4.2.1q) so that the ion chemistry equations described in Section 5.1.2 must be solved or an electron concentration profile assumed.)

The results of such a calculation are shown in Figure 4.22 for the case of an incident electron spectrum of form $Ee^{-E/5}$ for which the ionization rate profile was shown in Figure 4.19. It is quite similar within a factor of 2 to the results of Rees and Jones (1973) and to those of Banks *et al.* (1974) below 100 eV. Experimentally the spectrum of 10 to 100 eV electrons has been measured by Feldman *et al.* (1971) who found their results were fitted very well by a power law spectrum of form $E^{-3.17}$ at all heights between 100 and 150 km. This distribution is shown on Figure 4.22. This fits the theoretical results quite well since it does not appear that the irregularities between 10 to 100 eV in Figure 4.22 have any significance. The Rees *et al.* (1969) formula for $S(E_p, E_S)$ was used in calculating the solid curves of Figure 4.22 and this probably accounts for the larger slope towards 100 eV in comparison with the newer

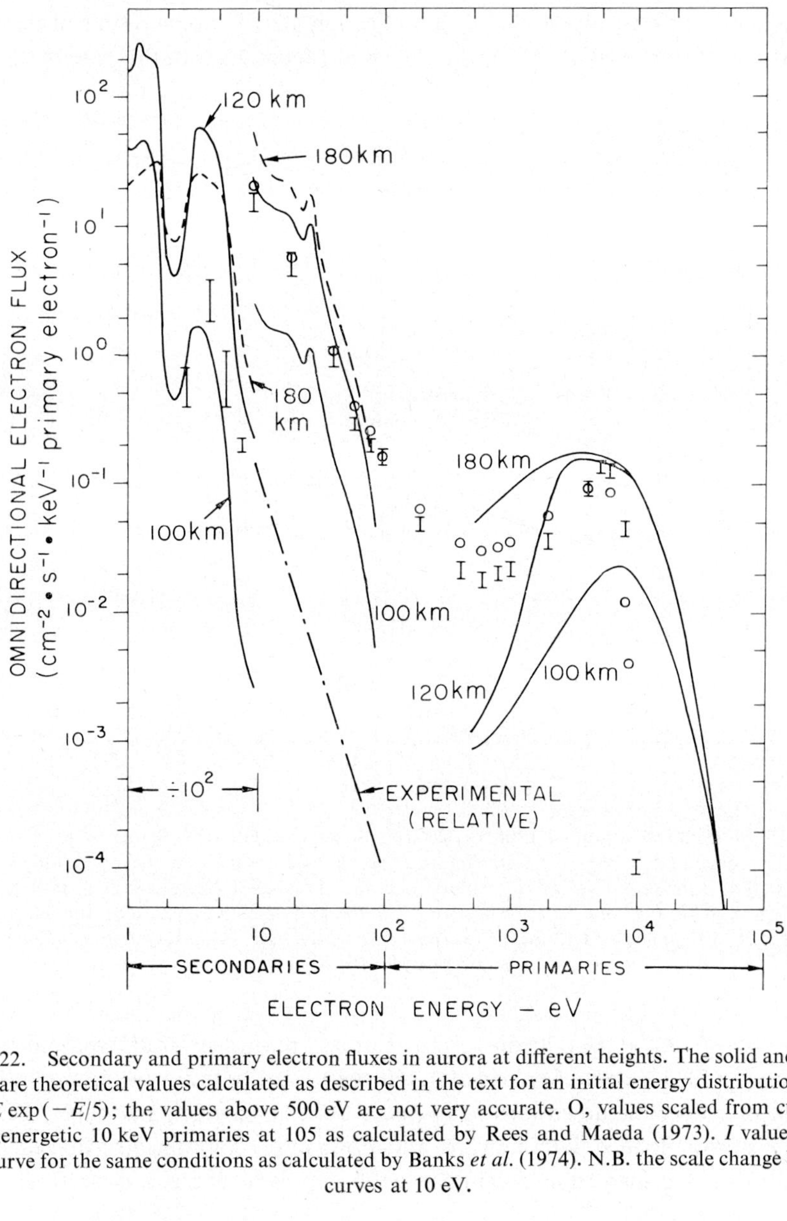

Fig. 4.22. Secondary and primary electron fluxes in aurora at different heights. The solid and dotted curves are theoretical values calculated as described in the text for an initial energy distribution of the form $E\exp(-E/5)$; the values above 500 eV are not very accurate. O, values scaled from curve for mononenergetic 10 keV primaries at 105 as calculated by Rees and Maeda (1973). *I* values scaled from curve for the same conditions as calculated by Banks *et al.* (1974). N.B. the scale change between curves at 10 eV.

results. Beyond 100 eV there is also an increasing contribution from degraded primaries. Thus the points for the newer results on Figure 4.22 fit a power law with an exponent closer to -2.

4.2.1.7. *Primary Electron Flux*

As outlined in Section 4.2.1 the primary flux may be calculated down to 1 keV by the Fokker-Planck method of Walt (1967) as extended by Donat *et al.* (1971) and by Banks *et al.* (1974). This method is accurate but cumbersome. The other approach is the Monte Carlo treatment discussed in Section 4.2.1.2. Some results of the newest calculations of Berger *et al.* (1974) are shown in Figure 4.23. In this case the quantity

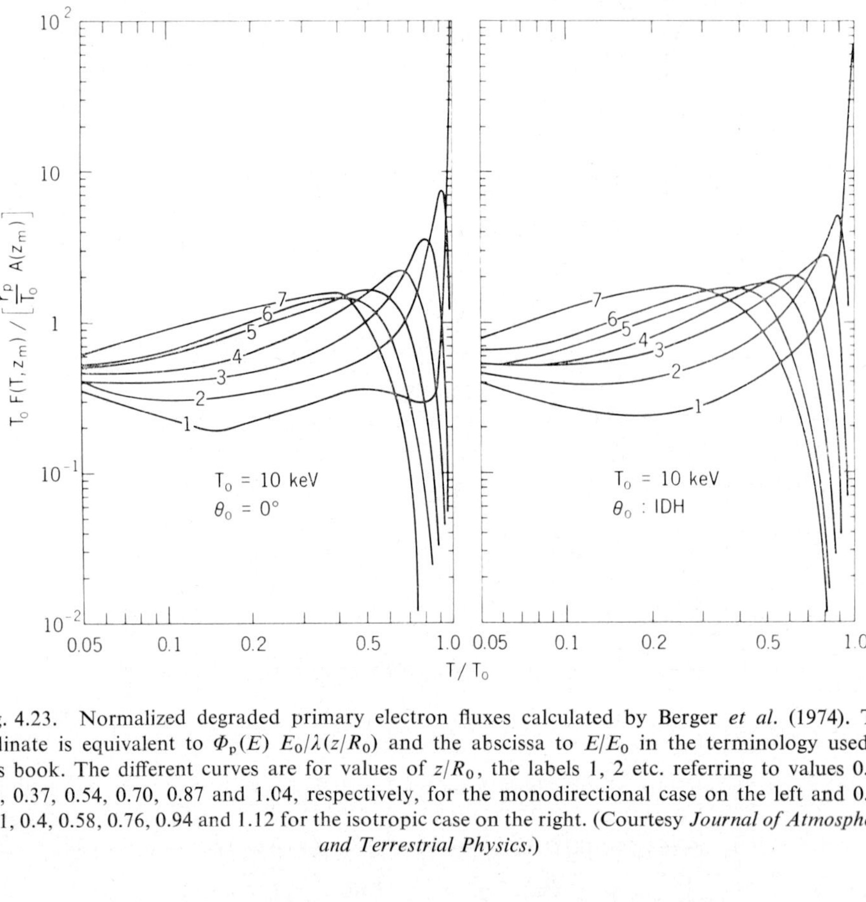

Fig. 4.23. Normalized degraded primary electron fluxes calculated by Berger *et al.* (1974). The ordinate is equivalent to $\Phi_p(E)\ E_0/\lambda(z/R_0)$ and the abscissa to E/E_0 in the terminology used in this book. The different curves are for values of z/R_0, the labels 1, 2 etc. referring to values 0.03, 0.2, 0.37, 0.54, 0.70, 0.87 and 1.04, respectively, for the monodirectional case on the left and 0.03, 0.21, 0.4, 0.58, 0.76, 0.94 and 1.12 for the isotropic case on the right. (Courtesy *Journal of Atmospheric and Terrestrial Physics.*)

plotted is $\Phi_p(E)\ E_0/\lambda(z/R_0)$ against E/E_0. Curves are shown for different normalized depths, z/R_0, for incident electrons of initial energy E_0, both for the monodirectional and isotropic-over-the-downward-hemisphere cases. In the original paper the fluxes are also shown divided into upward and downward components. In general the upward fluxes are important except for small depths at energies close to E_0. Rees

(1969) quoted and used an empirical formulae which had been fitted to the earlier Monte Carlo data by Maeda and Aikin (1968); this work is now superseded by that of Rees and Maeda (1973) who used the results of the latest Monte Carlo calculations.

4.2.1.8. *Total Electron Fluxes*

The total electron flux may be obtained by adding the primary and secondary components obtained as described in Sections 4.2.1.6 and 4.2.1.7 or it may be derived directly by the method of Banks *et al.* (1973). Points from an electron flux curve calculated by Banks *et al.* for 10 keV quasi-monoenergetic electrons are shown in Figure 4.22 together with points from curves calculated by Rees and Maeda (1973) by the addition method. (It should be noted that the Monte Carlo method does take account of high energy secondaries in the primary energy range.) The calculations of Banks *et al.* also give ratios of upward to downward fluxes and show there are considerable fluxes of lower energy particles backscattered from the atmosphere. Also shown on Figure 4.22 are some less accurate estimates of primary electron flux based on an energy deposition normalized version of earlier (now superseded) Monte Carlo data of Maeda (1965). The latter curves which are for an initial energy distribution of the form $E\exp(-E/5)$ show the hardening effect which occurs in such cases due to the loss of the lower energy primaries. With an initially monoenergetic beam the spectrum softens.

4.2.2. Excitation of Atmospheric Atoms by Primary and Secondary Electrons

If the primary and secondary electron fluxes, $\Phi_p(E)$ and $\Phi_S(E)$, are known and σ_{xj} is the cross section of the jth process involving particle x then

$$\eta_j(x) = n(x)\int_0^\infty [\Phi_S(E) + \Phi_p(E)]\sigma_{xj}(E)\,dE. \qquad (4.2.2a)$$

If the process is an ionization then the total cross section, integrated over all energies of the ejected electron, must be used, while for molecular bands the total cross section summed over all vibrational states is required.

The production rates of the principal excited species by secondary electrons for an incident flux proportional to $Ee^{-E/5}$ are shown in Figure 4.24. These are the rates calculated from the electron fluxes of Figure 4.22 using the cross sections of Figure 4.15. Integrated rates are given in Table 4.5 for three fluxes of primary electrons having different energy spectra which may correspond roughly to observed types of aurora. In addition results calculated by Rees and Jones (1973) and Green *et al.* (1969) are included. The 5 keV power exponential spectrum may most nearly correspond to normal bright aurora. The corresponding ion production vs height curve was given in Figure 4.19. The production rates have been normalized to 30 kR for the 0,1 4278 N_2^+ band since it has been found experimentally that the ratio of $I(5577, \text{O I})/I(4278)$ is about 3 for bright aurora (see also Section 4.2.4.1). (Since the

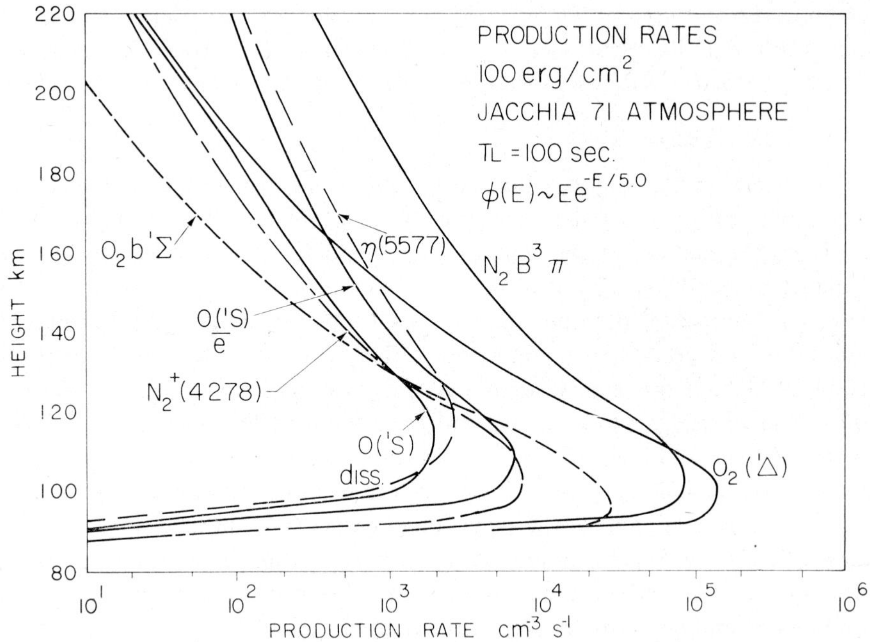

Fig. 4.24. Calculated height profiles of production rates for the excitation of some states in aurora.

excitation mechanism of the O(^{1}S) state is still uncertain it is unsatisfactory to normalize to the theoretical values of I(5577)). All the rates correspond to excitation by secondary electrons with $E \leq 100$ eV except for λ4278 N_2^+ which was scaled from the total ionization rate according to the branching ratios of Figure 4.20.

The different production rates displayed in Table 4.5 agree reasonably well, almost always within a factor of 2 or less. The differences mostly reflect uncertainties in the cross sections except for O(^{1}S) and O(^{1}D) for which the contributions of the indirect mechanisms (here included) are an additional source of uncertainty. The values in Table 4.5 will be discussed further and compared with observation in later parts of this section.

4.2.2.1. *Excitation and Ionization by Primary and Secondary Electrons*

The role of secondary electrons in producing ionization has proved difficult to assess. It was originally thought that perhaps 2/3 of the ionization was produced by secondaries (e.g. Omholt, 1967). Direct measurement of the secondary spectrum and of the 3914 Å N_2^+ emission led Feldman *et al.* (1971) to conclude that only 10% of the ionization is due to secondaries. The latest theoretical calculations of Rees and Maeda (1973) suggest that, for 5 keV electrons, the contribution of ionization by secondaries plus degraded primaries with energies below ~300 eV may be about 1/3. In fact calculations of the ionization due to secondary fluxes such as those of Figure 4.22 originally gave ratios of about 10% but it appears that the larger values

TABLE 4.5
Calculated production and emission rates (kR) of excited species for $I(4278) = 30$ kR ($\simeq$IBC3)

Emission or excited species	Electron spectrum $Ee^{-E/5}$ [a]	$e^{-E/1.0}$	562 eV	$Ee^{-E/2.3}$ [b]	1 KeV [c]	100 eV [d]	Observed [g]
N_2^+1N(1, 0)	(30)	(30)	(30)	(30)	(30)	(30)	(30)
O(^{1}S)	55	130	133	140	160		
λ5577[h]	24	110	130				(100)
O(^{1}D)	1700	2380	1800		3510		
λ6300[h]	8.0	9.4	200	21	18		2–100
$O_2(a^1\Delta)$	400	78	16	500	17		2500
$O_2(b^1\Sigma)$	72	15	3.4	110	7		1300
$N_2$1P	780	820	770	460	730	340	880
$N_2$2P	120	120	114	140	210	133	110
$N_2A^3\Sigma$	340	310	230	450	1750	480	55
$N_2a^1\Pi$				345		340	200–400
$N_2^+A^2\Pi$				690	520		630
$O_2^+b^4\Sigma$				70	38		26
N_2(b)[e]					244	760 }	2000 (b + h)
N_2(h)[e]					173	550 }	
O(3^3S^0)				99	193		30
O(3^5S^0)				21	36		2
[N I]5200 Å				2.3			0.1–2
[N I]3466 Å				2.6			1
[N I]10400 Å				38			6
[O II]7319 Å	2.9			5.6		3.9[f]	0.4–100
[O II]3727 Å	2–3 R		~200–300 R			4.1 R[f]	0–1
Energy flux (erg cm^{-2} s^{-1})	154	234	331	163			

[a] The first three columns refer to results from the model used in this book.
[b] Results of Rees and Jones (1973) normalized for $I(4278)$.
[c] Results of Green *et al.* (1969) similarly normalized.
[d] Monte Carlo result of Brinkmann and Trajmar (1970) for pure N_2 atmosphere.
[e] See footnote Figure 4.11 for meaning of b and h states.
[f] Derived from results of Gérard (1970) for electron spectrum $\sim \exp(-E/1.3)$.
[g] Data from Table 4.3.
[h] Includes quenching factors.

of the measured electron energy distribution function of Opal *et al.* (1971) produce an increase over the values based on the Born-approximation formula. Banks *et al.* (1973) show that at most heights about 30% of the ionization is due to electrons with less than 500 eV (secondaries plus degraded primaries) but that the fraction rapidly increases to 100% near and below the height of maximum ionization.

The converse question of the amount of excitation of optically forbidden states by primary electrons is more certain. Rough calculations using the electron spectrum shown in Figure 4.22 and cross sections as in Figure 4.15 show that, for the optically spin-forbidden type of excitation process, the excitation by electrons with energy greater than 150 eV is much less than 1% while for the spin-allowed forbidden transitions the fraction is less than 10%. Since most of the important band systems (except the N_2 LBH bands) are of the former type it is clear that most auroral electron impact excitation rates can be derived from a knowledge of the secondary spectrum alone.

4.2.2.2. *Excitation and Ionization by Secondary Photons*

Very recently Zipf (1973a) has suggested that the strong flux recently observed in the Birge-Hopfield bands (see Section 4.2.4.4) is an important source of ionization of O, O_2 and NO. This could lead to profound effects which have not been fully evaluated. At lower heights the photons may be absorbed locally whereas those produced higher would have longer mean free paths.

Zipf (1973b) has also made the important suggestion that $O(^1S)$ produced in photodissociation by the same extreme UV flux should be a major source of the λ5577 emission.

4.2.3. Indirect Excitation Processes and Quenching

In a number of cases excitation and de-excitation in secondary processes assume a prominent role. Probable examples of the most important of such processes are as follows.

(i) Dissociative recombination (DR)

$$O_2^+ + e \rightarrow \begin{matrix} O(^1D) + O \\ O(^1S) + O \end{matrix}$$

(ii) Energy transfer (ET)

$$O_2 + O(^1D) \rightarrow O_2(b^1\Sigma) + O \qquad (4.2.3a)$$

(iii) Quenching

$$O(^1D) + N_2 \rightarrow O(^3P) + N_2^\dagger \qquad (Q)$$

$$N_2(A^3\Sigma) + O \rightarrow N_2 + O$$

(† denotes vibrational excitation)

(iv) Thermal excitation

$$O(^3P) + e(T_e \lesssim 2000\ K) \rightarrow O(^1D) + e$$

(v) Cascading

$$N_2(C^3\Pi) \rightarrow N_2(B^3\Pi) + h\nu(2PN_2).$$

Processes thought to be significant are listed in Table 4.6 with values of their reaction rates where appropriate. The most important will now be discussed in turn.

TABLE 4.6

Rates of some reactions in auroral atmosphere

rocess	Type	Available energy (eV)	Rate constant[a] ($cm^3\ s^{-1}$)	Reference
$_2^+ + e \to O(^1D) + O$	DR	5.0	$k_{16a} \simeq k_{16} = 2.2(-7) \times (300/T_e)^{1/2}$	b
$+ O(^1S) + O$	DR	2.8	$k_{16b} \approx 0.1k_{16}$	
$O^+ + e \to N(^2D) + O$	DR	0.4	$k_{19a} = 0.5k_{19}$ $k_{19} = 6.1(-7)\ (300/T_e)^{1.5}$	d
$_2^+ + e \to N(^4S, ^2D, ^2P) + N$	DR	5.8(max)	$k_{12} = 2.9(-7)[300/T_e]^{1/3}$ Equal yields assumed	e
$(^1D) + N_2 \to O(^3P) + N_2^\dagger$	Q	1.97	$k_1 = 5(-11)$	f
$+ O_2 \to O(^3P) + O_2(?)$	Q, ET		$k_2 = 1.1k_1$ Yield $O_2(a^1\Sigma)$ 0.8 to 1.0 (k_{2a})	f, g
$+ O_2 \to O_2(b^1\Sigma, v' = j) + O(^3P)$	ET		$k_{2a}(j)$ $k_{2a}(1) \sim 0.4\ k_2$	p
$+ NO \to ?$	Q, (ET?)		$k_{65} = 1.5(-10)$	h
$(^1S) + O_2 \to O(^3P) + O_2(?)$	Q	4.2	$k_3 = 4.9(-12)\exp(-1700/RT)$, $3(-13)$	i, h
$+ O \to ?$	Q		$k_4 = 7.5(-12)$	f
$+ N_2 \to (?)$	Q		$k_5 < 5(-17)$	i
			$k_{53a} = 1.5(-11)$	j
$_2(A^3\Sigma) + O \to N_2(X^1\Sigma) + O(^1S)$	Q, ET(?)	6.1	$k_{53} = 3(-12)$ [$\sim 8(-11)$ aurora interp]	k, l
$+ O_2 \to ?$	Q, ET?		$k_{54} = 3(-12)$	l
$+ NO \to NO(X^2\Pi, A^2\Sigma) + N_2$	Q, ET		$k_{56} = 7(-11)$ Yield $NO(A^2\Sigma) \sim 0.5$	m, n, o
$_2(b^1\Sigma, v' > 1) + O_2 \to O_2(b^1\Sigma, v' = 0)$				
$+ O_2(v' > 1)$	Q		$k_{59} \simeq 9(-11)$	p
$_2(b^1\Sigma) + N_2 \to ?$	Q		$k_8 = 2(-15)$	f
$+ O_2 \to 2O_2$	Q		$k_7 = 1(-16)$	f
$_2(^1\Delta) + O_2 \to 2O_2$	Q		$k_{51} = 2.2(-18)$	f
$^+(^2D) + N_2 \to O + N_2^+$	Q		$k_{27} = 1(-9)$ or $3(-10)$	d, h
$+ N_2 \to O + N_2^+(A^2\Pi, v' = j)$	ET		$k_{27a}(j)$	q
$+ e \to O^+(^4S) + e$	Q		$k_{28} = 3(-8)$	b, d
$^+(^2P) + N_2 \to O + N_2^+(?)$	Q		$k_{25} = 1(-10)$?	r
$(^2D) + O_2 \to$	Q		$k_{44} = 6\text{–}14(-12)$	f, s
$+ NO \to$	Q		$k_{64} = 1(-10)$, $2.2(-11)$	f, r
$+ O \to O(^1D) + N$	ET		$k_{45} = 2(-13)$ (??)	t
$+ N_2$	Q		$k_{63} = 2(-14)$	f
$(^2P) + N_2 \to$	Q		$k_{58} < 3(-19)$	u
$+ O_2 \to$	Q		$k_{66} = k_{44}$ (?)	r
$+ NO \to$	Q		$= k_{64}$ (?)	r
$^+(^1S, ^1D) + O_2 \to NO^+ + O$	Q		$= k_{13}$ (?)	
$+ O_2 \to O_2^+ + N$	Q		$= k_{14}$ (?)	
$^+(^3P) + O_2 \to NO^+ + O$			$k_{13} = 5(-10)$	b
$+ O_2 \to O_2^+ + N$			$k_{14} = 4.5(-10)$	b
$_2^+ + O \to NO^+ + N(^2D)$	ET		$k_{11b} = 0.25k_{11}$ $k_{11} = 1.4(-10)$	v
$_2^+(a^4\Pi) + e \to O_2^+ + e\ (?)$	Q		$k_{23} = 1(-7)$	b
$+ N_2 \to N_2^+ + O_2$	Q		$k_{21} = 2(-10)$	w
$O \to O_2^+ + O$	Q		$k_{22} = 1(-10)$	b

Litteral subscripts are used to denote partial reaction rates; unlettered subscripts denote corresponding total rate; $2.2(-7) \equiv 2.2 \times 10^{-7}$.
Walker and Rees (1968).
Zipf (1970).
Strobel (1971).
Bardsley and Biondi (1970).
Schiff (1973).
Schiff (1972).
Zipf (1969).
Atkinson and Welge (1972).
Zipf (1973b).
Meyer *et al.* (1969).

[l] Shemansky *et al.* (1971).
[m] Young *et al.* (1969).
[n] Stuhl and Niki (1969).
[o] Dreyer and Perner (1973).
[p] Vallance Jones and Gattinger (1973).
[q] Gattinger and Vallance Jones (1973a).
[r] Jones and Rees (1973).
[s] Slanger *et al.* (1971).
[t] Weill (1969).
[u] Noxon (1962).
[v] Fehsenfeld *et al.* (1970).
[w] Ryan (1969).

4.2.3.1. *Dissociative Recombination*

The possible importance of dissociative recombination as an excitation process in aurora was pointed out by Bates *et al.* (1948) and later by Dalgarno and Khare (1967). The question of whether dissociative recombination or secondary electron impact excitation is dominant in the production of $\lambda 5577$ and its contribution to $\lambda 6300$–64 is discussed in detail in Section 4.2.4.9. A theoretical treatment requires the calculation of the concentrations of electrons and positive ions and consequently the solution of the ion chemistry equations. This is discussed in Section 5.1. The production rate of $O(^1S)$ and $O(^1D)$ can be calculated from $n(O_2^+)$ and $n(e)$ and the reaction rates of Table 4.6. The reaction rates are sensitive to electron temperature of which the height and time variations are discussed in Section 5.2. The calculated height profiles for $O(^1S)$ and $O(^1D)$ production by dissociative recombination for a 5 keV power exponential electron flux are shown in Figures 4.24 and 4.28. This process is also important in producing $N(^2D)$ and $N(^2P)$ from NO^+ and N_2^+.

4.2.3.2. *Energy Transfer*

This may be regarded as the reverse of quenching. The one well established example is the $O(^1D)-O_2(b^1\Sigma)$ process of Equation (4.2.3a) for which the reaction rate constants and yields are given in Table 4.6 (k_2). Since the rate constants for quenching of $O(^1D)$ by O_2 and N_2 are almost equal and $n(O_2)/n(N_2) \geq 0.1$ up to 150 km it follows that between 10 and 20% of the $O(^1D)$ production is transferred to the O_2 atmospheric system. (There could be some competition from NO quenching.) This possibility was investigated by Seaton (1958) with reference to the quenching $O(^1D)$. The reaction gives an almost exact energy balance for the $v' = 2$ vibrational level $O_2(b^1\Sigma)$. Originally it was thought that the reverse reaction would severely limit the quenching rate but Wallace and Chamberlain (1960) suggested that the vibrational energy of the $O_2(b^1\Sigma)$ could be lost by the vibrational quenching process,

$$O_2(b^1\Sigma,v') + O_2(X^3\Sigma_g^-) \rightarrow O_2(b^1\Sigma,v' = 0) + O_2(X^3\Sigma_g^-,v'). \qquad (4.2.3b)$$

This results in an interesting quenching effect since O_2 appears both in the energy transfer reaction and in the vibrational quenching reaction. Thus for the $v' = 1$ level we may equate the production and loss rates so that

$$k_{2a}(1)\, n(O^1D)\, n(O_2) = [k_{59}\, n(O_2) + A_{bX}]n(O_2^*,1), \qquad (4.2.3c)$$

where $n(O_2^*,1)$ represents the concentration of $O_2(b^1\Sigma)$ in the $v' = 1$ level and A_{bX} is the probability for the $b \rightarrow X$ radiative transition. Thus,

$$n(O_2^*,1)/n(O^1D) = \frac{k_{2a}(1)}{k_{59} + A_{bX}/n(O_2)}. \qquad (4.2.3d)$$

The height distribution for the bands arising from the $v' = 1$ level should thus be similar to that of $\lambda 6300$[OI] for heights somewhat below the level where $A_{bX} = k_{59}n(O_2)$. The value of k_{59} is unknown, but must be such that $k_{59}n(O_2) > A_{bX}$

up to 160 to 180 km to account for the observations by Omholt (1957) and Vallance Jones and Gattinger (1973) that $I(O_2, \text{Atm}; 1,1)$ follows $I(6300)$ in the upper parts of auroral forms. Wallace and Chamberlain estimated that $k_{59} \simeq 10^{-10}$ cm^3 s^{-1} while Vallance Jones and Gattinger derived a value of 8×10^{-11} from their height effect observations. The observed ratio of $I(1,1)/I(6300)$ fixes the value of $k_{2a}(1)/k_{59}$ at about 0.2 according to the recent data so that $k_{2a}(1)$ was estimated to be about 2×10^{-11} cm^3 s^{-1}. Since $k_2 \simeq 6 \times 10^{-11}$ and the efficiency of $O_2(b^1\Sigma)$ production in the quenching process is about 0.85 (see Table 4.6) it follows that up to 1/3 of the $O_2(b^1\Sigma)$ molecules produced in the energy transfer are in the $v'=1$ level. The observed weakness of the 2,2 O_2 atmospheric band indicates either that $k_{2a}(2) < k_{2a}(1)$ or that vibrational quenching is stronger for the $v'=2$ level of $O_2(b^1\Sigma)$. Very recently Gattinger and Vallance Jones (1974) have found bands in aurora with $v' \leq 5$. It is possible that these also arise from reaction (4.2.3a) but with $O(^1D)$ produced with excess kinetic energy from O_2^+ dissociative recombination. Since about 2.5 eV is needed for excitation to $O_2(b^1\Sigma, v'=5)$ there is enough K.E. produced in reaction 16a (Table 4.6) even if two $O(^1D)$ atoms are produced. Alternatively an energy transfer from $N(^2D)$ seems a possibility (Vallance Jones and Gattinger, 1974).

Several other energy transfer processes have been suggested and will be discussed in turn:

$$O^+(^2D) + N_2(^1\Sigma) \rightarrow O(^3P) + N_2^+(A^2\Pi, v' \leqslant 1). \tag{4.2.3e}$$

This was proposed by Omholt (1957) as a possible source of N_2^+ in the upper state of the Meinel bands. It should lead to enhancements in the $v'=1$ and possibly the $v'=0$ progressions. However Gattinger and Vallance Jones (1973a) recently found no evidence of abnormalities in the 0,1 and 1,2 bands compared to those of higher progressions at shorter wavelengths.

The corresponding process with $O^+(^2P)$ ions could populate a wide range of levels of the N_2^+A state but there is no evidence that this takes place even though this state is believed to be quenched by N_2.

$$N_2(A^3\Sigma) + O \rightarrow N_2(X^1\Sigma) + O(^1S). \tag{4.2.3f}$$

This process was suggested as a possible means of exciting part of the 5577 Å [OI] emission by Parkinson and Zipf (1970). As is clear from Table 4.5 the production rate of $N_2(A^3\Sigma)$ is sufficient to be an important source of $\lambda 5577$ if it is quenched via Equation (4.2.3f). The possible role of this mechanism is discussed in Section 4.2.4.9

$$O^+(^2D) + O(^3P) \rightarrow O^+(^4S) + O(^1D). \tag{4.2.3g}$$

This was suggested by Mahaderan and Roach (1968) as a possible mechanism for $O(^1D)$ excitation in SAR-arcs but has been shown to be unimportant by Gérard (1970). More generally by applying the ion production formulae of Section 4.2.1.4, the branching ratios of Figure 4.20 and the curve for $\eta(4278)$ from Figure 4.24 it can be calculated that $\eta(O^+, {}^2D) \leq 1/3\, \eta(O, {}^1D)$ as produced by secondary electrons (Figure 4.28). Thus even if Equation (4.2.3g) was the dominant quenching process, this reaction would be only a minor source of $O(^1D)$.

$$N(^2D) + O(^3P) \rightarrow N(^4S) + O(^1D).$$

This was suggested by Weill (1969) to be of some importance in the nightglow in the 250 km region. Our calculations indicate that it should be a minor source in aurora.

4.2.3.3. *Quenching*

This is the loss of excitation energy in collisional processes. At upper atmospheric pressures it is only of importance for metastable particles. Quenching reactions are of the type

$$X(n) + M_k \rightarrow X + M_k \text{ (Rate const } k_{Qk})$$

where M is the kth quenching particle and X the excited particle which is quenched from the nth excited state. Let us assume the particle emits by making a transition to the mth state. In the steady state, the excitation rate, $\eta(X,n)$, the quenching term and the radiation transition probability $A(n,m)$ are related by the equation

$$\eta(X,n) = [A(n,m) + k_Q\, n(M)]n(X,n)$$

so that the emission rate per unit volume becomes

$$\eta(X,n,m) = n(X,n)A(n,m) = \frac{\eta(X,n)}{1 + \dfrac{k_Q\, n(M)}{A(n,m)}}. \tag{4.2.3h}$$

If there are p lower states to which the particle can decay by radiating $A(n,m)$ must be replaced by $\sum_i^p A(n,i)$ and the right hand side of Equation (4.2.3h) must be multiplied by $b(n,m) = A(n,m)/(\sum_i^p A(n,i))$, the branching factor, for the transition to the mth state. Likewise if there were r competing quenching reactions $k_Q n(M)$ would be replaced by $\sum_k^r k_{Qk}\, n(M_k)$ where M_k represents the kth quenching species. When cascading from higher levels occurs the rate of the cascading transition must be added to direct production term $\eta(X,n)$.

The general expression for the emitted intensity of the transition from the nth to the mth state then becomes

$$\eta(X,n,m) = \eta'(X,n)\, b(n,m)q(n), \tag{4.2.3i}$$

where

$$q(n) = 1/(1 + \sum_{k}^{r} k_{Qk}\, n(M_k)/(\sum_{i}^{p} A(n,i))) \tag{4.2.3j}$$

and $\eta'(X,n)$ is the total production rate of particle X in the nth state including cascade from higher levels. $q(n)$, the quenching factor, determines the severity of the effect. At a height for which $q(n) = 1/2$ (which occurs for the simple case of Equation (4.2.3h) when $n(M) = A(n,m)/k_Q$) the emission rate is reduced to one half of its unquenched value; this height may be called the *quenching height*, h_Q. Above h_Q the quenching

TABLE 4.7

Quenching rates and heights for metastable emitters

Emission	Upper state	Lifetime	Dominant quenching particle, Q	Quenching[a] rate constant, k_Q cm^3 s^{-1}	Critical density, Q $n(Q)_c$	Quenching height h_Q km
λ5577	OI ^1S	0.7 s	O_2	3(−13)	5(+12)	~95[j]
λ2972			O	7.5(−12)	2(+11)	115
λλ6300–64	OI ^1D	110 s	N_2	5(−11)	2(+8)	250–350
λλ7319–30	OII ^2P	5 s	N_2	1(−10)	1(+8)	250–350
λλ3727–9	OII ^2D	2.9 h	N_2	1(−9)	1(+5)	~500
λλ10395–404	NI ^2P	12 s	N_2	<3(−19)	>2.8(+17)	<40
λ3466			O_2	~1(−11)	1(+10)	135
λλ5199–201	NI ^2D	26 h	O_2	6(−12)	2(+6)	250–300
			O	2(−13)	5(+7)	320
			NO ?[e]	1(−10)		
λ5755	NII ^1S	0.9 s	O_2, O	≃1(−9)[b]	1(+9)	180
λ6584	NII ^1D	246 s	O_2, O	?	4(+6)	250
N_2(LBH)	$a^1\Pi_g$	0.14 ms[c]	?	<3(−10)	>2.5(+13)	<95
N_2(VK)	$A^3\Sigma_u^+$	~2 s[d]	O	1.5(−11)	3(+10)	145
O_2(Atm.)	$b^1\Sigma_g^+$	12 s	N_2	2(−15)	4(+13)	95
O_2(IR Atm.)	$a^1\Delta_g$	60 mn	O_2	2.2(−18)	1.25(+14)	71
N_2^+(1N)	$B^2\Sigma$	70 ns[f]	N_2+O_2	4(−10)[g]	3.6(+16)	~48
N_2^+(M)	$A^2\Pi$	14 μs[h]	N_2	~5(−10)[k]	1.4(+14)	85–90[l]
N_2(1P)	$B^3\Sigma$	6 μs[f]	N_2	1(−11)[g]	1.7(+16)	~53
N_2(2P)	$C^3\Pi$	50 ns[f]	O_2	3(−10)[g]	7(+16)	~30
O_2^+(1N)	$b^4\Sigma$	1.2 μs[x]	N_2	2(−10)	4(+15)	~60

[a] Unless indicated otherwise the rate constants come from Table 4.6.

[b] Sum of k_{13} and k_{14}.

[c] Lifetime from Shemansky (1969b); value of h_Q is upper limit for gas kinetic cross section.

[d] Shemansky (1969a).

[e] According to Zipf *et al.* (1970) n(NO, 120 km) ≈ n(O_2, 120 km) for one 80 kR aurora at Churchill.

[f] Nicholls (1969).

[g] O'Neill and Davidson (1969).

[h] Peterson and Moseley (1973).

[i] From observation of Simpson and McConkey (1969) that quenching threshold is at 1.5×10^{-3} torr.

[j] Gault's (1970) analysis of meteor train λ5577 decay times indicates a greater value for h_Q in a meteor wake in which conditions may differ from the normal atmosphere.

[k] Cartwright *et al.* (1974).

[l] Borst and Zipf (1970b).

rapidly becomes negligible while below, its effects are severe. A tabulation of quenching heights is given in Table 4.7 while quenching reactions and their rates are given in Table 4.6. The most notable emissions affected by significant quenching will now be discussed in turn.

[*O I*] $\lambda 5577$, $\lambda 6300$ *and* $\lambda 2972$. It is convenient to discuss these together because of the additional cascading effect. The transitions are indicated on Figure 4.8. For brevity we will denote the energy levels 1, 2 and 3 in order of increasing energy. For the $\lambda 5577$ emission, by applying Equation (4.2.3i) one obtains

$$\eta(5577) = \eta(\mathrm{O}^1\mathrm{S}) \frac{A(3,2)}{A(3,2)+A(3,1)} \frac{1}{1+\dfrac{k_4[\mathrm{O}]+k_3[\mathrm{O}_2]}{A(3,2)+A(3,1)}} \tag{4.2.3k}$$

since quenching by O and O_2 are the important reactions. For the $\lambda 6300$ emission

$$\eta(6300) = (\eta(\mathrm{O}^1\mathrm{D}) + \eta(5577)) \frac{1}{1+\dfrac{k_1[\mathrm{N}_2]+k_2[\mathrm{O}_2]}{A(2,1)}} \tag{4.2.3l}$$

(Quenching by NO has been, perhaps wrongly, neglected.) Finally $\eta(2972) = (A(3,1)/A(3,2))\eta(5577)$. For $\lambda 5577$ the quenching height determined by k_4 is about 115 km so quenching will be important if the recent values for k_4 are correct. For $\lambda 6300$ the quenching height is 250 to 350 km depending on atmospheric composition above 200 km (which is sensitive to exospheric temperature, turbopause composition and possible dynamic effects). The severity of the effect can be seen from the production and emission curves of Figure 4.28.

The ratio $\eta(6300)/\eta(5577)$ is very sensitive to quenching effects and is readily seen to be

$$\eta(6300)/\eta(5577) = \left\{\frac{\eta(\mathrm{O}^1\mathrm{D})}{0.94\eta(\mathrm{O}^1\mathrm{S})\, q(\mathrm{O}^1\mathrm{S})} + 1\right\} q(\mathrm{O}^1\mathrm{D}).$$

As may be seen from Figures 4.24 and 4.28 $\eta(\mathrm{O}^1\mathrm{D})/\eta(\mathrm{O}^1\mathrm{S})$ is in the range 15 to 25 at most heights but $q(\mathrm{O}^1\mathrm{D})$ becomes extremely small below 150 km, so that the ratio $\eta(6300)/\eta(5577)$ should be very small.

$\lambda 5200$, $\lambda 10\,400$ *and* $\lambda 3466$ [*N I*]. The upper levels of these transitions (see Figure 4.10) have lifetimes of approximately 26 h for the first and 12 s for the last two. The quenching height (according to the data from Table 4.6) is about 300 km for the ^{2}D state and about 130 km for the ^{2}P if the value assumed for k_{66} is correct. Consequently $\lambda 5200$ is strongly quenched and has a height distribution similar to that of $\lambda 6300$ while $\lambda 3466$ and $\lambda 10\,400$ should have more normal profiles.

λ7325 and λ3728 [O II]. The upper level radiative lifetimes are 5 s and 3.6 h (see Figure 4.9). The quenching rate constant for $O^+(^2P)$ is merely a plausible value assumed by Dalgarno and McElroy (1965). The rate constant for $O^+(^2D)$ quenching has been measured to be 3×10^{-10} cm^3 s^{-1}. The corresponding quenching heights are 200 to 300 km and 500 to 600 km.

λ5755 and λ6584 [N II]. These transitions are similar spectroscopically to λ5577 and λ6300 but as ions are probably more readily deactivated. One can only assume that they react at the same rate as the ground state ion and this gives the quenching heights of Table 4.7.

N_2 Vegard-Kaplan Bands. The upper state, $N_2A^3\Sigma$ has a radiative lifetime of about 2 s. It is believed to be quenched by O and possibly O_2 and NO with rate constants as given in Table 4.6. It was suggested by Shemansky *et al.* (1971) that NO is the important quencher below 130 km. Zipf *et al.* (1970) have measured the NO and O_2 concentrations in an aurora of IBC = 2^+ and the NO concentration appears to exceed that of O_2 above 100 km. If this observation is confirmed the quenching factor for $N_2A^3\Sigma$ at lower levels would decrease for more intense aurora and consequently the height distribution of the VK N_2 bands would also be a function of the time-integrated auroral intensity.

According to Dreyer and Perner (1973), k_{56} is more than twice as fast for $v' = 1$. This could affect the relative band intensities. It also appears that there is a vibrational deactivation analogous to that of Equation (4.2.3b).

4.2.3.4. *Thermal Excitation*

As will be discussed in Section 5.3 electron temperatures increase rapidly in the presence of an energetic auroral electron flux. Figure 4.25 shows electron temperatures reached after 100 s for an incident electron flux having a Maxwellian energy distribution with a characteristic energy of 5 keV. The enhanced electron temperature gives rise to a significant excitation of the O^1D at heights above 200 km. The thermal electron contribution is indicated on Figure 4.28. It is calculated from an expression of the form

$$\eta(O^1D, \text{thermal}) = \alpha(T_e)n(e)n(O),$$

where

$$\alpha(T_e) = \int_0^\infty f(E)v(E)\sigma(O^1D, E)\,dE,$$

where $f(E)$ is the fraction of electrons in the range E to $E + dE$, $v(E)$ is the velocity of an electron of energy E and σ is the excitation cross section. $\alpha(T_e)$ has been evaluated and expressed in analytical form by Rees *et al.* (1967). The expression is equal to the coefficient of $n(e)n(O)$ in the $O(^1D)$ cooling rate of Table 5.3 divided by the $O(^1D)$ excitation energy.

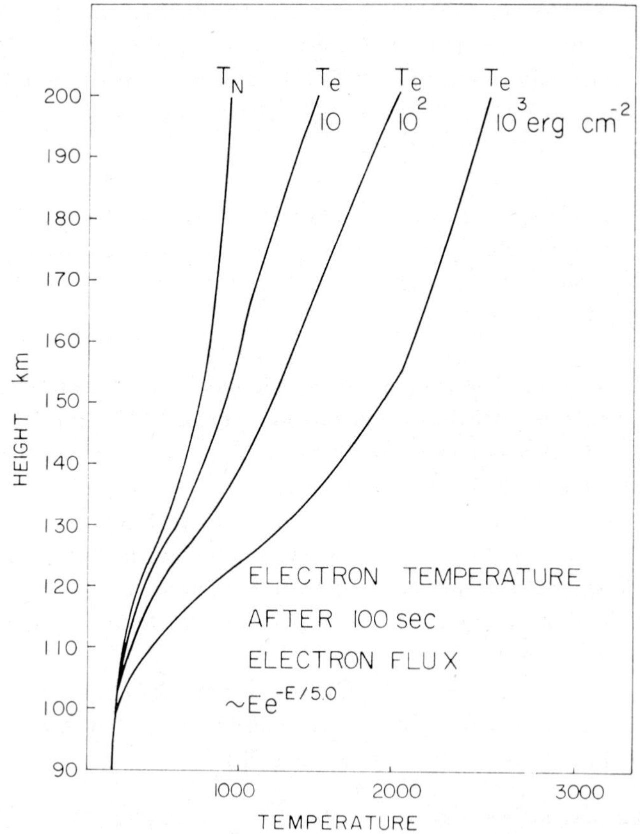

Fig. 4.25. Electron temperature profiles in aurora calculated for different energy fluxes according to the method described in Section 5.3.9.

Electron temperatures may also be driven up by the auroral electric fields described in Section 3.6.3. As discussed in Section 5.3.9 such fields are not very effective in raising electron temperatures within auroral forms at heights less than 250 km and consequently would not be expected to contribute to visible emissions closely associated with discrete forms. Outside discrete aurora the larger fields might be effective in exciting low lying states such as $O(^1D)$ or $O_2(a^1\Delta)$. (See also Section 4.2.4.5.)

Ion thermal excitation. In contrast to electrons, ions are strongly heated by electric fields. As discussed in Section 5.3 ion temperatures may be raised by 1000 K or more by fields of 50 mV m^{-1} as found outside auroral forms. This has been discussed by Walker (1971), and Cole (1971). Unfortunately the excitation cross sections for ion collisions are mostly unknown so no quantitative estimates of the real importance of this mode of excitation are possible. It may be important in the extensive glows seen outside auroral forms in strong displays.

4.2.3.5. *Cascading*

This is the population of an excited state by radiative or perhaps non-radiative transitions from a higher excited state. One example, noted in Section 4.2.3.3 is the contribution to $\eta(6300\text{–}64)$ entailed in the emission $\lambda 5577$. The importance of cascading depends on the population rate of the uppermost state, the branching ratio for the cascading transition and the relative population rate of the intermediate state by other paths. For example for the [OI] lines the branching ratio, b(3,2), is 0.94 while the population rate of $O(^1S)$ is high; however $\eta(O^1D) \gg \eta(O^1S)$ so that cascading is only a minor contributor to $\eta(6300\text{–}64)$. For the N_2^+ band systems (see Figure 4.12) cascading to the $A^2\Pi$ state is not important because the branching ratio to this state from $B^2\Sigma$ is small.

The most discussed example of cascading is that of the triplet systems of N_2 as shown in Figure 4.12. The potential curve energy level diagram showing the vibrational levels (Figure 4.26) gives a better idea of the processes involved. Clearly the cascading

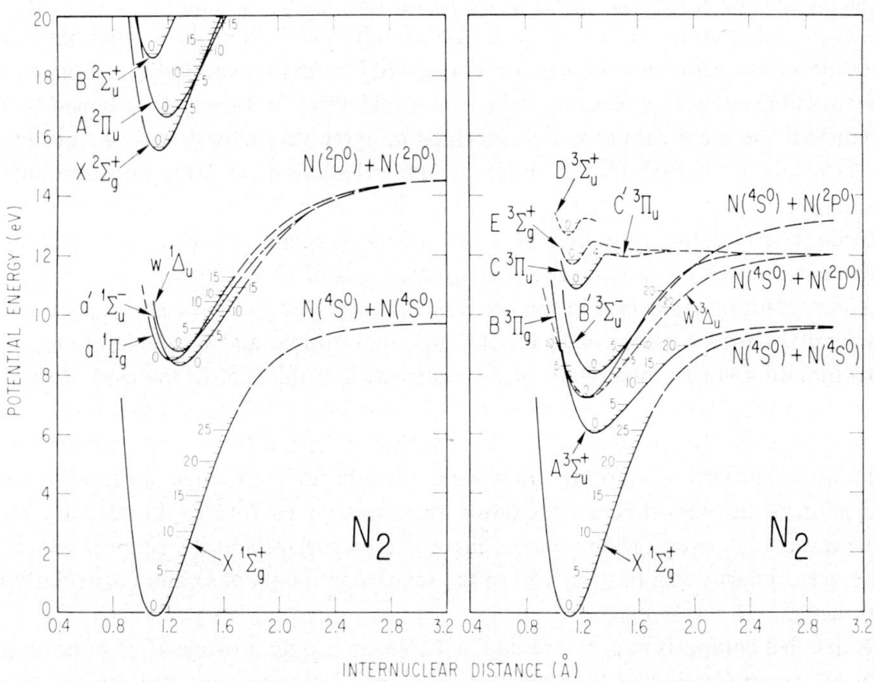

Fig. 4.26. Potential energy curves and vibrational levels for electronic states of N_2. The observed transitions are shown on Figure 4.12 which shows only the minimum energy levels for each state from Cartwright (1973b); the original curves for all except the $W^3\Delta_u$ state are from Gilmore (1965).

involves transitions between large numbers of vibrational levels rather than the simple three-level situation for the [OI] states. Equation (4.2.3i) can be applied in turn to the vibrational levels of each excited state. A full set of equations describing the cascading processes has been given by Cartwright *et al.* (1971). This involves consideration of the B→A, E→A, C→B, W→B, A→B, E→B, D→B, B→W, E→C, D→E, B′→B and A→X transitions. In the cases of A – B, W – B transitions reverse radiative transitions are allowed from the higher vibrational levels of the 'lower' energy state to the lower vibrational levels of the 'higher' energy state. This complicates the solution for these pairs of states since the equations, through the cascade terms, involve the vibrational populations of both members. The excitation rates of the vibrational levels can be determined from the total excitation rate divided according to Equation (4.1e). Cartwright *et al.* have described how the resulting equations may be solved numerically.

4.2.4. Predicted electron auroral spectrum and comparison with observation

It is somewhat optimistic to claim that the intensities of all the features in the auroral spectrum can yet be calculated from a given input flux of electrons. Some important cross sections have not been determined to within less than a factor of 2 while others are even less certain. In some cases the excitation mechanism is debatable. The question of atmospheric composition above 120 km and its variability is not settled. The possible effects of electric fields are difficult to evaluate. Consequently the 'predicted' spectrum cannot yet be expected to agree very closely with the observed. Nevertheless it is desirable to attempt the calculations, taking due account of reasonable variations in the uncertain quantities, since in the first place the results provide a test of the excitation processes assumed while secondly the calculations furnish the means of correlating many isolated sets of observations.

The procedures described in the preceding sections led to the height vs production curves of Figures 4.24 and 4.28 for the principal atomic and molecular states. The corresponding height integrated rates were given in Table 4.5. In the case of $O(^1D)$, both the total production*, $I(O^1D)$, and the emission rate after quenching, $I(6300)$, have been given. Quenching has not been included for the $I(N_2, A^3\Sigma)$ and $I(O_2b^1\Sigma)$ columns as implied by quoting the production rate for the excited state. With some exceptions it appears that the excitation rates relative to $I(4278)$ do not vary more than about 15% over a wide range of input energy spectra and the changes which do arise are primarily due to decreases in the relative amounts of O and particularly O_2 with height.

A detailed comparison of the results of Table 4.5 is only possible after some discussion of the application of the theory of Section 4.2.3 and of the distribution of the total production rates among the vibrational bands. Various features of the spectra will be considered in turn, singly or in groups as appropriate.

* We shall use I to denote a height integrated production rate. When the argument denotes an optical emission, I is proportional to emission intensity and $I/10^6$ equals the brightness in rayleighs when there is no absorption. Likewise we shall use N to indicate a height integrated concentration (or abundance).

tions (Gattinger and Vallance Jones, 1974) are given in column 4; they are in reasonable agreement with both sets of predictions except for $v' = 1$ and 2.

In setting up a table of predicted band intensities for the 1P system we have used Equation (4.2.4b) with the observed relative $N(v')$ values of column 4 of Table 4.8 a, together with A-values of Shemansky and Broadfoot (1971b). The intensities were normalized to the value $I(1\text{P}; 5{,}2) = 18.2$ kR observed by Gattinger and Vallance Jones (1974) and tabulated in Table 4.12. This results in a system intensity of 880 kR. The populations of column 2 give a total intensity of 590 kR when treated in the same way. These observed system intensity estimates are comparable to the calculated excitation rate of about 800 kR.

TABLE 4.9

IN N2+BAND INTENSITIES

	V'' 0	1	2	3	4	5
0	3914.4 0.110E 08 98.51	4278.1 0.335E 07 30.00	4709.2 0.675E 06 6.04	5228.3 0.111E 06 0.99	5864.7 0.161E 05 0.14	0.0 0.000 0.00
1	3582.1 0.670E 07 6.80	3884.3 0.387E 07 3.93	4236.5 0.381E 07 3.87	4651.8 0.132E 07 1.34	5148.8 0.312E 06 0.32	5754.4 0.590E 05 0.06
V' 2	3308.0 0.127E 07 0.02	3563.9 0.916E 07 0.15	3857.9 0.899E 06 0.01	4199.1 0.313E 07 0.05	4599.7 0.171E 07 0.03	5076.6 0.543E 06 0.01
SUM V''	105.33	34.08	9.92	2.38	0.49	0.07

```
            V''
       WAVELENGTH P-HEADS
V'     A(V',V'')
       INTENSITY(KR) IBC3 AURORA
       SYSTEM INTENSITY =  152. KR
       BANDS I< 0.0 KR  KR OMITTED
```

N_2 VK system. The relative contributions to the population rate of the $A^3\Sigma$ state have been calculated by Cartwright *et al.* (1971); the most recent results (Cartwright, 1973b) are not very different. As originally pointed out by Broadfoot and Hunten (1964), the rates for low v' are dominated by cascading through the 1P ($B \rightarrow A$) bands. According to Table 4.5 there should be about 350 kR of direct excitation mostly at higher values of v'. Table 4.8b, column 3, gives the relative population values calculated by Cartwright for a height of 170 km where quenching is less severe and also in column 2 the populations derived from the results of Shemansky *et al.* (1971) who used the simpler cascade scheme described above.

As noted in Section 4.2.3.3 quenching is severe for the A-state and as pointed out by Cartwright *et al.* (1971) there may also be a loss (for $v' \geqslant 8$) due to the $A \rightarrow B$ (reverse 1P) transition.

TABLE 4.10*

M N2+ BAND INTENSITIES

V' \ V''	0	1	2	3	4	5	6	7	8	9
0	11086.9 0.455E 05 122.69	14609.8 0.136E 05 36.67	21264.9 0.119E 04 3.21	38560.8 0.279E 02 0.08	0.0 0.170E-01 0.00	0.0 0.000 0.00	0.0 0.000 0.00	0.0 0.000 0.00	0.0 0.000 0.00	0.0 0.000 0.00
1	9181.5 0.553E 05 158.01	11472.4 0.352E 04 10.06	15210.5 0.112E 05 32.00	22395.7 0.202E 04 5.77	41860.3 0.675E 02 0.19	0.0 0.216E-01 0.00	0.0 0.000 0.00	0.0 0.000 0.00	0.0 0.000 0.00	0.0 0.000 0.00
2	7852.5 0.373E 05 65.49	9469.9 0.367E 05 64.43	11879.8 0.105E 04 1.84	15851.9 0.573E 04 10.06	23628.7 0.219E 04 3.84	45679.6 0.992E 02 0.17	0.0 0.767E-02 0.00	0.0 0.000 0.00	0.0 0.000 0.00	0.0 0.000 0.00
3	6873.6 0.189E 05 14.22	8081.8 0.517E 05 38.91	9774.0 0.128E 05 9.63	12312.2 0.621E 04 4.67	16540.5 0.186E 04 1.40	24982.6 0.186E 04 1.40	50168.0 0.112E 03 0.08	0.0 0.141E-49 0.00	0.0 0.000 0.00	0.0 0.000 0.00
4	6123.0 0.817E 04 2.11	7063.7 0.407E 05 10.50	8323.1 0.426E 05 10.99	10095.4 0.140E 04 0.36	12772.5 0.882E 04 2.28	17282.1 0.205E 03 0.05	26477.2 0.133E 04 0.34	55519.8 0.107E 03 0.03	0.0 0.000 0.00	0.0 0.000 0.00
5	5529.4 0.320E 04 0.33	6285.2 0.239E 05 2.43	7263.1 0.514E 05 5.23	8577.1 0.246E 05 2.50	10435.5 0.528E 03 0.05	13263.1 0.790E 04 0.80	18082.5 0.651E 02 0.01	28133.0 0.806E 03 0.08	61996.2 0.885E 02 0.01	0.0 0.000 0.00
6	5048.3 0.119E 04 0.05	5671.0 0.119E 05 0.46	6455.1 0.405E 05 1.55	7472.6 0.480E 05 1.84	8844.8 0.927E 04 0.85	10795.6 0.405E 04 0.15	13786.3 0.518E 04 0.20	18946.9 0.552E 03 0.02	29972.8 0.405E 03 0.02	69963.1 0.649E 02 0.00
7	4650.7 0.426E 03 0.01	5174.1 0.538E 04 0.08	5819.0 0.252E 05 0.37	6633.2 0.508E 05 0.74	7692.6 0.353E 05 0.51	9127.0 0.137E 04 0.02	11176.9 0.728E 04 0.11	14344.4 0.248E 04 0.04	19881.4 0.105E 04 0.02	32022.6 0.156E 03 0.00
8	4316.7 0.151E 03 0.00	4764.0 0.228E 04 0.01	5305.4 0.136E 05 0.08	5974.0 0.393E 05 0.22	6819.9 0.515E 05 0.29	7923.9 0.201E 05 0.11	9424.6 0.190E 03 0.00	11580.9 0.826E 04 0.05	14940.2 0.714E 03 0.00	20892.9 0.129E 04 0.01
9	4032.5 0.531E 02 0.00	4420.1 0.927E 03 0.00	4882.4 0.670E 04 0.02	5442.9 0.251E 05 0.06	6136.4 0.495E 05 0.12	7016.0 0.434E 05 0.10	8167.5 0.805E 04 0.02	9739.0 0.269E 04 0.01	12009.9 0.710E 04 0.02	15577.7 0.333E 02 0.00
10	3787.8 0.187E 02 0.00	4127.9 0.368E 03 0.00	4528.3 0.310E 04 0.00	5006.5 0.143E 05 0.01	5587.3 0.376E 05 0.02	6307.2 0.529E 05 0.03	7222.6 0.306E 05 0.02	8424.8 0.149E 04 0.00	10072.3 0.528E 04 0.00	12467.1 0.481E 04 0.00
SUM V''	450.05	196.88	76.26	29.54	9.86	3.02	0.81	0.22	0.06	0.0'

V' \ V''	10	11	12	13
0	0.0 0.000 0.00	0.0 0.000 0.00	0.0 0.000 0.00	0.0 0.000 0.00
1	0.0 0.000 0.00	0.0 0.000 0.00	0.0 0.000 0.00	0.0 0.000 0.00
2	0.0 0.000 0.00	0.0 0.000 0.00	0.0 0.000 0.00	0.0 0.000 0.00
3	0.0 0.000 0.00	0.0 0.000 0.00	0.0 0.000 0.00	0.0 0.000 0.00
4	0.0 0.000 0.00	0.0 0.000 0.00	0.0 0.000 0.00	0.0 0.000 0.00
5	0.0 0.000 0.00	0.0 0.000 0.00	0.0 0.000 0.00	0.0 0.000 0.00
6	0.0 0.000 0.00	0.0 0.000 0.00	0.0 0.000 0.00	0.0 0.000 0.00
7	79957.8 0.424E 02 0.00	0.0 0.000 0.00	0.0 0.000 0.00	0.0 0.000 0.00
8	34315.0 0.361E 02 0.00	92818.8 0.245E 02 0.00	0.0 0.000 0.00	0.0 0.000 0.00
9	21991.1 0.123E 04 0.00	36893.7 0.765E 00 0.00	0.0 0.124E 02 0.00	0.0 0.000 0.00
10	16262.8 0.128E 03 0.00	23190.1 0.988E 03 0.00	39822.5 0.814E 01 0.00	0.0 0.535E 01 0.00
SUM V''	0.00	0.00	0.00	0.00

```
         V''
    WAVELENGTH ORIGIN
V'  A(V',V'')
    INTENSITY(KR) IBC3 AURORA
    SYSTEM INTENSITY =  631. KR
    BANDS I< 0.0 KR  KR OMITTED
```

* In this and following similar tables the entry 'sum v''' is in error and should be ignored.

TABLE 4.11

2P N2 BAND INTENSITIES

V' \ V''	0	1	2	3	4	5	6	7	8	9
0	3370.0	3576.0	3804.0	4058.0	4343.0	4665.0	0.0	0.0	0.0	0.0
	0.110E 08	0.733E 07	0.294E 07	0.923E 06	0.247E 06	0.606E 05	0.000	0.000	0.000	0.000
	30.68	20.44	8.20	2.57	0.69	0.17	0.00	0.00	0.00	0.00
1	3158.0	3338.0	3536.0	3754.0	3997.0	4268.0	4573.0	4917.0	0.0	0.0
	0.102E 08	0.528E 06	0.461E 07	0.410E 07	0.249E 07	0.769E 06	0.235E 06	0.674E 05	0.000	0.000
	14.99	0.78	6.77	6.02	3.66	1.13	0.35	0.10	0.00	0.00
2	2976.0	3135.0	3309.0	3499.0	3709.0	3942.0	4200.0	4489.0	4813.0	0.0
	0.349E 07	0.884E 07	0.589E 06	0.146E 07	0.337E 07	0.263E 07	0.132E 07	0.526E 06	0.182E 06	0.000
	1.88	4.76	0.32	0.79	1.82	1.42	0.71	0.28	0.10	0.00
3	2818.0	2961.0	3115.0	3284.0	3468.0	3671.0	3894.0	4140.0	4415.0	4722.0
	0.507E 06	0.661E 07	0.548E 07	0.218E 07	0.112E 06	0.201E 07	0.248E 07	0.169E 07	0.807E 06	0.333E 06
	0.08	1.04	0.86	0.34	0.02	0.32	0.39	0.27	0.13	0.05
4	2684.0	2812.0	2952.0	3102.0	3266.0	3445.0	3641.0	3856.0	4093.0	4355.0
	0.184E 05	0.133E 07	0.864E 07	0.324E 07	0.201E 07	0.912E 05	0.911E 06	0.195E 07	0.173E 07	0.101E 07
	0.00	0.06	0.40	0.15	0.09	0.00	0.04	0.09	0.08	0.05
SUM V''	497.67	223.96	92.82	39.42	16.13	6.06	2.30	0.96	0.37	0.11

```
        V''
        WAVELENGTH ORIGINS
V'      A(V',V'')
        INTENSITY(KR) IBC3 AURORA
        SYSTEM INTENSITY =   113. KR
        BANDS I< 0.0 KR   KR OMITTED
```

As a guide to the possible intensities of the auroral bands Table 4.13 was prepared. The relative populations of Cartwright (1973b) were used with the transition probabilities of Shemansky (1969a) and the table was normalized to $I(1,11) = 1.5$ kR (cf., Vallance Jones, 1971). Because of the severe and variable quenching of the system the absolute intensities in Table 4.16 might be expected to vary by a factor of at least 3. The system emission rate is about 55 kR, a value consistent with considerable quenching since the total production rate including the B → A cascade from Table 4.5 is about 1000 kR.

Other triplet systems. Some bands of the Goldstein-Kaplan system ($C'^3\Pi_u - B^3\Pi_g$) have been suspected at 2863 and 2890 Å by Crosswhite and Fastie (1962). The 0,10 band at 4432 Å was reported by Dick and Fastie (1970) in high resolution spectra.

Transitions in the IR beyond 1 μm for the A → B, W → B, B → W, E → C have been predicted and their intensities calculated by Cartwright *et al.* (1972a). None of these bands have yet been observed in aurora. A number of the bands beyond 1.5 μm should have IBC3 intensities greater than 10 kR.

The D − B (4th positive) and E − A (Herman-Kaplan) systems are observed in the laboratory and their intensities were computed but not given by Cartwright *et al.* (1971). Presumably they are weak since the excitation cross sections are an order of magnitude smaller than those of the A, B and C states.

The B′ − B system (IR afterglow system or Y Bands) is observed in the laboratory and very recently has been detected in aurora by Gattinger and Vallance Jones (1974). Its electron impact cross section according to Cartwright (1973b) is larger than originally supposed (Cartwright, 1970).

TABLE 4.12

1P N2 BAND INTENSITIES

V''

V'	0	1	2	3	4	5	6	7	8	9
0	10508.3 0.625E 05 96.36	12373.0 0.356E 05 54.88	14977.0 0.112E 05 17.27	18874.0 0.247E 04 3.81	0.0 0.397E 03 0.61	0.0 0.424E 02 0.07	0.0 0.000 0.00	0.0 0.000 0.00	0.0 0.000 0.00	0.0 0.000 0.00
1	8912.4 0.872E 05 124.50	0.0 0.412E 03 0.59	11925.0 0.185E 05 26.41	14269.0 0.148E 05 21.13	17687.0 0.569E 04 8.12	23060.0 0.140E 04 2.00	0.0 0.227E 03 0.32	0.0 0.000 0.00	0.0 0.000 0.00	0.0 0.000 0.00
V' 2	7753.7 0.444E C5 53.61	8723.0 0.617E 05 74.50	9939.9 0.125E 05 15.09	11516.0 0.268E 04 3.24	13635.0 0.105E 05 12.68	16643.0 0.729E 04 8.80	21210.0 0.271E 04 3.27	0.0 0.639E 03 0.77	0.0 0.887E 02 0.11	0.0 0.000 0.00
3	6875.2 0.107E 05 8.80	7626.8 0.773E 05 63.56	8542.2 0.217E 05 17.84	9680.4 0.285E 05 23.43	0.0 0.797E 03 0.66	13061.0 0.388E 04 3.19	15717.0 0.633E 04 5.20	19619.0 0.368E 04 3.03	0.0 0.124E 04 1.02	0.0 0.257E 03 0.21
4	6186.7 0.129E 04 0.75	6788.6 0.302E 05 17.59	7504.7 0.836E 05 48.70	8370.1 0.154E 04 0.90	9436.4 0.294E 05 17.13	10781.4 0.784E 04 4.57	0.0 0.193E 03 0.11	14890.0 0.152E 04 0.89	0.0 0.374E 04 2.18	0.0 0.184E 04 1.07
5	0.0 0.742E 02 0.03	6127.3 0.508E 04 1.76	6704.8 0.526E 05 18.20	7387.2 0.686E 05 23.74	8205.5 0.302E 04 1.04	9203.9 0.191E 05 6.61	10448.2 0.152E 05 5.26	0.0 0.110E 04 0.38	0.0 0.102E 04 0.35	17063.0 0.281E 04 0.97
6	0.0 0.000 0.00	0.0 0.371E 03 0.07	6069.7 0.119E 05 2.27	6623.6 0.709E 05 13.53	7274.0 0.438E 05 8.36	8047.9 0.158E 05 3.01	8983.4 0.710E 04 1.35	10133.3 0.176E 05 3.36	11588.5 0.511E 04 0.97	0.0 0.000 0.00
7	0.0 0.000 0.00	0.0 0.000 0.00	0.0 0.107E 04 0.12	6013.5 0.214E 05 2.36	6544.9 0.810E 05 8.93	7164.8 0.203E 05 2.24	7896.9 0.288E 05 3.17	0.0 0.509E 03 0.06	9841.8 0.143E 05 1.58	11169.7 0.939E 04 1.03
8	0.0 0.000 0.00	0.0 0.000 0.00	0.0 0.000 0.00	0.0 0.234E 04 0.13	5959.0 0.325E 05 1.80	6468.6 0.815E 05 4.52	7060.0 0.494E 04 0.27	7752.0 0.350E 05 1.94	0.0 0.138E 04 0.08	0.0 0.811E 04 0.45
9	0.0 0.000 0.00	0.0 0.000 0.00	0.0 0.000 0.00	0.0 0.777E 02 0.00	0.0 0.429E 04 0.13	5906.0 0.438E 05 1.28	6394.7 0.731E 05 2.14	0.0 0.000 0.00	7612.9 0.328E 05 0.96	0.0 0.736E 04 0.21
10	0.0 0.000 0.00	0.0 0.000 0.00	0.0 0.000 0.00	0.0 0.000 0.00	0.0 0.159E 03 0.00	0.0 0.692E 04 0.10	5854.4 0.538E 05 0.75	6322.8 0.586E 05 0.81	0.0 0.399E 04 0.06	0.0 0.244E 05 0.34
11	0.0 0.000 0.00	0.0 0.000 0.00	0.0 0.000 0.00	0.0 0.000 0.00	0.0 0.000 0.00	0.0 0.282E 03 0.00	0.0 0.101E 05 0.07	5804.2 0.611E 05 0.43	6253.1 0.414E 05 0.29	0.0 0.133E 05 0.09
12	0.0 0.000 0.80	0.0 0.000 0.00	0.0 0.000 0.00	0.0 0.000 0.00	0.0 0.000 0.00	0.0 0.000 0.10	0.0 0.449E 03 0.00	0.0 0.138E 05 0.04	5755.2 0.649E 05 0.21	6185.2 0.248E 05 0.08
SUM V''	781.72	436.92	238.73	131.68	75.59	42.45	24.23	12.67	8.16	4.58

V'	10	11	12
0	0.0 0.000 0.00	0.0 0.000 0.00	0.0 0.000 0.00
1	0.0 0.000 0.00	0.0 0.000 0.00	0.0 0.000 0.00
V' 2	0.0 0.000 0.00	0.0 0.000 0.00	0.0 0.000 0.00
3	0.0 0.254E 02 0.02	0.0 0.000 0.00	0.0 0.000 0.00
4	0.0 0.544E 03 0.32	0.0 0.903E 02 0.05	0.0 0.000 0.00
5	0.0 0.213E 04 0.74	0.0 0.895E 03 0.31	0.0 0.224E 03 0.08
6	0.0 0.140E 04 0.27	0.0 0.193E 04 0.37	0.0 0.117E 04 0.22
7	0.0 0.112E 04 0.12	0.0 0.231E 03 0.03	0.0 0.126E 04 0.14
8	10779.9 0.115E 05 0.64	0.0 0.374E 04 0.21	0.0 0.933E 02 0.01
9	0.0 0.244E 04 0.07	0.0 0.104E 05 0.30	0.0 0.647E 04 0.19
10	0.0 0.154E 05 0.21	0.0 0.000 0.00	0.0 0.696E 04 0.10
11	0.0 0.138E 05 0.10	0.0 0.209E 05 0.15	0.0 0.158E 04 0.01
12	0.0 0.237E 05 0.08	0.0 0.486E 04 0.02	0.0 0.219E 05 0.07
SUM V''	2.56	1.43	0.81

V''
V' WAVELENGTH P11 HEAD
A(V',V'')
INTENSITY(KR) IBC3 AURORA
SYSTEM INTENSITY = 882. KR
BANDS I< 0.6 KR KR OMITTED

TABLE 4.13

VK N2 BAND INTENSITIES

V' \ V''	0	1	2	3	4	5	6	7	8	9
0	2009.5 0.410E-05 0.00	2108.3 0.600E-03 0.01	2215.8 0.651E-01 1.15	2333.3 0.292E-01 0.52	2462.2 0.757E-01 1.34	2604.3 0.131E 00 2.31	2761.5 0.163E 00 2.88	2936.4 0.154E 00 2.72	3132.2 0.113E 00 2.00	3352.7 0.653E-01 1.15
1	1953.2 0.310E-04 0.00	2046.5 0.650E-03 0.01	2147.6 0.979E-02 0.14	2257.8 0.329E-01 0.48	2378.3 0.453E-01 0.67	2510.6 0.223E-01 0.33	2656.4 0.000 0.00	2817.9 0.353E-01 0.52	2997.7 0.114E 00 1.68	3199.0 0.166E 00 2.44
V' 2	1901.1 0.750E-03 0.01	1989.3 0.000 0.00	2084.7 0.514E-02 0.06	2188.4 0.123E-01 0.14	2301.4 0.283E-02 0.03	2425.0 0.830E-02 0.09	2560.8 0.521E-01 0.59	2710.6 0.649E-01 0.74	2876.6 0.188E-01 0.21	3061.5 0.487E-02 0.06
3	1852.6 0.433E-02 0.04	1936.2 0.169E-02 0.01	2026.5 0.590E-03 0.00	2124.4 0.870E-03 0.01	2230.7 0.396E-02 0.03	2346.7 0.275E-01 0.23	2473.6 0.233E-01 0.20	2613.0 0.000 0.00	2767.0 0.381E-01 0.32	2937.6 0.798E-01 0.67
4	1807.4 0.141E-01 0.09	1886.9 0.792E-02 0.05	1972.6 0.160E-03 0.00	2065.2 0.220E-03 0.00	2165.5 0.798E-02 0.05	2274.7 0.103E-01 0.06	2393.7 0.430E-03 0.00	2524.1 0.313E-01 0.19	2667.4 0.395E-01 0.25	2825.6 0.142E-02 0.01
5	1765.3 0.337E-01 0.17	1841.0 0.162E-01 0.08	1922.5 0.370E-03 0.00	2010.3 0.160E-03 0.00	2105.3 0.307E-02 0.02	2208.3 0.330E-04 0.00	2320.3 0.139E-01 0.07	2442.6 0.214E-01 0.11	2576.5 0.000 0.00	2723.9 0.357E-01 0.18
6	1725.9 0.623E-01 0.30	1798.2 0.205E-01 0.10	1875.8 0.570E-04 0.00	1959.4 0.290E-03 0.00	2049.5 0.140E-03 0.00	2147.0 0.232E-02 0.01	2252.7 0.113E-01 0.05	2367.8 0.370E-03 0.00	2493.5 0.205E-01 0.10	2631.2 0.280E-01 0.13
7	1689.0 0.996E-01 0.48	1758.2 0.171E-01 0.08	1832.3 0.338E-02 0.02	1912.0 0.220E-02 0.01	1997.7 0.000 0.00	2090.2 0.229E-02 0.01	2190.3 0.170E-02 0.01	2298.9 0.596E-02 0.03	2417.2 0.191E-01 0.09	2546.5 0.480E-04 0.00
8	1654.4 0.136E 00 0.29	1720.8 0.820E-02 0.02	1791.8 0.136E-01 0.03	1867.8 0.312E-02 0.01	1949.5 0.500E-04 0.00	2037.5 0.363E-03 0.00	2132.5 0.210E-03 0.00	2235.4 0.906E-02 0.02	2347.1 0.206E-02 0.00	2468.7 0.124E-01 0.03
9	1622.0 0.179E 00 0.14	1685.8 0.750E-03 0.00	1753.8 0.277E-01 0.02	1826.6 0.124E-02 0.00	1904.7 0.155E-02 0.00	1988.6 0.320E-04 0.00	2079.0 0.105E-02 0.00	2176.6 0.331E-02 0.00	2282.4 0.204E-02 0.00	2397.3 0.166E-01 0.01
10	1591.6 0.211E 00 0.05	1652.9 0.266E-02 0.00	1718.3 0.383E-01 0.01	1788.1 0.130E-03 0.00	1862.9 0.497E-02 0.00	1943.1 0.000 0.00	2029.3 0.340E-03 0.00	2122.2 0.930E-04 0.00	2222.6 0.603E-02 0.00	2331.4 0.407E-02 0.00
11	1563.0 0.232E 00 0.02	1622.1 0.191E-01 0.00	1685.0 0.394E-01 0.00	1752.1 0.539E-02 0.00	1823.8 0.698E-02 0.00	1900.6 0.220E-03 0.00	1983.1 0.280E-04 0.00	2071.7 0.290E-03 0.00	2167.3 0.387E-02 0.00	2270.6 0.170E-03 0.00
12	1536.2 0.243E 00 0.01	1593.2 0.504E-01 0.00	1653.8 0.302E-01 0.00	1718.4 0.187E-01 0.00	1787.4 0.500E-02 0.00	1861.1 0.250E-02 0.00	1940.0 0.330E-03 0.00	2024.8 0.250E-03 0.00	2116.0 0.680E-03 0.00	2214.3 0.330E-02 0.00
13	1510.9 0.241E 00 0.00	1566.0 0.925E-01 0.00	1624.6 0.156E-01 0.00	1686.9 0.357E-01 0.00	1753.3 0.108E-02 0.00	1824.1 0.697E-02 0.00	1899.9 0.120E-03 0.00	1981.1 0.000 0.00	2068.3 0.100E-04 0.00	2162.2 0.355E-02 0.00
SUM V''	783.32	437.29	240.17	132.85	77.73	45.50	28.03	17.00	12.82	9.27

V' \ V''	10	11	12	13	14	15	16	17	18	19
0	3602.8 0.305E-01 0.54	3888.8 0.115E-01 0.20	4218.9 0.360E-02 0.06	4604.0 0.890E-03 0.02	5058.9 0.180E-03 0.00	5604.4 0.300E-04 0.00	6269.9 0.000 0.00	0.0 0.000 0.00	0.0 0.000 0.00	0.0 0.000 0.00
1	3426.0 0.154E 00 2.26	3683.6 0.102E 00 1.50	3978.5 0.512E-01 0.75	4319.2 0.199E-01 0.29	4717.1 0.610E-02 0.09	5187.9 0.150E-02 0.02	5753.3 0.290E-03 0.00	6444.4 0.450E-04 0.00	0.0 0.000 0.00	0.0 0.000 0.00
V' 2	3267.9 0.725E-01 0.82	3502.3 0.148E 00 1.68	3767.9 0.158E 00 1.79	4072.1 0.111E 00 1.26	4424.0 0.558E-01 0.63	4835.5 0.212E-01 0.24	5323.1 0.618E-02 0.07	5909.5 0.140E-02 0.02	6627.9 0.140E-02 0.02	7527.9 0.340E-04 0.00
3	3127.9 0.372E-01 0.31	3341.2 0.340E-03 0.00	3582.0 0.560E-01 0.47	3855.9 0.135E 00 1.13	4169.9 0.148E 00 1.24	4533.7 0.102E 00 0.85	4959.6 0.491E-01 0.41	5464.9 0.174E-01 0.15	6073.7 0.466E-02 0.04	6821.0 0.467E-02 0.04
4	3001.2 0.335E-01 0.21	3197.1 0.841E-01 0.52	3416.9 0.412E-01 0.26	3665.2 0.240E-03 0.00	3947.9 0.553E-01 0.34	4272.4 0.128E 00 0.80	4648.6 0.132E 00 0.82	5089.6 0.839E-01 0.52	5613.7 0.369E-01 0.23	6246.3 0.118E-01 0.07
5	2886.7 0.470E-01 0.24	3067.5 0.148E-02 0.01	3269.3 0.379E-01 0.19	3495.9 0.846E-01 0.43	3752.1 0.334E-01 0.17	4044.1 0.202E-02 0.01	4379.6 0.636E-01 0.32	4768.9 0.122E 00 0.62	5226.1 0.620E-01 0.32	5770.1 0.240E-01 0.12
6	2782.8 0.000 0.00	2950.4 0.443E-01 0.21	3136.7 0.454E-01 0.22	3344.7 0.000 0.00	3578.5 0.494E-01 0.24	3843.1 0.790E-01 0.38	4144.8 0.189E-01 0.09	4491.9 0.894E-02 0.04	4895.3 0.753E-01 0.36	5369.5 0.112E 00 0.54
7	2688.2 0.305E-01 0.15	2844.3 0.272E-01 0.13	3017.0 0.256E-02 0.01	3208.9 0.553E-01 0.27	3423.5 0.344E-01 0.17	3664.9 0.257E-02 0.01	3938.3 0.633E-01 0.31	4250.4 0.637E-01 0.31	4609.9 0.507E-02 0.02	5028.1 0.230E-01 0.11
8	2601.7 0.221E-01 0.05	2747.7 0.112E-02 0.00	2908.5 0.421E-01 0.09	3086.5 0.179E-01 0.04	3284.5 0.121E-01 0.03	3506.1 0.616E-01 0.13	3755.5 0.168E-01 0.04	4038.2 0.145E-01 0.03	4361.3 0.713E-01 0.15	4733.8 0.396E-01 0.08
9	2522.5 0.310E-03 0.00	2659.4 0.254E-01 0.02	2809.8 0.168E-01 0.01	2975.6 0.908E-02 0.01	3159.2 0.481E-01 0.04	3363.7 0.508E-02 0.00	3592.6 0.304E-01 0.02	3850.5 0.440E-01 0.04	4143.1 0.235E-02 0.00	4477.9 0.346E-01 0.03
10	2449.7 0.831E-02 0.00	2578.6 0.190E-01 0.00	2719.7 0.158E-02 0.00	2874.8 0.363E-01 0.01	3045.8 0.565E-02 0.00	3235.4 0.263E-01 0.01	3446.7 0.401E-01 0.01	3683.4 0.000 0.00	3950.3 0.494E-01 0.01	4253.5 0.336E-01 0.01
11	2382.6 0.139E-01 0.00	2504.4 0.990E-03 0.00	2637.3 0.207E-01 0.00	2782.9 0.119E-01 0.00	2942.9 0.129E-01 0.00	3119.5 0.356E-01 0.00	3315.4 0.000 0.00	3533.9 0.361E-01 0.00	3778.8 0.194E-01 0.00	4055.3 0.121E-01 0.00
12	2320.7 0.590E-02 0.00	2436.2 0.440E-02 0.00	2561.8 0.173E-01 0.00	2698.9 0.120E-02 0.00	2849.1 0.307E-01 0.00	3014.3 0.160E-02 0.00	3196.8 0.314E-01 0.00	3399.5 0.162E-01 0.00	3625.6 0.111E-01 0.00	3879.3 0.190E-02 0.00
13	2263.6 0.170E-03 0.00	2373.2 0.109E-01 0.00	2492.3 0.223E-02 0.00	2621.8 0.162E-01 0.00	2763.4 0.101E-01 0.00	2918.6 0.133E-01 0.00	3089.4 0.260E-01 0.00	3278.2 0.261E-02 0.00	3488.0 0.408E-01 0.00	3722.2 0.242E-02 0.00
SUM V''	7.15	5.72	4.67	3.45	2.95	2.46	2.10	1.73	1.16	1.01

```
         V''
      WAVELENGTH  R2 HEAD
V'    A(V',V'')
      INTENSITY(KR) IBC3 AURORA
      SYSTEM INTENSITY =    55. KR
      BANDS I< 0.0 KR  KR OMITTED
```

TABLE 4.14

LBH N2 BAND INTENSITIES

V' \ V''	0	1	2	3	4	5	6	7	8	9
0	1450.0 0.412E 03 2.27	1501.0 0.132E 04 7.26	1555.0 0.193E 04 10.63	1621.0 0.171E 04 9.42	1672.0 0.108E 04 5.94	1736.0 0.483E 03 2.66	1805.0 0.164E 03 0.90	1878.0 0.425E 02 0.23	1956.0 0.860E 01 0.05	2040.0 0.135E 01 0.01
1	1416.0 0.119E 04 9.20	1464.0 0.181E 04 13.97	1515.0 0.676E 03 5.21	1570.0 0.432E 01 0.03	1627.0 0.619E 03 4.77	1688.0 0.115E 04 8.88	1752.0 0.959E 03 7.39	1821.0 0.493E 03 3.80	1895.0 0.178E 03 1.37	1973.0 0.460E 02 0.35
2	1384.0 0.192E 04 17.00	1430.0 0.992E 03 8.80	1479.0 0.315E 02 0.28	1530.0 0.901E 03 7.99	1585.0 0.637E 03 5.65	1642.0 0.333E 01 0.03	1703.0 0.418E 03 3.71	1768.0 0.908E 03 8.06	1838.0 0.775E 03 6.87	1911.0 0.389E 03 3.45
3	1354.0 0.223E 04 19.75	1398.0 0.136E 03 1.21	1444.0 0.763E 03 6.74	1493.0 0.625E 03 5.52	1545.0 0.343E 02 0.30	1600.0 0.716E 03 6.33	1658.0 0.423E 03 3.74	1719.0 0.330E 01 0.03	1785.0 0.454E 03 4.02	1854.0 0.794E 03 7.02
4	1325.0 0.212E 04 15.91	1368.0 0.745E 02 0.56	1412.0 0.105E 04 7.92	1459.0 0.476E 01 0.04	1508.0 0.703E 03 5.28	1560.0 0.288E 03 2.16	1616.0 0.139E 03 1.05	1674.0 0.642E 03 4.82	1736.0 0.193E 03 1.45	1801.0 0.653E 02 0.49
5	1299.0 0.174E 04 10.03	1339.0 0.605E 03 3.49	1382.0 0.551E 03 3.18	1427.0 0.370E 03 2.13	1474.0 0.545E 03 3.14	1523.0 0.831E 02 0.48	1576.0 0.630E 03 3.64	1631.0 0.453E 02 0.26	1690.0 0.332E 03 1.91	1752.0 0.485E 03 2.80
6	1273.0 0.129E 04 5.24	1312.0 0.120E 04 4.88	1353.0 0.583E 02 0.24	1396.0 0.854E 03 3.48	1441.0 0.264E 02 0.11	1489.0 0.620E 03 2.53	1539.0 0.116E 03 0.47	1592.0 0.342E 03 1.39	1648.0 0.376E 03 1.53	1707.0 0.230E 02 0.09
7	1249.0 0.909E 03 2.44	1287.0 0.157E 04 4.21	1326.0 0.831E 02 0.22	1367.0 0.741E 03 1.99	1411.0 0.213E 03 0.57	1456.0 0.500E 03 1.34	1504.0 0.141E 03 0.38	1555.0 0.501E 03 1.34	1608.0 0.154E 02 0.04	1664.0 0.504E 03 1.35
8	1227.0 0.641E 03 1.08	1263.0 0.172E 04 2.91	1301.0 0.576E 03 0.97	1340.0 0.281E 03 0.47	1382.0 0.770E 03 1.30	1426.0 0.473E 02 0.08	1472.0 0.655E 03 1.10	1520.0 0.400E 02 0.07	1571.0 0.509E 03 0.86	1624.0 0.129E 03 0.22
9	1206.0 0.410E 03 0.41	1240.0 0.155E 04 1.56	1277.0 0.118E 04 1.18	1315.0 0.194E 01 0.00	1355.0 0.884E 03 0.89	1397.0 0.148E 03 0.15	1441.0 0.519E 03 0.52	1487.0 0.217E 03 0.22	1536.0 0.409E 03 0.41	1587.0 0.145E 03 0.15
SUM V''	83.33	48.84	36.58	31.08	27.96	24.64	22.90	20.23	18.52	15.93

V' \ V''	10	11	12	13	14	15	16
0	2130.0 0.166E 00 0.00	2227.0 0.185E-01 0.00	2331.0 0.165E-02 0.00	2444.0 0.277E-06 0.00	2567.0 0.590E-05 0.00	2700.0 0.173E-04 0.00	2846.0 0.176E-04 0.00
1	2057.0 0.904E 01 0.07	2147.0 0.134E 01 0.01	2244.0 0.152E 00 0.00	2349.0 0.164E-01 0.00	2462.0 0.119E-02 0.00	2584.0 0.190E-03 0.00	2717.0 0.626E-05 0.00
2	1990.0 0.132E 03 1.17	2074.0 0.322E 02 0.29	2165.0 0.577E 01 0.05	2262.0 0.757E 00 0.01	2366.0 0.751E-01 0.00	2479.0 0.563E-02 0.00	2601.0 0.391E-03 0.00
3	1928.0 0.592E 03 5.23	2007.0 0.267E 03 2.36	2092.0 0.804E 02 0.71	2182.0 0.174E 02 0.15	2279.0 0.271E 01 0.02	2384.0 0.306E 00 0.00	2497.0 0.278E-01 0.00
4	1871.0 0.545E 03 4.10	1945.0 0.686E 03 5.16	2025.0 0.422E 03 3.18	2109.0 0.161E 03 1.21	2200.0 0.413E 02 0.31	2297.0 0.759E 01 0.06	2402.0 0.988E 00 0.01
5	1818.0 0.275E 02 0.16	1888.0 0.218E 03 1.26	1963.0 0.607E 03 3.50	2042.0 0.543E 03 3.13	2127.0 0.264E 03 1.52	2218.0 0.824E 02 0.48	2315.0 0.177E 02 0.10
6	1769.0 0.475E 03 1.93	1835.0 0.240E 03 0.98	1906.0 0.183E 02 0.07	1981.0 0.405E 03 1.65	2060.0 0.585E 03 2.38	2145.0 0.376E 03 1.53	2236.0 0.145E 03 0.59
7	1723.0 0.784E 02 0.21	1786.0 0.225E 03 0.60	1853.0 0.434E 03 1.16	1923.0 0.359E 02 0.10	1998.0 0.176E 03 0.47	2078.0 0.537E 03 1.44	2164.0 0.482E 03 1.29
8	1681.0 0.283E 03 0.48	1740.0 0.357E 03 0.60	1804.0 0.185E 02 0.03	1870.0 0.451E 03 0.76	1941.0 0.217E 03 0.37	2017.0 0.253E 02 0.04	2097.0 0.420E 03 0.71
9	1641.0 0.442E 03 0.44	1698.0 0.233E 02 0.02	1758.0 0.480E 03 0.48	1821.0 0.551E 02 0.06	1888.0 0.255E 03 0.26	1960.0 0.402E 03 0.40	2035.0 0.152E 02 0.02
SUM V''	13.79	11.27	9.19	7.07	5.34	3.95	2.72

V''
V' WAVELENGTH ORIGIN
A(V',V'')
INTENSITY(KR) IBC3 AURORA
SYSTEM INTENSITY = 383. KR
BANDS I< 0.0 KR KR OMITTED

4.2.4.4. *N_2 Singlet Systems*

N_2 LBH system: ($a^1\Pi_g - X^1\Sigma_g^+$). This system has an upper state with a 0.14 ms lifetime so that its quenching height should be no more than 95 km, the value for quenching by N_2 at the gas kinetic rate. Some calculated values of the total production rate, $I(a^1\Pi\ N_2)$, are entered in Table 4.5. Recently Cartwright (1973 b) has examined the cascade processes which should populate the $a^1\Pi$ state and concluded that cascade from the $a'^1\Sigma_u^-$ state (see Figure 4.12) is dominant for $v' < 2$ and that cascade from this state and the $w^1\Delta_u$ state provides about half the excitation out to $v' = 12$. Table 4.14 gives band intensities computed from Equations (4.2.3b) and (4.11) on the assumption that R_e is constant. The excitation rates, $I(v')$, were obtained from Equation (4.2.4c) divided by the fraction of direct excitation from Cartwright (1973 b). The Franck-Condon factors of Benesch *et al.* (1966 a) were used.

An intensity of 17 kR for the 2,0 band has been chosen to normalize Table 4.14 in accordance with the measurements of Vreux (1971). This leads to a system intensity of 380 kR. The results of Opal *et al.* (1970) and especially of Peek (1970) would suggest that lower values are also possible although there is no obvious reason for variability. There could be significant cascade contributions to the intensity of this system from the higher singlet levels of N_2. These could double the effective population rate of the $a^1\Pi$ state according to the estimates of Green *et al.* (1969). The cascading contributions to $a^1\Pi_g$ are uncertain since the branching ratios are unknown for the transitions (see Figure 4.12) from b, b′, c, c′ and o upper states to the ground state (Birge-Hopfield systems) compared to those to the $a^1\Pi_g$ state (Gaydon-Herman singlet systems). The latter systems lie in the 2200–3400 Å region so that the fact that they have not as yet been reported in the auroral spectrum probably indicates that the far UV transitions are favored and that the LBH system is not enhanced by cascading from these higher levels.

Other singlet systems of N_2. There are many other singlet systems of which only a few are shown on Figure 4.12. Their cross sections are not known in detail although energy loss measurements made by Brinkmann and Trajmar (1970) give the total cross sections for unresolved groups of transitions. Stolarski *et al.* (1967) have also analysed electron loss spectra to obtain estimates of the higher level cross sections.

Calculations of auroral intensities using the above cross sections have been made by Green *et al.* (1969) and Brinkmann and Trajmar. Some of their results are given in Table 4.5. There are considerable differences in the estimated intensities of the bands from the higher states but it would appear that the upper singlet levels are populated at rates of at least 1000 kR in IBC3 aurora. Dick (1970 a) has tentatively identified the 0,0 and 0,2 bands of the $c'^1\Sigma_u^+ - a^1\Pi_g$ Gaydon-Herman system at 2827 and 3119 Å while Miller *et al.* (1968) have a tentative identification of features ascribed to the Birge-Hopfield and Watson-Koontz systems. It has recently been reported by Paresce *et al.* (1972) that emissions have been observed in the Birge-Hopfield region (850–1000 Å) with IBC3 intensities of the order of 2000 kR. This is of the same order of magnitude as the predictions of Brinkmann and Trajmar.

4.2.4.5. O_2 *Systems*

The energy levels for these are shown on Figure 4.13. No excitation of the states above the $b^1\Sigma$ has been reported in aurora, no doubt because the A and B states have comparatively modest cross sections and because $n(O_2) \ll n(N_2)$. (The cross sections to discrete levels are small because the Franck-Condon principle favors dissociation.) Electron impact produces a significant excitation of the $b^1\Sigma$ and $a^1\Delta$ states. The former is shown in Table 4.5 to be about 70 kR for normal height IBC3 aurora while the latter is 400 kR. The electron impact excitation of these systems has also been discussed by Cartwright *et al.* (1972b).

TABLE 4.15

O2 ATM BAND INTENSITIES

V' \ V''	0	1	2	3	4	5	6	7	8	9
0	7619.0 0.793E-01 1176.08	8645.0 0.391E-02 58.00	9967.6 0.971E-04 1.44	11730.6 0.135E-05 0.02	14202.3 0.125E-07 0.00	17915.7 0.702E-10 0.00	24116.7 0.115E-11 0.00	36556.4 0.197E-13 0.00	74145.6 0.972E-15 0.00	-7193789.0 -0.223E-21 -0.00
1	6882.0 0.799E-02 1.66	7708.0 0.673E-01 13.96	8742.0 0.775E-02 1.61	10070.9 0.319E-03 0.07	11839.9 0.650E-05 0.00	14313.1 0.845E-07 0.00	18013.5 0.608E-09 0.00	24152.3 0.106E-10 0.00	36316.1 0.984E-13 0.00	71890.8 0.161E-13 0.00
2	6288.2 0.434E-03 0.05	6970.3 0.157E-01 1.93	7802.0 0.552E-01 6.79	8844.3 0.113E-01 1.40	10180.1 0.694E-03 0.09	11956.4 0.194E-04 0.00	14433.1 0.337E-06 0.00	18124.2 0.317E-08 0.00	24209.0 0.678E-10 0.00	36125.9 0.424E-12 0.00
3	5796.4 0.153E-04 0.00	6371.1 0.143E-02 0.38	7060.3 0.228E-01 6.09	7901.4 0.432E-01 11.53	8950.6 0.146E-01 3.89	10295.5 0.124E-02 0.33	12080.5 0.463E-04 0.01	14562.8 0.105E-05 0.00	18248.2 0.126E-07 0.00	24287.2 0.299E-09 0.00
4	5384.0 0.293E-06 0.00	5876.4 0.766E-04 0.01	6457.7 0.310E-02 0.29	7154.4 0.289E-01 2.75	8003.9 0.320E-01 3.04	9062.5 0.171E-01 1.63	10417.4 0.198E-02 0.19	12212.6 0.955E-04 0.01	14702.6 0.278E-05 0.00	18386.1 0.422E-07 0.00
5	5033.4 0.985E-10 0.00	5461.2 0.225E-05 0.00	5959.8 0.234E-03 0.02	6548.3 0.552E-02 0.56	7252.9 0.336E-01 3.44	8111.5 0.220E-01 2.25	9180.2 0.189E-01 1.93	10546.3 0.291E-02 0.30	12353.0 0.179E-03 0.02	14853.1 0.659E-05 0.00
6	4731.9 0.136E-08 0.00	5108.1 0.710E-08 0.00	5541.7 0.995E-05 0.00	6047.0 0.561E-03 0.00	6643.0 0.869E-02 0.00	7356.1 0.366E-01 0.00	8224.4 0.134E-01 0.00	9304.1 0.197E-01 0.00	10682.5 0.401E-02 0.00	12502.3 0.312E-03 0.00
7	4470.0 0.738E-09 0.00	4804.2 0.880E-08 0.00	5185.9 0.903E-07 0.00	5625.8 0.330E-04 0.00	6138.1 0.116E-02 0.00	6742.1 0.126E-01 0.00	7464.4 0.378E-01 0.00	8343.1 0.691E-02 0.00	9434.7 0.195E-01 0.00	10826.5 0.527E-02 0.00

V' V''
WAVELENGTH ORIGIN
A(V',V'')
INTENSITY(KR) IBC3 AURORA
SYSTEM INTENSITY = 1302. KR
BANDS I< 0.0 KR KR OMITTED

The $b^1\Sigma$ state is excited both by electron impact and by energy transfer as discussed in detail in Section 4.2.3.2. As noted about 10% of $I(O^1D)$ could be converted to $O_2(^1\Sigma)$. The total unquenched production of O^1D, corresponding to the 5 keV power-exponential spectrum of Table 4.5, is about 2000 kR suggesting an energy transfer production rate of perhaps 200 kR. The total theoretical $O_2(^1\Sigma)$ production rate of $\sim$300 kR would divide in a 50:1 ratio (Noxon, 1961), between the O_2 atmospheric (At) system and the $b^1\Sigma - a^1\Delta$ Noxon system of which the 0,0 band lies at 1.908 μm. However the observed O_2 At system intensity is about 1300 kR. The Noxon band has not yet been observed in aurora; it should be possible to detect 26 kR in this region. The Franck-Condon factors (Nicholls, 1965a) indicate that for the electron impact component the production will be mostly in the $v'=0$ level. The transfer component according to Section 4.2.3.2 produces an important fraction in the $v'=1$ and possibly in the $v' \leqslant 5$ levels. There is evidence that the $v'=1$ level is vibrationally quenched below 150 km. In Table 4.15 an attempt is made to organize the data on the band intensities for the system. The A-values are calculated from the Franck-Condon factors (Nicholls, 1974) with R_e assumed constant. The vibrational population rates cannot be predicted theoretically and so were adjusted so as to reproduce the intensities observed by Gattinger and Vallance Jones (1974). The 0,0 band is not observable from the ground because of self-absorption by ground state O_2. These data lead to a system intensity of about 1300 kR. This is considerably higher than the predicted value.

The $a^1\Delta$ state, giving rise to the IR atmospheric bands, should be populated in normal IBC3 aurora as mentioned above at a rate of about 400 kR by electron impact. Since the radiative lifetime is about 60 min the steady state intensity is unlikely to be approached. Even a continuous IBC3 excitation of 10 min in the same atmospheric volume would produce only about 75 kR, a value comparable with the nightglow intensity. Aircraft observations by Noxon (1970) however show that on rare occasions there are enhancements of up to 500 kR. These appear always to be associated with large magnetic disturbances although the converse is not necessarily true.

Recently Gattinger and Vallance Jones (1973c) succeeded in recording simultaneously the time variation of 1.27 μm 0,0 O_2 and of the 5577 Å [OI] from the ground during several enhancements of the emission associated with auroral substorms. It was found that the O_2 emission time variation corresponded to that predicted if $I(O_2\,{}^1\Delta)/I(4278)$ had a value of about 85 and $\tau(O_2\,{}^1\Delta) \simeq 60$ min. The ratio from Table 4.5 is about 15 but a considerable increase should occur for an aurora with a lower border about 95 km because of the increase in $n(O_2)/n(N_2)$ and the decrease $n(O)/n(O_2)$. The latter change is particularly important since O is the principal competitor of O_2 in the important 4 to 6 eV secondary electron energy range (Figure 4.21). It thus appears likely that at least one class of $O_2(^1\Delta)$ enhancements may be due to electron impact excitation in energetic substorms.

One of the problems in observing auroral $O_2(^1\Delta)$ emission is the fluctuations which occur in the nightglow level. As shown by Gattinger and Vallance Jones (1973b) enhancements of 200 to 300 kR can occur and appear to be associated with λ5577 and OH nightglow fluctuations. It is thus difficult to study $O_2(^1\Delta)$ enhancements associated with any but the strongest aurora.

Noxon (1970) and Walker (1970) have considered electric field excitation of $O_2(^1\Delta)$ by way of increased electron or ion temperatures. Noxon has also suggested catalytic chemical processes. It must be emphasized that some unrecognized process may be involved.

Band intensities for the O_2 IR atmospheric system in the steady state are given in Table 4.16. This is based on the production ratio measured by Gattinger and Vallance Jones and may be applicable only to low level aurora. Non steady state intensities would usually be much smaller. The Franck-Condon factors are from Nicholls (1965a)

TABLE 4.16

O_2 IR Atmospheric Band Intensities

v' \ v''	0	1	2	3
0	12686.7	15807.7	20859.2	30429.8
	0.276E-03	0.186E-05	0.255E-02	0.500E-11
	2500.00	16.89	0.02	0.00
1	10677.2	12804.2	15929.8	20965.5
	0.625E-05	0.267E-03	0.403E-05	0.287E-07
	0.00	0.00	0.00	0.00

v' \ v''
Wavelength origin
A(v',v'')
Intensity (kR) IBC3 aurora
System intensity = 2517.kR

and R_e is assumed constant. (Laboratory measurements give $A(0,0)/A(0,1) \sim 50$.) It is also assumed that all $O_2(^1\Delta)$ is relaxed to $v' = 0$ before radiating.

4.2.4.6. *O_2^+ Systems*

First negative (1N). The relative production rate of this system should be a definite fraction of $\eta(O_2^+)$ as given by Equation (4.2.1j). Rees and Jones (1973) adopt a branching ratio of 0.24 for the $b^4\Sigma_g^-$ upper state.

The band intensities for the system are given in Table 4.17. The relative transition moment of Druetta and Dufay (1966) was used in conjunction with the Franck-Condon factors of Nicholls (1965b). The population rates $I(O_2^+ b, v')$ were assumed proportional to the Franck-Condon factors for the $v'' = 0$ sequence of the $O_2^+ b^4\Sigma - O_2 X^3\Sigma$ transition as calculated by Wacks and Krauss (1961). Somewhat different relative intensities are obtained in the table of Shemansky and Vallance Jones (1968) by using different data for the A-values. Table 4.17 has been normalized for IBC3 aurora by taking $I(1,0) = 5.9$ kR as measured by Gattinger and Vallance Jones (1974). From Table 4.12 this leads to a value of $I(O_2^+;0,0)$ of about 4 kR and a system intensity of about 26 kR. The relative system intensity should be sensitive to height effects which will be further considered in Section 4.2.5. The theoretical value from Table 4.5 is about 50 kR.

O_2^+ second negative system. This has been predicted by Rees and Jones (1973) to have an intensity almost twice that of the O_2^+ 1N system. Since there are many bands between 4000 Å and 2500 Å the individual bands may be quite weak.

4.2.4.7. *NO γ System*

The detection of NO in the auroral atmosphere by Zipf *et al.* (1970) has rendered more reasonable the observations (see Figure 4.1) reported by Duysinx and Monfils (1972) and by Sharp and Rees (1972) of the occurrence of a peak at 2160 Å

TABLE 4.17

IN 02+BAND INTENSITIES

V' \ V''	0	1	2	3	4	5	6	7	8	9
0	5999.9 0.332E 06 3.98	6389.0 0.289E 06 3.47	6822.3 0.158E 06 1.90	7307.6 0.706E 05 0.85	7854.4 0.279E 05 0.33	8474.6 0.101E 05 0.12	9183.6 0.349E 04 0.04	10001.2 0.115E 04 0.01	10953.7 0.365E 03 0.00	12076.4 0.114E 03 0.00
1	5608.7 0.592E 06 5.90	5947.2 0.234E 05 0.23	6320.9 0.378E 05 0.38	6735.3 0.861E 05 0.86	7197.2 0.738E 05 0.74	7714.5 0.434E 05 0.43	8297.7 0.214E 05 0.21	8959.5 0.932E 04 0.09	9716.4 0.374E 04 0.04	10589.6 0.141E 04 0.01
V' 2	5274.8 0.383E 06 1.84	5573.2 0.226E 06 1.08	5900.1 0.135E 06 0.65	6259.6 0.402E 04 0.02	6656.6 0.173E 05 0.08	7096.7 0.405E 05 0.19	7587.3 0.384E 05 0.18	8136.9 0.255E 05 0.12	8756.3 0.140E 05 0.07	9459.3 0.682E 04 0.03
3	4987.0 0.110E 06 0.20	5252.9 0.499E 06 0.90	5542.3 0.321E 05 0.06	5858.3 0.129E 06 0.23	6204.6 0.393E 05 0.07	6585.3 0.382E 02 0.00	7005.6 0.122E 05 0.02	7471.6 0.234E 05 0.04	7990.7 0.222E 05 0.04	8572.0 0.155E 05 0.03
SUM V''	7640.06	2583.20	2619.77	1882.26	1558.43	1542.77	1220.14	8404.16	4665.50	2595.53

V' \ V''	10
0	13417.9 0.339E 02 0.00
1	11607.2 0.506E 03 0.01
V' 2	10263.0 0.303E 04 0.01
3	9226.8 0.909E 04 0.02
SUM V''	1636.42

V''
V' WAVELENGTH ORIGIN
A(V',V'')
INTENSITY(KR) IBC3 AURORA
SYSTEM INTENSITY = 26. KR
BANDS I< 0.0 KR KR OMITTED

in rocket and satellite UV auroral spectra tentatively identified as the 1,0 NO γ band. Sharp and Rees argue that excitation by energy transfer from $N_2A^3\Sigma$ is the most likely excitation mechanism. The $v' = 0$ level of the $A^3\Sigma$ state could excite NO $A^2\Sigma$ up to the $v' = 2$ level.

Such an energy transfer, if it were the dominant quenching process in the region of peak production could give an excitation rate of $NO(A^2\Sigma)$ of the same order of magnitude as the excitation rate of $N_2(A^3\Sigma)$. According to Table 4.5 this could amount to about 500 kR including cascade contributions and assuming a conversion efficiency of 50%.

The direct electron impact excitation rate would be much less if the cross section for exciting the 1,0 band has a peak value of $\sim 10^{-18}$ cm^2 as quoted by Sharp and Rees. In the most favorable case with a peak cross section comparable to that for the $N_2(B^3\Pi)$ state ($\sim 10^{-16}$ cm^2) and a relative concentration $n(NO)/n(O_2) \simeq 1$ as reported by Zipf *et al.* (1970) the direct excitation rate might approach 100 kR according to Table 4.5.

Sharp and Rees estimated the observed intensity of the 1,0 band to be ~ 0.9 kR for $I(3914) = 7.5$ kR or about 10 kR for $I(4278) = 33$ kR as in Table 4.5. Since we do not know the relative population rates of the v' levels we cannot derive a system intensity. As a starting point we have arbitrarily assumed $I(0):I(1)$:etc. to be in the ratio 4:4:3:1:0.5:0.25:0.12:0.06 and used the Franck-Condon factors of Nicholls (1964b) and the $R_e(\bar{r})$ relation of Jeunehomme (1966) to compile Table 4.18 with $I(\text{NO}\gamma;\,1{,}0) = 10$ kR. With these population rate ratios the system intensity is about 60 kR but it would be less if the $v' = 1$ level were populated preferentially. One can conclude that the energy transfer process is a plausible mechanism and that excitation by electron impact is excluded if the cross section has the small value quoted above.

TABLE 4.18

NO GAMBAND INTENSITIES

V' \ V''	0	1	2	3	4	5	6	7
0	2262.0 0.1C5E 07 3.99	2362.0 0.150E 07 5.71	2470.0 0.121E 07 4.58	2586.0 0.728E 06 2.76	2712.0 0.365E 06 1.38	2848.0 0.161E 06 0.61	2998.0 0.648E 05 0.25	3171.0 0.243E 05 0.09
1	2148.0 0.263E 07 10.00	2238.0 0.728E 06 2.77	2338.0 0.740E 04 0.03	2439.0 0.430E 06 1.63	2550.0 0.707E 05 0.27	2670.0 0.617E 06 2.34	2800.0 0.401E 06 1.52	2941.0 0.216E 06 0.82
2	2047.0 0.257E 07 7.32	2128.0 0.143E 06 0.41	2215.0 0.847E 06 2.41	2309.0 0.463E 06 1.32	2410.0 0.104E 04 0.00	2516.0 0.198E 06 0.56	2630.0 0.432E 06 1.23	2754.0 0.450E 06 1.28
3	1956.0 0.144E 07 1.37	2030.0 0.182E 07 1.72	2109.0 0.345E 06 0.33	2193.0 0.322E 06 0.31	2283.0 0.762E 06 0.72	2379.0 0.282E 06 0.27	2481.0 0.541E 03 0.00	2580.0 0.133E 06 0.13
4	1878.0 0.4C5E 06 0.19	1740.0 0.265E 07 1.26	2019.0 0.339E 06 0.16	2096.0 0.805E 06 0.38	2171.1 0.420E 04 0.00	2257.6 0.342E 06 0.16	2349.6 0.430E 06 0.20	2447.9 0.130E 06 0.06
5	1803.0 0.111E 06 0.03	1866.0 0.132E 07 0.31	1933.0 0.202E 07 0.48	2003.0 0.490E 04 0.00	2078.0 0.877E 06 0.21	2160.0 0.297E 06 0.07	2232.8 0.545E 05 0.01	2321.4 0.418E 06 0.10
6	1735.0 0.173E 05 0.00	1795.0 0.426E 06 0.05	1855.0 0.204E 07 0.23	1920.0 0.119E 07 0.14	1965.0 0.386E 06 0.04	2060.0 0.392E 06 0.04	2128.4 0.624E 06 0.07	2208.7 0.339E 05 0.00
7	1673.0 0.171E 04 0.00	1727.0 0.827E 05 0.00	1784.0 0.906E 06 0.05	1840.0 0.227E 07 0.13	1907.0 0.368E 06 0.02	1965.1 0.829E 06 0.05	2034.5 0.380E 05 0.00	2107.8 0.602E 06 0.03
SUM V''	7662.94	2595.43	2628.04	1888.93	1561.08	1546.88	1223.43	8406.66

V''
V' WAVELENGTH P2 HEAD
A(V',V'')
INTENSITY(KR) IBC3 AURORA
SYSTEM INTENSITY = 63. KR
BANDS I< 0.0KR KR OMITTED

4.2.4.8. *OH Vibration-Rotation Bands*

The strong rapid enhancements of these bands in type-b aurora mentioned in a review by Vallance Jones (1971) are now thought to be due to some other unidentified emission (Brown, 1971). Slow enhancements of perhaps 2× above the average nightglow level have been reported by Shefov (1969) and Stair *et al.* (1971) in association with magnetic disturbances and in the auroral oval.

The relatively low concentration of OH ($\sim 10^7$ cm^{-3} at 85 km) and the long time constant (hours) associated with the observed 'auroral' enhancements rule out direct electron impact as the excitation mechanism. Gattinger (1969) has considered the photochemical response of the atmosphere to an auroral production of atomic oxygen down to 70 km. A production rate of 3×10^8 cm^{-3} s^{-1} would produce a 10 × enhancement of OH if maintained for about 7 min. According to the calculations of Maeda and Aikin (1968) this would require electrons of an average energy of 100 keV of flux 1.6×10^3 erg cm^{-2} s^{-1} which is near the value for IBC4. The results of Shefov (1969), indicating a traveling enhancement propagating equatorwards after a magnetic disturbance, suggest some dynamic effect possibly involving increased vertical mixing or heating.

4.2.4.9. [*O I*] *Forbidden Transitions*

λ5577

Although they are the most prominent features of the visible spectrum, it has been difficult to predict the intensities of these lines. In the section on quenching (Section 4.2.3.3), the equations connecting the production rates of the ^{1}S and ^{1}D levels and the line intensities were discussed while the production rates by electron impact, photodissociation, dissociative recombination, energy transfer and thermal excitation have also been considered.

We must now attempt to assess the relative importance of these processes. The recent discovery of the importance of quenching of O(^{1}S) by O(^{3}P) (see k_4 in Table 4.6) makes a drastic change in theoretical values of I(5577). As may be seen from Table 4.5 the theoretical value of I(5577) is cut by about a factor of 2.

Secondary electron excitation of O was for long considered dominant. Figure 4.24 shows the production rates of O(^{1}S) for the model developed here. This would suggest that electron impact could be several times more important than dissociative recombination. This conclusion is dependent on NO^+ being the dominant ion in the 100 to 110 km region. As may be seen from Figure 5.3 a the ion concentrations are highly time dependent and sensitive to NO concentration. The change between Figures 5.5 b and 5.5 a illustrates this possibility. Experimentally Donahue *et al.* (1970) and Parkinson *et al.* (1970) measured the ion concentrations reproduced in Figure 5.8 in an IBC2 aurora and concluded that the dissociative recombination mechanism could not be important. The recent work of Rees and Jones (1973) however shows that the role of this mechanism is very time dependent.

As discussed in Section 4.2.4.1 the observed value of $I(5577)/I(3914)$ in bright aurora seems close to unity so that $I(5577)/I(4278)$ should be about 3.3. Even without any quenching the calculated values from Table 4.5 are lower for aurora of normal height except where there is an important contribution from dissociative recombination. There have been, two reasons, however, for suspecting that dissociative recombination may not play an important role. The first is the measured ion concentrations of Figure 5.8. The second problem concerns the time variation effects observed in pulsing aurora. We shall examine these effects at this point.

Analysis of pulsing aurora. This provides a time variation of the exciting flux in the volume of observation. For O(^{1}S) we can assume that the line-of-sight integrated electron impact production rate equals $K\,I(4278)$ where K is a constant and $I(4278)$ is the observed $\lambda 4278$ N_2^+ intensity. K will be constant if the energy spectrum of the exciting flux does not vary strongly with time. (This assumption may be a weak point if pulsations are normally associated with the kind of energy variations suggested by Figure 4.34.) Then from the time-varying form of Equation (4.2.3i) it follows that

$$\frac{dI(5577)}{dt} + \frac{I(5577)}{\tau'} = K_1 I(4278) + \int K_2 n(e)\, n(O_2^+)\, ds, \qquad (4.2.4e)$$

where τ' is the effective lifetime. ($1/\tau'$ is the line-of-sight averaged value of $A_{32}+A_{31}+\Sigma k_{Qk}\, n(M_k)$.) K_1 and K_2 are constants equal to KA_{32} and $k_{16a}A_{32}$ respectively. The second term allows for a contribution to $O(^1S)$ production from dissociative recombination and is integrated along the line-of-sight.

If the dissociative recombination term is omitted Equation (4.2.4e) indicates that the $\lambda 5577$ signal should be proportional to that of $\lambda 4278$ as it would be modified by a simple low pass filter of time constant τ'. This is certainly the case qualitatively. Various quantitative methods for estimating τ' are described by Omholt (1971) who reproduces a number of examples of original photometric data. In general the effective lifetimes found have been close to 0.7 s, the theoretical lifetime for the $O(^1S)$, or shorter in the case of type-b aurora where quenching becomes important. These results reinforced the belief that the dissociative recombination production term in Equation (4.2.4e) could not be very important because of the extra time lag introduced by the production and decay of O_2^+.

Thus Parkinson (1971) found that an 18% contribution from dissociative recombination gave an excellent match to his observed pulsation records. Vreux and Marette (1971), from the analyses of a 25 kR giant pulse, concluded that only direct excitation was significant.

Recent results by Brekke and Pettersen (1972), based on a more refined analysis of many digital photometer records indicate that the dissociative recombination term of Equation (4.2.4e) must be taken into account and that the time constant deduced for the decay of O_2^+ can be related to the concentration profile of NO. It thus appears that the experimental data do not in fact exclude an important contribution from dissociative recombination and that NO plays an important part in limiting this contribution and in reducing its time constant. This possibility was pointed out by Parkinson.

Photodissociation of O_2 by extreme UV auroral emissions of N_2 has recently been proposed by Zipf (1973b) as a major source of $O(^1S)$. This mechanism would have much in common with secondary electron excitation and would not involve any additional time delay. Since the EUV band intensity is about 20 times the observed value of $I(5577)$ it is clear that this process could be an important source of $O(^1S)$.

Two other possible contributors to $\eta(O^1S)$ are dissociative excitation,

$$O_2 + e \rightarrow O + O(^1S) + e,$$

and energy transfer,

$$N_2(A^3\Sigma) + O \rightarrow O(^1S) + N_2 .$$

The first requires a rather large cross section to be important. This can be estimated from that for $\lambda 3914$ excitation; since $n(O_2)$ is no more than $0.2n(N_2)$ while $I(5577) \geqslant I(3914)$ then $\sigma(O_2 \rightarrow O(^1S)+O) > 5\sigma(N_2 \rightarrow \lambda 3914) = 7.5 \times 10^{-17}$ cm^{-2}. The cross section for dissociation of O_2 by electron impact is about 4×10^{-16} cm^{-2} but recent measurements (Zipf, 1973c) indicate that $\sigma(O_2, O^1S) \leqslant 2 \times 10^{-17}$ cm^2 so that this source cannot be dominant. The case for this process has been presented by Parkinson *et al.* (1970).

The energy transfer process has been considered by Parkinson and Zipf (1970) and by Parkinson (1971). As pointed out in Section 4.2.4.3 the total production rate for the $N_2(A^3\Sigma)$ state is about 1000 kR for IBC3 aurora while less than 10% appears as emission. Thus with an efficiency of 0.1 for $O(^1S)$ production, the quenching reaction could produce the observed $\lambda 5577$ emission if O were the only quencher. Meyer *et al.* (1969) have measured an efficiency near this value while the recent measurement of Zipf (1973b) implies an efficiency approaching unity (see Table 4.6). As pointed out by Parkinson a serious problem arises since the observed $\lambda 5577 - N_2^+(1N)$ time lags in fluctuating aurora could only be consistent with a major role for energy transfer if the radiative lifetime for $O(^1S)$ were much shorter than the accepted 0.7 s. Henriksen (1973), in another analysis of pulsating aurora, concluded that his observations are consistent with a minor contribution from this source.

In summary it appears that there are important contributions to $O(^1S)$ production

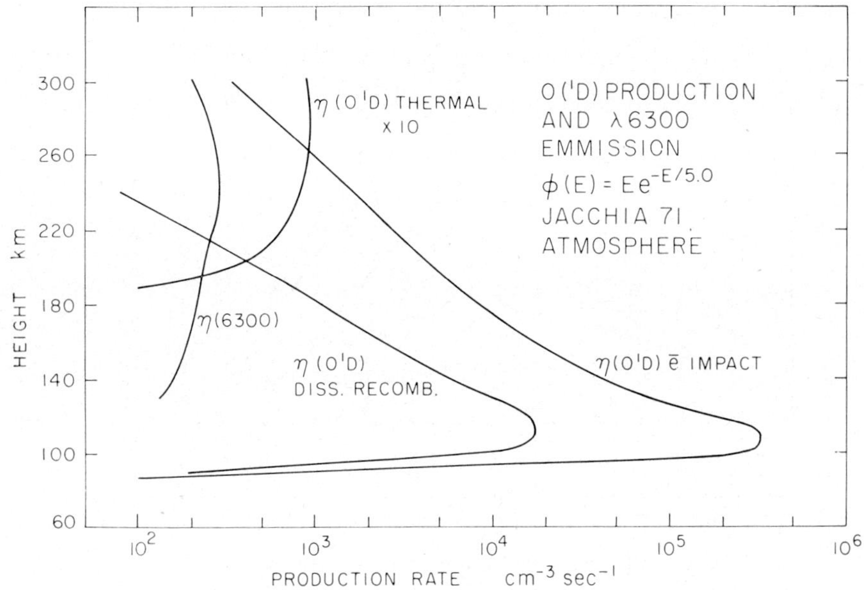

Fig. 4.28. Model calculation of excitation rates of $O(^1D)$ by electron impact, dissociative recombination of O_2^+ and by thermal electrons. The emission rate curve takes quenching into account.

from secondary electron and photodissociative excitation. There is also now evidence for a substantial contribution from indirect processes which may have a complicated time and composition dependence. The time constant effects in pulsating aurora must arise from a combination of the effect of the direct production term of Equation (4.2.4e), an indirect production term involving a time delay and an effective lifetime τ' which should be appreciably shorter than the radiative lifetime. It would appear that the two effects may largely cancel one another. The final resolution of this matter remains an intriguing problem.

λ6300

As may be seen from Table 4.5 the relative intensity of λ6300 is strongly dependent on average incident particle energy because of the variation of quenching factor with

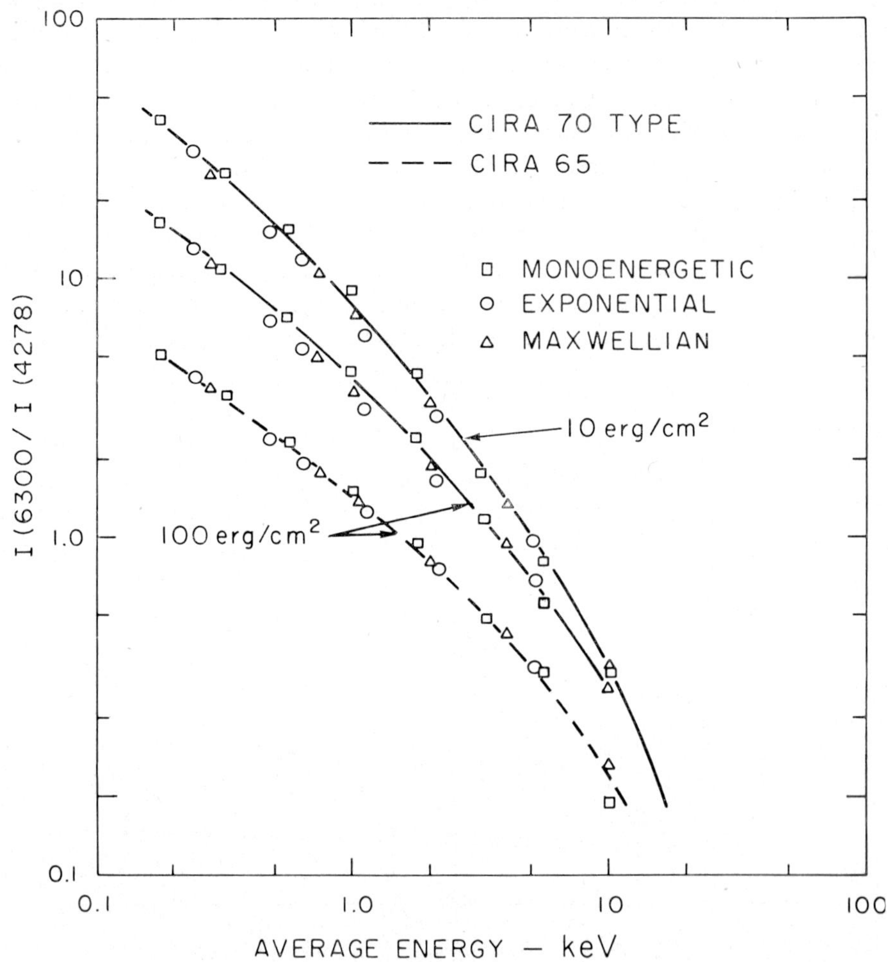

Fig. 4.29. Theoretical ratios $I(6300)/I(4278)$ for various energy fluxes, average electron energies and electron energy spectra.

height. Figure 4.28 shows a comparison between the production and emission height profiles, while Figure 4.29 shows how the ratio $I(6300)/I(N_2^+, \lambda 4278)$ varies with mean particle energy for a variety of particle energies, input energy fluxes and energy spectra. The details of these ratios and the height distributions are rather sensitive to the quenching rate constant k_1 and to the height profile $n(N_2, h)$ and consequently the observed intensity ratios can be taken as rough guides only to the incident particle energy. The observed zenith ratio $I(6300)/I(4278)$ is certainly variable with values from 0.3 to 5 occurring frequently according to the results of Gattinger and Vallance Jones (1972). Eather and Mende (1971a, 1972) have used the observed values of this ratio to infer time latitude distributions of particle energy across the auroral oval.

The relative height distributions of $\lambda 6300$ and $\lambda 5577$ (or $\lambda 4278$) in auroral arcs are undoubtedly qualitatively similar to the theoretical curves in Figures 4.28 and 4.24 as may be seen from the two-station photometer observations of Romick and Belon (1967b) from which Figure 4.30 is taken. These curves show clearly the strong displacement towards the zenith of the $\lambda 6300$ intensity for the arc for which the $\lambda 5577$ maximum is about 35° above the horizon as observed from College. It is difficult, however, to obtain an accurate height distribution for $\lambda 6300$ from these excellent data because of the uncertainties introduced by the latitude extent, non-localized background effects and possible non-equilibrium effects associated with this emission.

Because of uncertainties in the quenching rate and the large probability of quenching at low altitudes it is difficult to make a good test of the theoretical production rates.

Attempts have also been made to derive effective lifetimes for $O(^1D)$ by application of Equation (4.2.4e) but as reviewed by Omholt (1971), these have been unsatisfactory probably because of gross failures in the constant particle energy spectrum assumption and because of indirect excitation processes. Eather (1969) has been able to interpret low amplitude pulsations with periods in the 2 to 20 s range in terms of Equation (4.2.4e). These were ascribed to fluctuation in the higher energy component of the electron flux at heights where the effective lifetime is shortened by quenching.

4.2.4.10. [*NI*] *Forbidden Transitions*

The $\lambda 5200$, $\lambda 10\,400$ and $\lambda 3466$ transitions are easily observed in aurora. Qualitatively the height distribution of $\lambda 5200$ is similar to that of $\lambda 6300$. Prasad and Green (1971) have suggested that the ^2D state is produced mostly by dissociative recombination of N_2^+ and NO^+ while direct electron impact on N could possibly be important for $N(^2P)$. Brown (1968) discussed the growth of $n(N)$ in the auroral atmosphere and concluded that dissociative recombination would be dominant. Rees and Jones (1973) in their model calculation assumed that the ^2P, ^2D and ^4S states are produced with equal probability in the dissociative ionization of N_2. They also included the contributions from dissociative recombination of N_2^+ and NO^+, and obtained the total intensities listed in Table 4.5.

Gérard and Harang (1973) have made a quantitative study of $I(5200)$ and find it to co-vary with $I(6300)$. Over a wide range of brightnesses $I(6300)/I(5200)$ fell in the

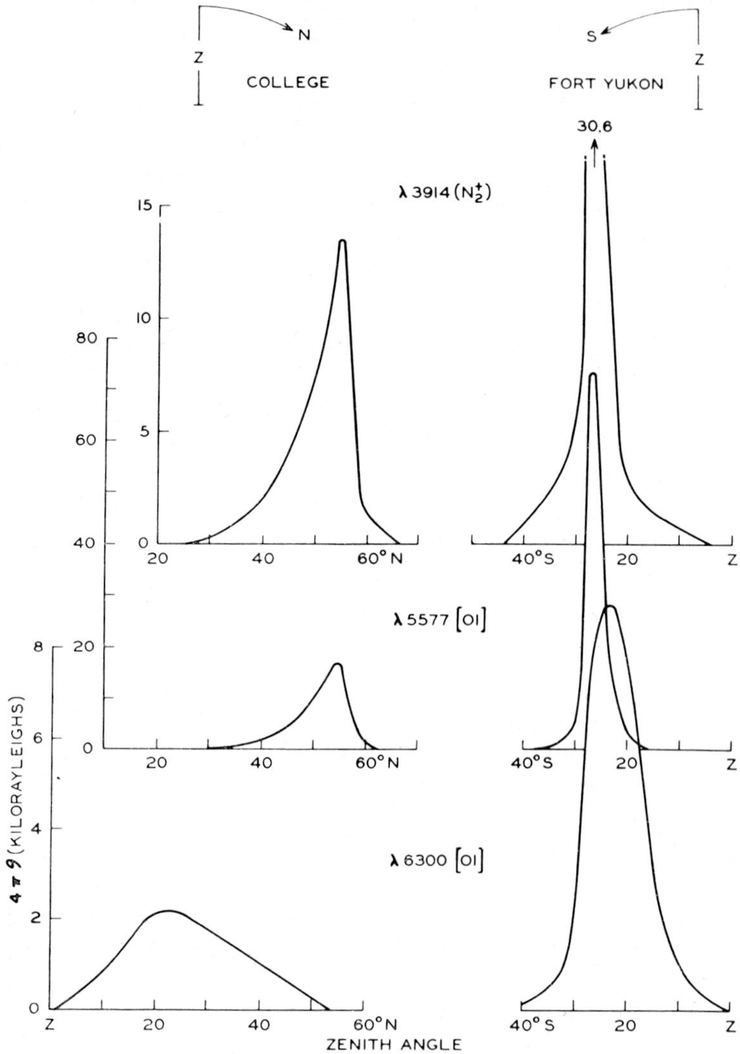

Fig. 4.30. Observed zenith scans of a simple auroral arc from two stations. The curves, obtained by Romick and Belon (1967b), are corrected for extinction, scattering and background. (Courtesy *Planetary and Space Science.*)

range 20 to 40, compared to the calculated value of about 10. This is a reasonable agreement considering that the yields for N(^{2}D) are little more than guesses. Gérard observed good correlation between λ5200 and λ6300 in intensity variations having characteristic times of 5 to 10 s. As with the short period pulsations in λ6300 this probably indicates that the time variation is predominantly in the higher energy parts of the electron energy spectrum.

4.2.4.11. *[O II] and [N II] Forbidden Transitions*

O^+ ions are probably produced directly in the ionization of O and O_2. Equation (4.2.1m) gives assumed values for the production rate for the ^2P, ^2D and ^4P ions. The ion chemistry Equation of (5.1.2) allows height profiles and volume emission rates for the $O^+(^2P)$ and (^2D) ions and their emissions to be calculated.

Table 4.5 gives values derived for emission from $O^+(^2P)$ by Gérard (1970), Rees and Jones (1973), and from Figure 5.5a. These fall in the range 3 to 6 kR; they are highly dependent on certain assumed values for ion yields and quenching reactions. For $O^+(^2D)$ the production rate is similar but because of the long 3.6 h radiative lifetime, quenching by electrons as well as N_2 must be considered. The calculated height profile is given in Figure 5.5a. Calculated emission rates given in Table 4.5 are in the 2 to 5 R range.

The observed intensity of $\lambda\lambda$7319–7330 is highly variable. It is normally undetectable within the N_2 1P 5,3 band of which the IBC3 intensity is about 20 kR, but becomes observable in low energy excited forms. This suggests its IBC3 intensity varies between 1 to 2 kR up to 20 kR or more. This is certainly in accord with predicted values. The IBC3 intensity of $\lambda\lambda$3726–29 Å obtained by Wallace (1969) in a great type-d aurora was 500 R according to Chamberlain (1961). The calculated value in Table 4.5 for 560 eV electrons is at least 180 R so the observed value does not appear unreasonable although this feature would normally be much weaker.

The N^+ forbidden transitions from the ^1S and ^1D levels (λ5755 and $\lambda\lambda$6548–84) are also very weak features. Brown (1969a) has made some calculations similar to those of Gérard for O^+ to give an estimate of the height profile of $N^+(^1S)$. Unfortunately the yield of $N^+(^1S)$ in the ionization of N_2 is unknown, as is the exact value of the quenching rate constants although these may be similar to those for $N^+(^3P)$. The intensity of λ5755 in normal aurora is small although possibly just detectable according to Gattinger and Vallance Jones (1974) (see also Figure 4.4). Both the λ5755 and λ6584 lines have been observed in type-d auroras, for example by Belon and Clark (1959) and Shefov and Yurchenko (1970). λ6584 was also observed by Harang and Pettersen (1967) in soft proton aurora.

4.2.4.12. *O I Higher Level Transitions*

Figure 4.8 shows these transitions for O, the solid lines indicating those certainly observed in the auroral spectrum. The 1304 Å (i.e. the 1302-5-6 triplet) is a very strong feature. The excitation cross sections for the $3s^3S°$ state have been measured by Stone and Zipf (1974) and Mumma and Zipf (1971a) and that for excitation from O is given in Figure 4.15b. It includes cascading to $3s^3S°$ via the $3p^3P$, $4p^3P$ and higher levels. The cross section for dissociative excitation of O_2 is also large, and again includes a major cascade contribution via the $3p^3P$ state. The λ1304 emission cannot be measured directly since it is strongly reabsorbed by ground state $O(^3P)$. A radiative transfer analysis of rocket measurements such as carried out by Donahue and Strickland (1970) is necessary. The intensity of the λ1304 feature scaled to

IBC3 ($I(3914) = 80$ kR) has been variously reported between 20 and 90 kR by Sharp and Rees (1972) and the sources quoted by Vallance Jones (1971). The $\lambda 8446$ line has an observed intensity of about 15 kR while the higher triplet transitions are much weaker.

The quintet states are also excited by electron impact on O but with a cross section about half of that for $3s^3S$. Moreover the cross section for the $3s^5S^\circ$ state by dissociative excitation of O_2 is about $2\times$ that for $3s^3S^\circ$. The observed IBC3 intensity of $\lambda 1356$ is about 11 kR according to the results of Sharp and Rees although less in the earlier results reviewed. They have computed the relative intensity for a normal auroral arc and have obtained an intensity in good agreement with that observed. $\lambda 7774(3s^5S_0 - 3p^5P)$ is a prominent spectral feature and it is reasonable to suppose that it is excited both directly from O as part of cascades down through $\lambda\lambda 1356$–9 and by dissociative excitation of O_2. The ratio $I(8466)/I(7774)$ is observed in bright aurora to be about 2. Since this is smaller than most theoretical estimates for excitation from $O(^3P)$ it has been argued that a substantial contribution from dissociative excitation must be involved. According to the measurements of Vallance Jones and Gattinger (1973) it appears likely that this ratio increases with height, an effect which would accord with the idea of dissociative excitation being more important for $\lambda 7774$. There is a strong increase in $I(8446)/I(N_2^+ \text{M})$ with height but this is probably due to the change in $n(O)/n(N_2)$. The generally reported absence of such an effect for $I(5577)/I(N_2^+ \text{1N})$ may be another indication of additional excitation processes for $O(^1S)$ at lower heights.

The other triplet, quintet and intercombination transitions shown on Figure 4.8 are rather weak and their excitation has not been systematically investigated since the pioneering study of Petrie and Small (1953).

4.2.4.13. *N I Higher Level Transitions*

The theoretical interpretation of the transitions of Figure 4.10 is rudimentary. Cross sections are given in Table 4.4 for the excitation the $3s^4P$ state for electron impact on N and from N_2. Prasad and Green (1971) suggest that dissociative excitation is important for the higher levels although direct excitation of N may be significant if $n(N)$ rises sufficiently under auroral conditions.

In the latter case $\lambda 1200$–01 suffers self-absorption and a radiative transport analysis of the rocket observations is necessary. Such an analysis by Strickland (1970) gives an observed production rate consistent with the calculations of Prasad and Green.

4.2.4.14. *O II and N II Higher Level Transitions*

Some of the multiplets associated with excited configurations of these ions appear in the spectrum but little has been done to elucidate their excitation mechanisms.

4.2.4.15. *Other Atomic Lines*

Na I. The NaD doublet appears in most auroral spectra (e.g. Figure 4.4) and in recent years there seemed to be good observational evidence that it was excited in aurora.

Theoretically this presented difficulties since the ratio $n(\mathrm{Na})/n(\mathrm{N_2})<10^{-9}$ so that even with a cross section for electron impact excitation of $10\times$ that for $\mathrm{N_2(B^3\Pi)}$ the IBC3 intensity should be less than $I(1\mathrm{PN_2})\times 10^{-8}\leq 10^{-2}$ R compared to an observed value of the order of 1 kR. Rees (1974) has carried out a more exact estimate including the process of energy transfer from vibrationally excited N_2 and obtained quite negligible excitation rates. He has also shown that the best documented observations (Derblom, 1964) were probably contaminated by twilight glow sodium emission.

He I. The 10830 Å He I line appears in the twilight airglow and its twilight intensity is much enhanced in the presence of aurora, particularly type-d aurora. This is presumably due to population of the metastable ^{3}P lower state through the excitation of atmospheric He by auroral electron fluxes.

The 5876 Å He I line has long been sought as evidence for a component of α-particle precipitation analogous to the H lines of proton aurora. In this case there should be a Doppler effect. The emission was not found by Eather (1968c) in a careful study with a sensitive filter tilting photometer system. Stoffregen (1969) presented evidence for the appearance of λ5876 in bright break-up aurora but the identification cannot yet be regarded as certain. As may be seen from Figure 4.4 the line occurs in a very complex region of the spectrum.

4.2.5. Variations in the Auroral Spectrum

4.2.5.1. *Introduction*

As mentioned in Section 3.3 aurora takes on many different forms and several different color classes can be distinguished visually. The variations in color seem largely due to variations in the energy spectrum of the exciting electrons coupled in some cases with variations in atmospheric composition and neutral and electron temperature profiles. Changes in form reflect similar factors together with spatial inhomogeneities in the electron precipitation patterns. Evans (1971) has argued that much of the finer structure discussed in Section 3.3.2 is not present in the exciting particle fluxes and must consequently correspond to some localized aeronomic process. As yet no process capable of generating such structure for λ5577 or the N_2^+ bands has been proposed and there is now much evidence that the structure of the particle fluxes and optical features do correspond (see Whalen and McDiarmid, 1974; and Evans, 1972).

In considering variations of the auroral spectrum one must distinguish between height effects due entirely to variations in the average height of the emission when an auroral form is viewed from the side, and variations in the integrated emission seen in the magnetic zenith due to variations in energy spectrum of the exciting particles. We will call the former *simple height effects* and the latter *simple energy spectrum effects*. There are probably additional variations due to modifications of the atmospheric parameters by the particle flux and we shall refer to these as *complex energy spectrum effects.*

4.2.5.2. *Simple Height Effects*

In the case of an auroral form having the ideal spatial distribution of a thin sheet aligned parallel to the terrestrial magnetic field the spectrum would vary with height according to the production rates of the different emissions as shown in Figures 4.24 and 4.28. The results indicate that most emissions have height distributions similar to those of the ionization rate with some modifications caused by atmospheric composition. The notable exceptions are the quenched emissions including [*O* I] λ6300, [N I] λ5200, O_2 (Atm.; 1,1), [O^+] $\lambda\lambda$7320–30. Such effects measured either by ground or rocket photometers are valuable in the study of the quenched emissions.

The range of variation of simple height effects is of course dependent on the exciting particle flux. In particular hard fluxes penetrating across the O/O_2 transition region probably produce the visually observable type-b effect while fluxes with a large component of soft electrons probably produce type-a height effects. Since in these color classes the color transition with height in the auroral form is essential they will be discussed here.

Type-b aurora. The morphology has been discussed in Section 3.3 where it is pointed out that several lines of evidence suggest that the transition between the red lower border and the green upper part lies in the neighborhood of 80 km. The spectroscopic changes reported are summarized in Table 4.19 an expanded version of that in the review of Vallance Jones (1971). There should theoretically be some change in the relative intensity of O_2^+(1N) between the higher weaker aurora with a maximum intensity at 140 km and type-b with a maximum about 90 km since $n(O_2)/n(N_2)$ can change by a factor of 2 or more in this range. According to the discussion of Hanna and Anger (1971) such enhancements of O_2^+ 1N could not account for the reddening. This leaves variations in $I(5577)$ relative to $I(1P\ N_2)$ as the most likely cause, with most or all of the variation being due to changes in $I(5577)/I(N_2^+\ 1N)$ rather than enhancement of $I(1P\ N_2)/I(N_2^+\ 1N)$. The most recent observations by Gattinger and Vallance Jones with a 1° field digital photometer indicate an increase of up to $4\times$ in the $I(1P\ N_2)/I(5577)$ ratio coupled with a small ($\sim$25%) increase in $I(1P\ N_2)/I(M\ N_2^+)$. The latter may be due to a decrease in the Meinel N_2^+ band intensity relative to the N_2^+ 1N bands. $I(O_2^+\ 1N)/I(1P\ N_2)$ did not change.

The simplest hypothesis to explain the changes in $I(5577)/I(N_2^+\ 1N)$ is the variation in the ratio $n(O)/n(N_2)$ which decreases rapidly between 100 and 80 km. This would be quite satisfactory if $O(^1S)$ is indeed excited primarily from O (by electron impact or energy transfer). Brown (1969b) has suggested that electron attachment could suppress O_2^+ dissociative recombination between 75 and 80 km. Likewise if significant concentrations of NO are formed enhanced quenching of $O(^1S)$ could occur since the quenching coefficient with NO is 10^3 larger than for O_2.

Our model calculations based on $O(^1S)$ excitation by electron impact and disso-

TABLE 4.19

Spectroscopic characteristics of type-b aurora

Observer	Spectral changes[a] $I(5577)$	$I(1P\ N_2)$	$I(O_2^+)$	Remarks
Vegard and Tönsberg (1937)		+		
Hunten (1955)		+	+	
Evans and Vallance Jones (1965)	−	+	Small	Relative to $I(N_2^+\ 1N)$
Shemansky and Vallance Jones (1968)		Small	+	Relative to $I(N_2^+\ M)$
Harris *et al.* (1970)			Small	Relative to $I(1P\ N_2)$
Gattinger and Vallance Jones (unpublished)	−	+	Small	Relative to $I(N_2^+\ M)$ which may decrease a little relative to $I(N_2^+\ 1N)$

[a] +(−) denotes a significant enhancement (decrease) in intensity relative to other features as specified in last column.

ciative recombination predict that $I(5577)/I(1P\ N_2)$ should decrease by a factor of at least 50 between 110 and 90 km. The fact that such large effects are not observed and that the type-b limit appears to occur below 90 km may support the hypothesis that photodissociative excitation from O_2 is important. In this case the onset of severe quenching of $O(^1S)$ by O_2 between 70 and 80 km could be of importance. The onset of quenching by $O(^3P)$ could produce the observed effect if excitation from O_2 is unimportant and if the height measurements for type-b borders are in error. On the the whole excitation from and quenching by O_2 accords better with the observations.

Type-a aurora. As defined this has red upper parts above green. As shown in Figures 4.24 and 4.28 any sufficiently soft power or exponential distribution of electrons should have this characteristic. Such forms might be fairly common visually if it were not for the color threshold of the eye since photometer scans such as Figure 4.30 show a strong red enhancement above most auroral forms. The development of significant red emission at higher levels may involve other factors which will be discussed below under type-d aurora.

4.2.5.3. *Simple Energy Spectrum Effects*

Variations of the average energy of the exciting electrons produces strong effects on the relative height integrated intensity of $\lambda 6300$. Some theoretical values of the ratio $I(6300)/I(4278)$ are given in Figure 4.29. In this figure the ratio is plotted against average electron energy for a variety of energy spectra and incident energy fluxes. The dependence on electron energy distribution is weak while that on energy flux is somewhat stronger, as is that on atmospheric composition. Eather and Mende (1972) have reproduced a figure calculated by M. H. Rees showing the ratio plotted as a function of $I(4278)$ for various incident electron energy spectra. Experimentally, wide variations of this ratio are observed in aurora viewed in the zenith as is shown, for example, in the work of Gattinger and Vallance Jones (1972). There appears to be a significant negative correlation between the $I(6300)/I(4278)$ ratio and $I(4278)$ indicating that softer electron fluxes are generally associated with fainter aurora and vice versa. Studies of $I(6300)/I(4278)$ provide a powerful method of detecting variations in the average energy of the incoming particle fluxes. Eather and Mende (1972) have made extensive use of this in the analysis of aircraft observations across the auroral oval.

The ratio $I(5577)/I(4278)$ should also be an index of particle energy for energies above 10 keV. Unfortunately the uncertainties in the excitation mechanism of $\lambda 5577$ render uncertain predictions of the variation of the ratio with energy. For electron impact excitation one expects a decrease in the ratio with increased particle energy because of the decrease of $n(\mathrm{O})/n(\mathrm{N_2})$ at lower heights and also a decrease for higher intensities because of increased competition from $\bar{e}-\bar{e}$ collisions for the energy of the secondaries. The contribution of the mechanisms involving O_2^+ and O_2 as discussed above tends to offset these effects. At sufficiently high particle energies quenching should produce a decrease in the ratio but at present we do not have a quantitative theory of the effect.

It has been reported that significant decreases in the ratio $I(5577)/I(4278)$ (or $I(5577)/I(3914)$) are observed in the 'mantle aurora' as discussed by Sandford (1964, 1968). This is a zone of diffuse emission in the form of a ring of radius 20° about the geomagnetic pole. This corresponds to the 'drizzle' particle precipitation zone of Hartz and Brice (1967) (see also Hartz, 1971). Sandford obtained $I(5577)/I(3914)$ as low as 0.4 in the morning period for this zone. Direct evidence for the mantle zone in the midday sector is given by the aircraft observations of Whalen *et al.* (1971c).

The O_2^+(1N) bands should also be enhanced relative to N_2^+ emissions with increasing particle energy. This effect has not yet been successfully exploited. Observations of the 0,0 and 0,1 O_2^+ bands are ambiguous because of other strong emissions nearby. The weaker 2,0 band at 5275 Å may give more reliable results but as yet no significant enhancement has been observed.

4.2.5.4. *Complex Energy Spectrum Effects*

The bright, often completely red, aurora sometimes seen particularly at low latitudes during great magnetic storms seems to involve more complex effects than a simple softening of the energy spectrum of the incoming particles. Figure 4.31 shows photometer results for the evolution of the display of March 23, 1969 as observed by Noxon and Evans (1974) from Boston. The ratio $I(6300)/I(5577)$ appears to reach a peak value of ~ 4 with $I(6300)$ having a maximum value of 150 kR. Several such displays have been studied by Noxon *et al.* using ground based photometry in conjunction with incoherent scatter measurements of electron temperature and concentration. Another type-d event of lesser intensity has been reported by McEwen and Sivjee (1972). The caculated value of $I(6300)/I(5577)$ for the 562 eV case in

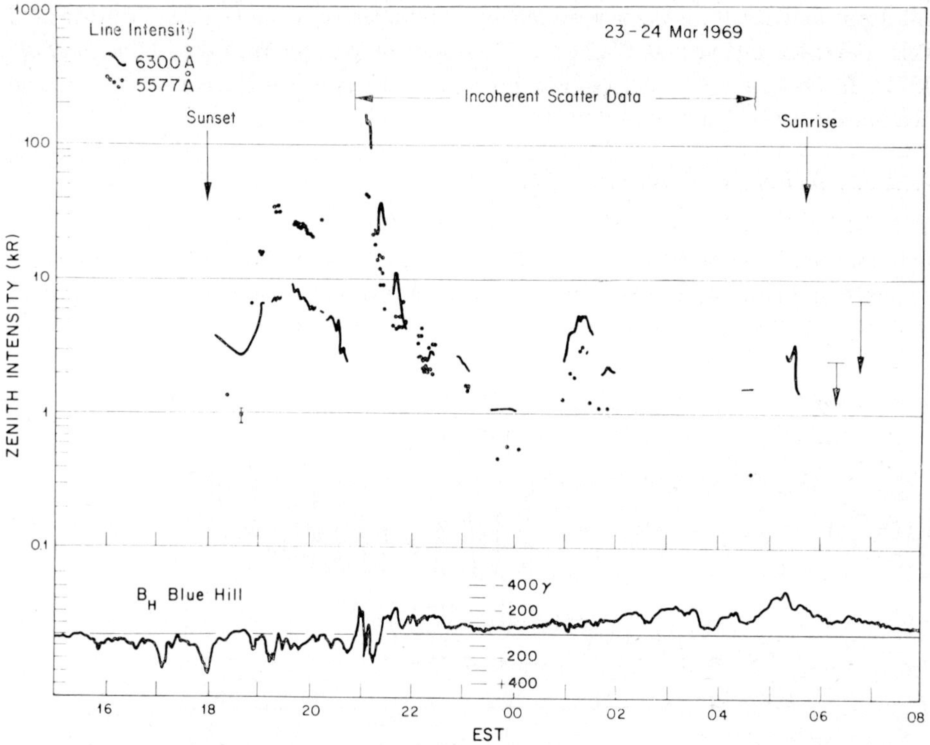

Fig. 4.31. Absolute values of $I(6300)$ and $I(5577)$ observed by Noxon and Evans (1974) from Blue Hill Observatory (geomagnetic latitude about 56° N) during the type-d display of March 23–24, 1969. The gaps are at times when the photometer was directed away from the zenith. The lower trace is the magnetometer H-component. I am very much indebted to Dr J. F. Noxon for providing these data in advance of publication.

Table 4.5 is 1.5 while the corresponding value for 180 eV electrons is 3.5. The model however is not very reliable at low energies because of the problems discussed in Section 4.2.1.3 and also because of the neglect of ion transport and heat conduction in the electron gas above 300 km. Noxon and Evans conclude from the direct measurements that downward thermal conduction raises the temperature of the electron gas below 350 km sufficiently to excite λ6300 by the thermal mechanism. In their analysis, this downward heat flux exceeds the height integrated electron cooling rate below 350 km by a factor of about 5 for a normal atmospheric composition. There are now indications (Mayr and Volland, 1972, 1973) that $n(N_2)$ above 200 km increases strongly as a result of heating (probably Joule heating) in strong magnetic storms. Noxon and Evans show that this effect could account for the additional electron cooling. Such composition effects induced by the external causes producing the auroral display would clearly produce important modifications to the spectrum through changes in quenching, electron temperature, and relative particle concentrations. Variations in density and composition associated with the auroral oval or substorm activity have been reported by a number of authors (e.g., Newton and Pelz, 1969; King-Hele and Walker, 1971; Anderson and Sharp, 1972; Taeusch *et al.*, 1971). In Section 5.3.11 it is shown how such effects are predicted by the auroral heating theory of Hays *et al.* (1973).

4.2.5.5. *Atmospheric Temperature Effects*

The distribution of intensity within spectral lines, bands and band systems is affected by atmospheric temperatures. Some temperature effects are very striking such as the example of Figure 4.32 showing changes in the rotational intensity distribution of the 0,0 1N N_2^+ band. The observable effects may be classified as follows.

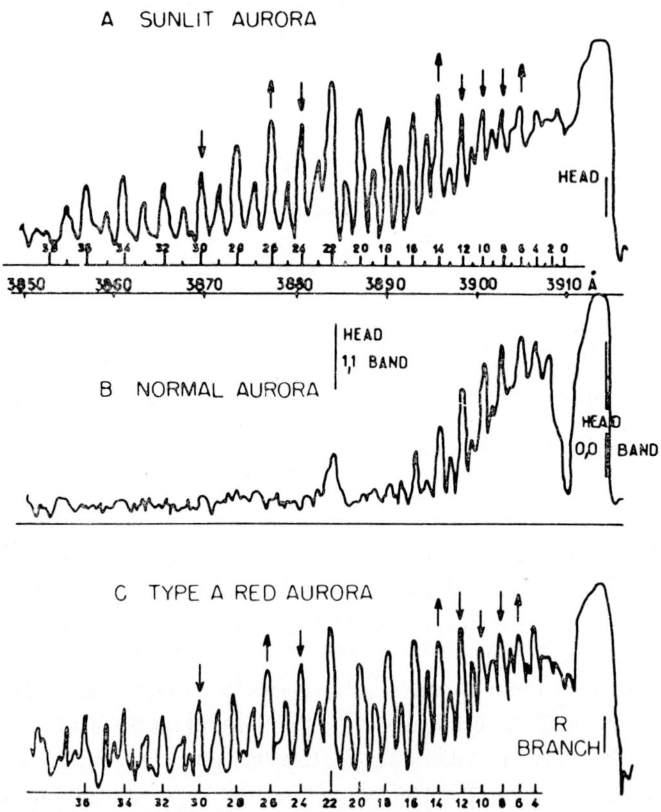

Fig. 4.32. High resolution spectra of 0, 0 N_2^+ 1N band showing changes in rotational distribution in (a) sunlit aurora (b) normal aurora and (c) type-d red aurora (Vallance Jones, 1967). (Courtesy Van Nostrand Reinhold Company.)

(a) *Doppler broadening of lines.* Translational motion of the emitting particles gives Doppler broadening or shifts of emission frequencies. The full width at half intensity for a spectral line broadened by the Doppler effect alone (Chamberlain, 1961, p. 7) is

$$\Delta\lambda = 2(\ln 2)^{\frac{1}{2}} \frac{\lambda}{c} \sqrt{\frac{2kT}{M}} \simeq 7.2 \times 10^{-7}\lambda \sqrt{\frac{T}{M}} \tag{4.2.5a}$$

λ is the wavelength in Å; M the molecular weight. Thus for O at 320 K, for λ6300 the Doppler width is about 20 mÅ. Table 2.1 shows that such a line width may be conveniently measured using a Fabry-Perot interferometer; it is also readily measurable by Fourier spectroscopy. Most work on aurora has been done on the λ5577 and λ6300 [OI] lines. As may be seen from Figures 4.24 and 4.28 temperatures from λ5577 should be typical of the height of maximum intensity of the visual aurora while λ6300 temperatures refer to a higher level, the height of which is somewhat dependent on the energy distribution of the particle flux.

In principle temperature distributions with height should be obtainable by combining λ5577 Doppler measurements with height-finding techniques. By assuming a lower border height of 105 km Turgeon and Shepherd (1962) obtained several sets of temperature values vs height as shown in Figure 4.33. They also obtained distri-

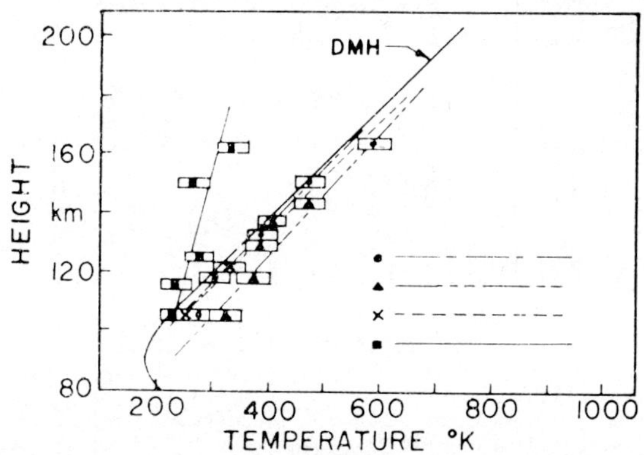

Fig. 4.33. Variation in measured λ5577 Doppler temperature with height calculated by assuming a lower border at 105 km from the study of Turgeon and Shepherd (1962). The solid line labeled DMH is the optical temperature height profile adopted by Hunten (1961). (Courtesy *Planetary and Space Science.*)

butions of temperatures for various auroral forms. Similar distribution measurements by the λ5577 Doppler method were made by Karandikar (1968, 1970) for diffuse glows and sub-visual aurora. Turgeon and Shepherd also showed a distribution of measured temperatures vs local time which indicated a decrease in median temperature from about 370 K to 250 K during the night. Other measurements of λ5577 Doppler temperatures have been reported by Jeannet and Vassy (1963), Weller (1961), Mylyarchik (1960), Armstrong (1959), Hilliard and Shepherd (1966b) and Zwick and

Shepherd (1973). The latter two studies were done using a field widened Michelson interferometer and have a time resolution down to about 0.6 s.

As a consequence of the large number of temperatures measured by the Michelson technique, some important relationships were demonstrated. The first is a very definite connection between brightness and temperature. The observed points on a $T(5577)$ vs $I(5577)$ plot were found to lie on a particular hyperbola-like curve for various intervals of time; subsequently a shift to another curve of the same family could occur.

The second striking result from the Michelson studies was a short period cyclic relationship between intensity and temperature in auroral pulsations. This is illustrated in Figure 4.34.

The significance of the observed temperature effects will be discussed after the description of rotational and vibrational temperatures.

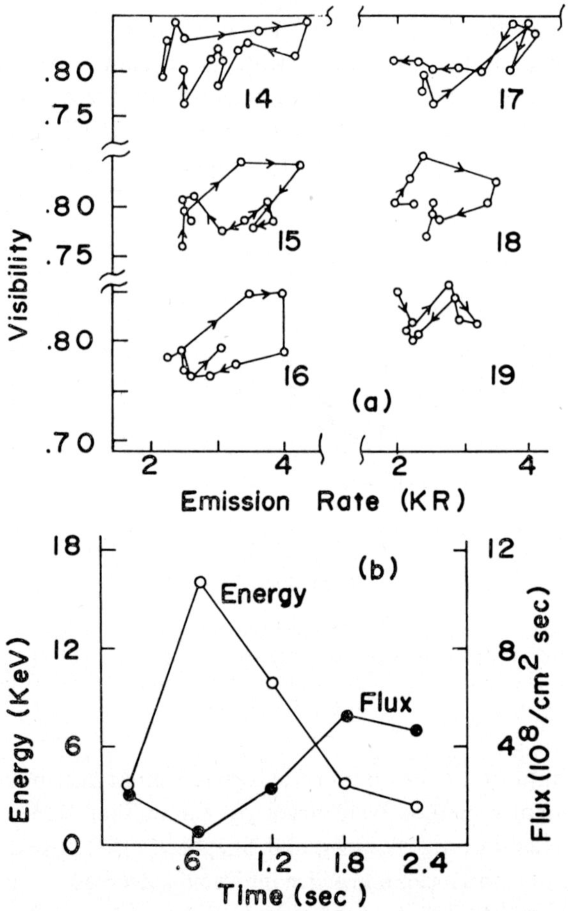

Fig. 4.34. (a) Variations of fringe visibility and intensity in auroral pulsations. (Visibility decreases with increasing temperature.) (b) Average energy and flux time variations which could generate cycles like 15. From Shepherd (1971). (Courtesy D. Reidel Publishing Co.)

(b) *Rotational intensity distribution of molecular bands.* In principle the rotational fine structure intensity distribution of molecular bands depends in most cases on the atmospheric temperature. The detailed theory involved is in general beyond the scope of this work, and reference is made to Herzberg (1950) for a full treatment and also for the definition of the spectroscopic symbols and quantum numbers to be used. We will confine ourselves as an example to a description of the situation in the simplest and most important case in auroral spectra. This concerns the 1N N_2^+ system.

It has already been pointed out that the system consists of bands arising from transitions between vibrational levels, v' of the upper state $B^2\Sigma_u^+$ to vibrational levels v'' of the lower $X^2\Sigma_g^+$ state. Each vibrational level has a fine structure due to rotation. The fine structure of two typical levels of the system is shown in Figure 4.35, together with the possible transitions between them. The single and double primes refer to quantities for the upper and lower states. K is the rotational quantum number.

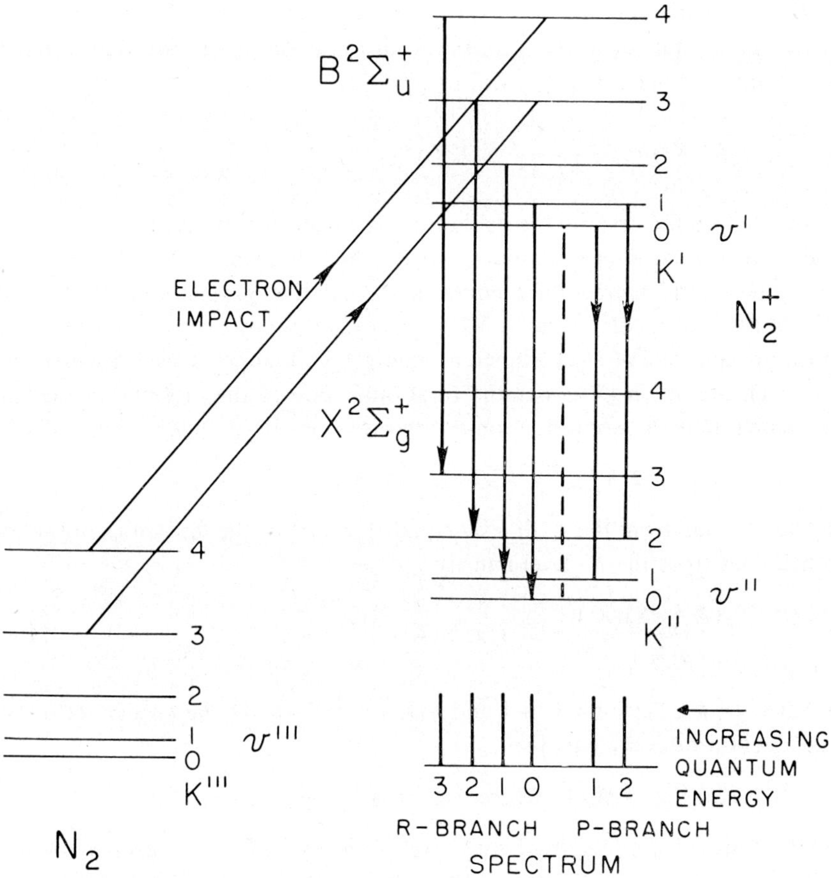

Fig. 4.35. Simplified energy level diagram illustrating typical excitation relations for an N_2^+ 1N rotation-vibration band v''', v' and v'' indicate respectively typical vibrational levels of the ground state of N_2 and the B and X states of N_2^+.

The rotational fine structure results from the allowed transitions for which $\Delta K = K' - K'' = \pm 1$. The $\Delta K = +1$ and -1 transitions give a series of lines called the R- and P-branches distributed in energy and hence wavenumber as shown. A higher resolution spectrum of the 0,0 band from aurora is shown in Figure 4.32. The lines resolved belong to the R-branch and are labeled by the value of K''.

The intensities of the rotational lines of the v', v'' band is given by the usual relation, Equation (4.1g),

$$I(v', v''; K', K'') = N(v', K')\, A(v', v'', K', K''). \tag{4.2.5b}$$

To a good approximation the rotational parts of the wavefunctions can be factored off so that (cf. Equations (4.1h) and (4.1l)

$$A(v', v'', K', K'') = A(v', v'')\, S(K', K'')/g(K'), \tag{4.2.5c}$$

where $A(v', v'')$ is the vibrational transition probability as defined in Equation (4.1l), $S(K', K'')$ is the line strength factor for the rotational wave functions (Equation (4.1i)) and $g(K')$ is the statistical weight of the K'th rotational level. For a given band, the relative rotational line intensities will be given by

$$I(K', K'') \propto \frac{S(K', K'')}{g(K')} N(K'). \tag{4.2.5d}$$

The line strength factors are calculable from the wave functions and in this case take the values K' for the R-branch and $K'+1$ for the P-branch.

The atmospheric temperature controls the relative populations, $N(K')$, of the rotational levels of the $B^2\Sigma$ state of N_2^+ via its excitation by electron impact ionization from the ground state of N_2 as shown in Figure 4.35. It appears, as is discussed below, that in such excitation processes the rotational state of the molecule is unchanged. If we assume that the excitation-ionization cross section is independent of K''' then

$$I(K') \propto N(K'''), \qquad K' = K'''$$

while the loss rate from the K'th level equals the sum of the intensities of the R- and P-branch lines from the K' level namely

$$\frac{S_R(K') + S_P(K')}{g(K')} N(K') = N(K')$$

since $S_R(K') = K'$, $S_P(K') = K'+1$ and $g(K') = 2K'+1$. In the steady state the loss and production rates are equal so

$$N(K') \propto N(K''') \quad \text{where} \quad K' = K'''.$$

In other words the population of each level of the excited state is proportional to the population of the level of the ground state of N_2 with the same quantum number. The latter is given simply by the Boltzmann relation

$$N(K''') \propto g(K''') e^{-hcB'''K'''(K'''+1)/kT}$$

since the rotational energy levels of the ground state are given approximately by the relation

$$E_{rot} = hcB'''K'''(K''' + 1).$$

The statistical weight $g(K''')$ of the ground state rotational levels, apart from a nuclear spin factor, is given by $2K'''+1$. Recalling that $K' = K'''$, we finally have after substituting into Equation (4.2.5d) that

$$I(K',K'') \propto S(K',K'')e^{-hcB'''K'(K'+1)/kT}. \tag{4.2.5e}$$

This is the same expression as would be obtained if the upper level itself had been in thermal equilibrium, except that B' is replaced by B'''.

In practice only the R-branch of the, 1N N_2^+ bands is easily measured (see Figure 4.32) and so Equation (4.2.5e) can be taken with $S(K',K'') = K' = K''+1$ and transformed to give

$$\ln(I_R/K') = -\frac{hcB'''}{kT}K'(K'+1). \tag{4.2.5f}$$

Thus the temperature can conveniently be obtained by plotting $\log(I_R/K')$ against $K'(K'+1)$ which should give a straight line of slope $-\dfrac{hc\,B'''}{2.303kT}$. Examples of the use of such plots are given by Vallance Jones and Harrison (1955).

This example has been treated at some length to show clearly the assumptions on which it is based. The important one is that the rotational angular momentum is unchanged during excitation. There is considerable evidence from laboratory tests that the method does give good values for the gas temperature for a gas excited by electric discharge, electron impact or even proton impact.* There is some laboratory evidence (Culp and Stair, 1967) for a breakdown of the $\Delta K \simeq 0$ approximation for electrons of low energy (<25 eV) but this would have a negligible effect on the 1N N_2^+ bands which are mostly excited by higher energy electrons (see Section 4.2.2.1).

Often it is desirable to derive temperatures from incompletely resolved spectra. This can conveniently be done by the synthetic spectrum method, i.e., by calculating the theoretical band profiles for the instrumental function used for a range of temperatures. The observed spectrum can be compared to the synthetic spectra and the temperature adopted to give the best fit. In other cases, fixed frequency photometers have been used to measure intensity ratios for pass bands at two or more wavelengths and from these ratios temperatures can be derived (Hunten *et al.*, 1963; Shepherd, 1971).

This method has been applied to other band systems. In most cases the synthetic spectrum method must be employed since the more complex bands consist of many partially overlapping branches. The individual branches have an intensity distribution

* See Section 4.3.3.2 however.

similar to that given by Equation (4.2.5e) but the form of the line strength factors is usually less simple.

(c) *Effects of temperature on vibrational band ratios.* In principle the relative intensities of the vibrational bands of electronic systems are affected by the initial state vibrational distribution of the target molecules. This has been noted in Section 4.2.4.1 where the excitation rate of the v'th level (see e.g. Figure 4.35) was shown to be given by Equation (4.2.4c). In that section $N(N_2, v''')$ was neglected for $v''' > 0$ but if this approximation is not made we must obtain $N(v''')$ from the Boltzmann distribution,

$$N(v''') \propto g(v''')e^{-hcG(v''')/kT}. \tag{4.2.5g}$$

$G(v''')$ is the vibrational term value. The vibrational energy levels of N_2 and like molecules are spaced at intervals corresponding to about 2000 cm^{-1}, i.e., $G(1) - G(0) \simeq 2000$ cm^{-1} while $kT/hc \simeq 200$ cm^{-1} for $T = 300$ K. Hence $N(1)/N(0) < 10^{-4}$ for normal auroral temperatures around 300 K. At 1300 K $N(1)/N(0) \simeq 0.1$. Clearly terms higher than the first will have little effect on the vibrational population rates until temperatures of at least 1200 K are reached. For the normal neutral atmosphere such temperatures are rarely if ever found below 200 km and consequently vibrational band ratio measurements are not in general useful for temperature measurements. Under exceptional conditions higher temperatures may occur in aurora and such ratios may then become significant.

(d) *Observed Doppler and rotational temperature effects in normal aurora.* Some observed Doppler temperature effects have already been noted in (a). Studies of rotational temperatures for the 1N N_2^+ bands have given results very similar to the Doppler measurements of λ5577. Most of the earlier systematic work on this has been reviewed by Hunten (1961). One long-term study has been reported by Brandy (1964) who attempted the simultaneous measurement of height by ground triangulation and rotational temperature of N_2^+ λ3914. The results are similar to the mean curve adopted by Hunten (1961) and shown in Figure 4.33. Other similar ground based results are given by Zwick and Shepherd (1973). Both the Doppler and N_2^+ rotational methods give similar temperature gradients in the 100 to 150 km regions. The temperatures inferred for 150 km seem however to be somewhat low compared to standard atmospheres, eg., CIRA 1965 for which the midnight 160 km value is 690 K even for values of solar activity lower than the mean. This difference is probably a result of the difficulty of estimating the height of the region from which the emission arises. Other optical methods of measuring temperatures at auroral latitudes namely AlO and BaO rotational and vibrational temperatures from rocket releases show temperatures higher if anything than the CIRA 65 values at 165 km and sensitive to the general level of geomagnetic activity (Rees, 1971). As in the auroral observations of Hilliard and Shepherd (1966 b) no definite heating effects in individual forms could be found. For the low level auroral emissions the Doppler and rotational intensity distributions seem consistent with our knowledge of the height variations in aurora together with the temperature profile of the neutral atmosphere.

(e) *Doppler and rotational temperature effects in unusual aurorae.* For λ6300 emission many rather high temperatures have been reported. Since the maximum emission rate of this line in normal aurora is about 250 km (where the temperature of the mean CIRA 65 atmosphere is about 1000K) it is not surprising that values of 1000 to 1500K are encountered, as reported by Turgeon and Shepherd (1962) or in some of the data of Mylyarchik (1960). However Mylyarchik obtained some values in the 1700 to 2000K range as well as a high temperature of 3400K in the November 17–18 1958 type-A red aurora. Ignatyev *et al.* (1971) studied λ6300 line widths in aurora at 65°N and obtained temperatures varying from 700K under quiet conditions to 1700K under disturbed conditions. Blamont and Luton (1972) have obtained measurements of temperatures up to 1700K from a satellite-borne λ6300 spherical Fabry-Perot interferometer. Truttse (1972) has noted a relation between $T(6300)$ and K_p with $T(6300)$ reaching 2000K for $K_p \sim 7$. It appears that the atmosphere above 250 km is heated during strong disturbances which are normally associated with strong aurora. Whether this heating is the direct result of the energy deposited by the same particles which produce the aurora or occurs through some other mechanism remains undecided.

Enhanced 1N N_2^+ rotational temperatures have also been observed during type-A red (type-d) auroral displays by Clark and Belon (1959) and by Vallance Jones (1967). It seems likely that these correspond to real increases in atmospheric temperature above 200 km.

(f) *Vibrational temperature effects.* As noted above, vibrational band ratios could show effects for high temperatures above 1500K. This temperature need only apply to the vibrational degree of freedom. It was suggested several years ago by Walker (1968) that T_v for N_2 might indeed be increased much above the neutral temperature as a result of collisions with thermal electrons in the quenching of $O(^1D)$, and in the reaction of N and NO. The problem is complicated by loss processes. These include an accidental resonance between N_2 and CO_2 vibrational levels, the possibility of an $O-N_2^\dagger$ reaction and the diffusion of $N_2^\dagger$. (The dagger indicates vibrational excitation.) These effects have been analyzed by Walker *et al.* (1969) who concluded that in bright aurora $T_v(N_2)$ might increase by 20K min^{-1} at 100 km. This would probably not lead to any observable effect and this indeed appears to be the case since none of the studies of the auroral band ratios have really shown any sign of the changes in ratios for the N_2^+ 1N or N_2 2P bands which should arise if $T_v(N_2)$ were enhanced to values above 1500K. Some such effects have been claimed from time to time but as pointed out by Broadfoot (1968) these claims are dubious.

For high level aurora such as type-d and sunlit rays, in which high rotational and Doppler temperatures are observed, the vibrational band ratios likewise indicate high temperatures. For the sunlit rays ($\sim$500 km) discussed by Vallance Jones and

Hunten (1960) and Shuiskaya (1961) this effect appears to exist but there are some difficulties in distinguishing between the effect of high T_v for the initial N_2 molecules and the optical pumping effect suggested by Bates (1949). This effect can produce high vibrational temperatures among the N_2^+ ions undergoing multiple fluorescent excitations. Clark and Belon (1959) studied lower resolution (~10 Å) spectra of the February 10–11, 1958, great type-A aurora and were able to fit the $\Delta v = -2$ sequence of 1N N_2^+ bands with a synthetic spectrum with $T_v(N_2) \simeq 4000$ K and $T_R(N_2) \simeq 2500$ K. While there have been some questions raised concerning possible contamination of these spectra by atomic lines (Harang and Gérard, 1970) it is possible that under the unusual conditions of this display that the vibrational temperature of the N_2 was indeed enhanced above the atmospheric temperature. Walker (1973) has reviewed recent theoretical ideas on this question.

In the case of rotational temperatures for sunlit auroral rays the analysis of Vallance Jones and Hunten (1960) favors the conclusion that the rotational intensity distribution arose from a high atmospheric temperature. The rotational structure (Figure 4.32a) shows irregularities because of the structure in the solar continuum exciting the fluorescence. Conceivably there could also be an optical pumping effect among the rotational levels but the evidence was against this being important.

Further high resolution studies of the structure of these bands in sunlit and type-d aurora would be of value especially since Broadfoot (1968) has shown that there are some problems in the interpretation of the spectrogram studied by Vallance Jones and Hunten. The increase in N_2 concentrations above 200 km described in Section 5.3.11 is likely to be important here.

In the case of the type-d aurora in the midday sector of the oval (Equation (3.3.6b)) Sivjee (1974) has observed both high vibrational and rotational temperatures. The former can be understood in terms of the processes suggested by Walker but the high T_R exceeding 3500 K is puzzling.

4.3. Proton Aurora

4.3.1. Interaction of a Proton Beam with the Atmosphere

The behavior of protons penetrating the atmosphere differs fundamentally from that of electrons. First the probability of the deflection of protons in collisions with atmospheric particles is small enough to be neglected. Secondly the possibility of a fast proton capturing an electron to become a fast neutral atom introduces the need to consider the collisional interactions of such neutral particles. In the neutral state the particle is free from control by the magnetic field. Thus in penetrating the atmosphere, protons alternate between spiral paths (as bare protons) around the field lines and straight paths (as H atoms) at an almost constant angle to the field direction. The typical path is somewhat as shown in Figure 4.36. Because of the smallness of collisional deflections the analysis of proton paths is simpler than for electrons. The possibility of partial neutralization of the beam is a complication which must be taken into account. To consider this and other processes the inelastic cross sections for protons and H atoms must first be considered.

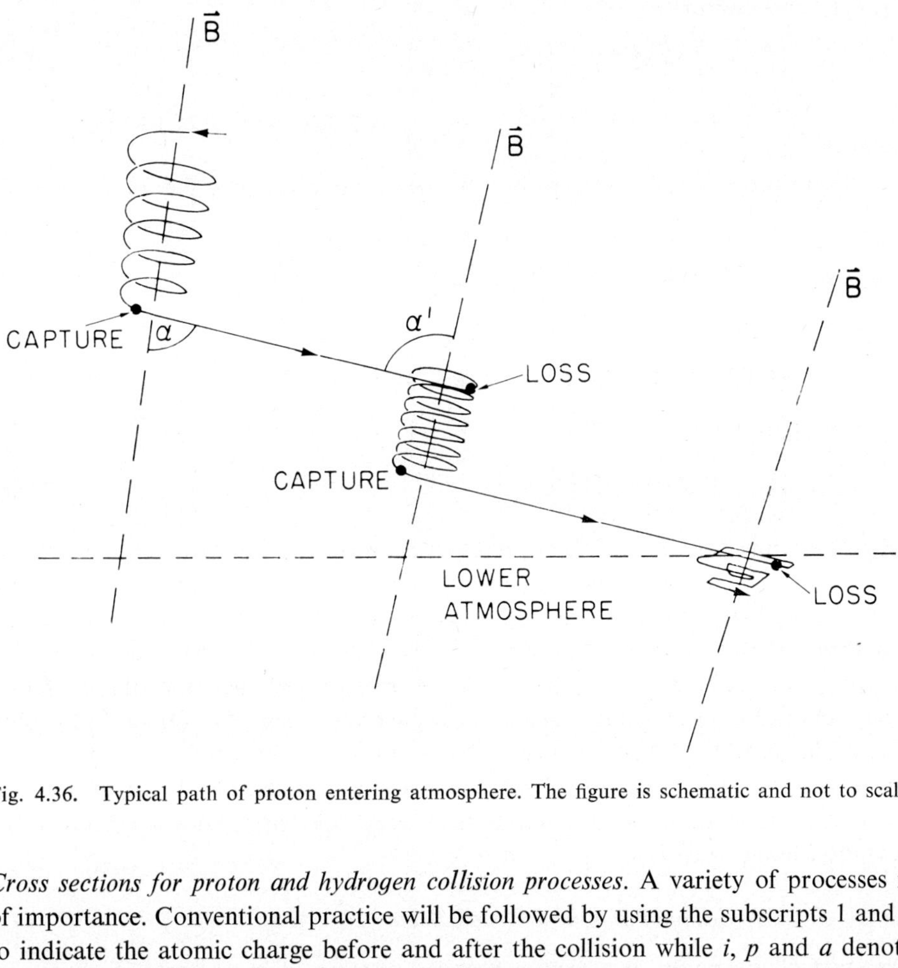

Fig. 4.36. Typical path of proton entering atmosphere. The figure is schematic and not to scale.

Cross sections for proton and hydrogen collision processes. A variety of processes is of importance. Conventional practice will be followed by using the subscripts 1 and 0 to indicate the atomic charge before and after the collision while *i*, *p* and *a* denote ionization, proton and neutral atom, respectively. Where excitation to a particular state or causing a particular optical emission is considered, this is indicated as an argument in brackets.

Collision processes involving protons include the following.

(i) Ionization

$$H^+ + M \rightarrow H^+ + M^{+(*)} + \bar{e} \quad \sigma_{ip}. \tag{4.3.1a}$$

The asterisk in brackets indicates that the ion may be left in an excited state from which a photon may be emitted while σ_{ip} (3914) denotes ionization to the $N_2^+ B^2\Sigma$ state with subsequent emission of a $\lambda 3914$ photon.

(ii) Excitation

$$H^+ + M \rightarrow H^+ + M^* \quad \sigma_p(e), \tag{4.3.1b}$$

where the parameter e identifies the excited state M* or a photon emitted.

(iii) Charge capture

$$H^+ + M \rightarrow M^{+(*)} + H^{(*)} \quad \sigma_{10}(e), \tag{4.3.1c}$$

The total cross section may be subdivided into partial ones for the excitation of various states.

Collision processes involving neutral hydrogen atoms include the following.

(iv) Charge loss

$$H + M \rightarrow H^+ + M^{(*)} + \bar{e} \quad \sigma_{01}. \tag{4.3.1d}$$

(v) Ionization

$$H + M \rightarrow M^{+(*)} + H^{(+)} + (2)\bar{e} \quad \sigma_{ia}. \tag{4.3.1e}$$

(vi) Excitation

$$H + M \rightarrow M^{(*)} + H \quad \sigma_a(e). \tag{4.3.1f}$$

Cross sections for some of these processes have been measured and typical results are given in Figure 4.37. Processes involving H^- do not seem important in aurora and have been omitted. The data in the figure are taken from the review of McNeal and Birely (1973) in which the present knowledge of these cross sections is fully discussed with references to earlier work. In brief, measurements of the excitation cross sections by charge capture and atom impact have been made for Lα and Hα but not satisfactorily for Hβ. Excitation cross sections for λ3914 for atom and proton impact are available, as are those of the ionization cross sections. Cross sections for the excitation of most other band systems are imperfectly known especially for excitation by neutral H.

4.3.1.1. *Ionization Equilibrium of a Proton Beam*

The state of ionization of a proton beam is determined by the rate of change of ionization per unit path (usually expressed in atmosphere-cm i.e. cm at N.T.P.). Suppose at a given point the beam contains n(p) protons and n(H) H atoms per unit volume and that σ_{10} and σ_{01} are the capture and ionization cross sections, respectively, then

$$-\,\mathrm{d}n(\mathrm{p}) = n(\mathrm{p})\sigma_{10}\,\mathrm{d}N - n(\mathrm{H})\sigma_{01}\,\mathrm{d}N, \tag{4.3.1g}$$

where dN is the number per cm^2 of atmospheric particles traversed. Now if α is the fraction of incident particles ionized then

$$-\,\mathrm{d}\alpha = \alpha\sigma_{10}\,\mathrm{d}N - (1-\alpha)\sigma_{10}\,\mathrm{d}N$$

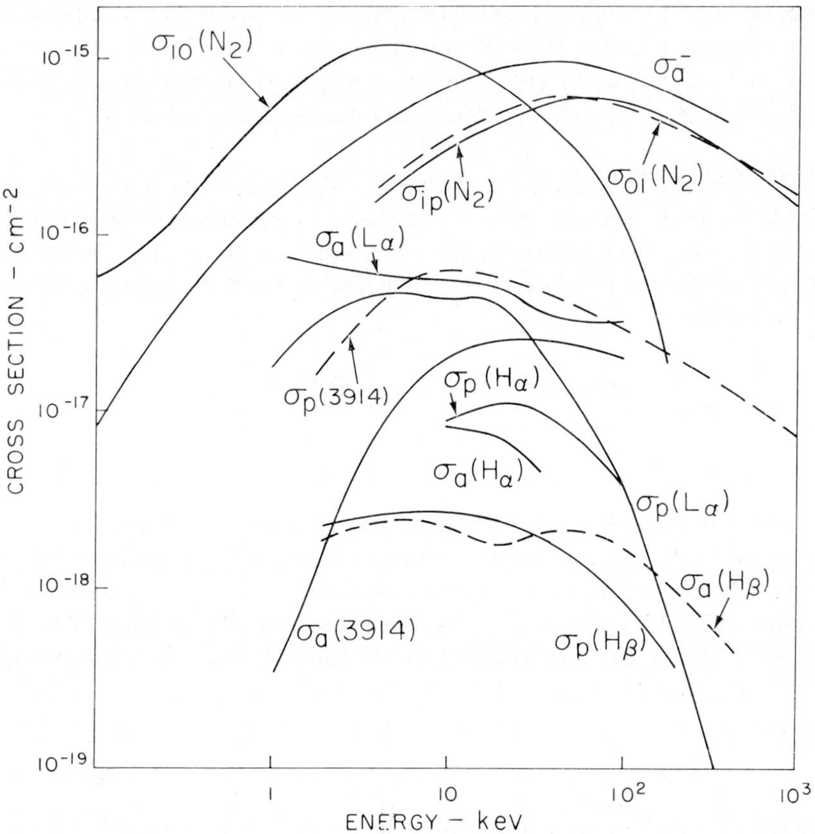

Fig. 4.37. Cross sections per target *molecule* for inelastic processes involving H^+ and H. $\sigma_{10}(N_2)$ is the capture cross section in N_2, $\sigma_a^- = \sigma_{ia} + \sigma_{01}$ is the total ionization cross section in N_2 for H atoms, $\sigma_{01}(N_2)$ is the loss cross section in N_2, $\sigma_{ip}(N_2)$ is the total ionization cross section in N_2 for protons, $\sigma_a(L_\alpha)$ and $\sigma_p(L_\alpha)$ are, respectively, the excitation cross section for H and the capture cross section for H^+ to give L_α in N_2, $\sigma_p(3914)$ and $\sigma_a(3914)$ are cross sections for production of $\lambda 3914$ by H^+ and H, respectively, in N_2. $\sigma_a(H_\alpha)$, $\sigma_a(H_\beta)$, $\sigma_p(H_\alpha)$ and $\sigma_p(H_\beta)$ are analogous to the L_α cross sections. All results come from the review of McNeal and Birely (1973) except for those for H_β which are theoretical estimates taken from Eather (1966a).

hence

$$-\frac{d\alpha}{dN} = \alpha(\sigma_{10} + \sigma_{01}) - \sigma_{01}$$

or

$$\alpha = \frac{\sigma_{10}}{\sigma_{10}+\sigma_{01}}\, e^{-N(\sigma_{10}+\sigma_{01})} + \frac{\sigma_{01}}{\sigma_{10}+\sigma_{01}}. \tag{4.3.1h}$$

Thus α decreases to a steady state value characteristic of ionization 'equilibrium'. α approaches its steady state value, $\sigma_{01}/(\sigma_{10}+\sigma_{01})$ after $1/(\sigma_{10}+\sigma_{01})$ particles cm^{-2} have been traversed. For example for 30 keV protons $\sigma_{01} \simeq \sigma_{10} \simeq 5\times10^{-16}$ cm^2 so that $N \simeq 10^{15}$ cm^2 for approach to charge equilibrium. Below 250 km equilibrium is virtually complete even for 300 keV protons. Tables of equilibrium values of α are given by McNeal and Birely (1973). Since at heights of interest the steady state is established α will be used below to denote the steady state value.

4.3.1.2. *Ion Production Rate*

The ionization of atmospheric particles produced by protons is a starting point for understanding other phenomena. This may be estimated by using an empirical relation for air giving the ion-pair production rate per atmosphere-cm per proton. The relation (Chamberlain, 1961) is

$$-\frac{dN_i}{dR} = CR^b \exp(-aR) = J(\text{ionization}, R) \qquad (4.3.1i)$$

$$C = 2.2 \times 10^5, \qquad b = 0.74, \qquad a = 4.63$$

and R is *residual* range in atmosphere-cm. This expression can be used to derive the ionization rate cm^{-3} for an initial proton flux having a given angular and energy distribution. Such calculations have been carried out by Eather and Burrows (1966) and Eather (1970) who give plots of the ionization rate as a function of height for various initial energy and angular distributions. Some of these results are reproduced in Figure 4.38. From the measured heights discussed in Section 3.4 this suggests that proton aurora is caused by particles with energies in the 10 to 100 keV range.

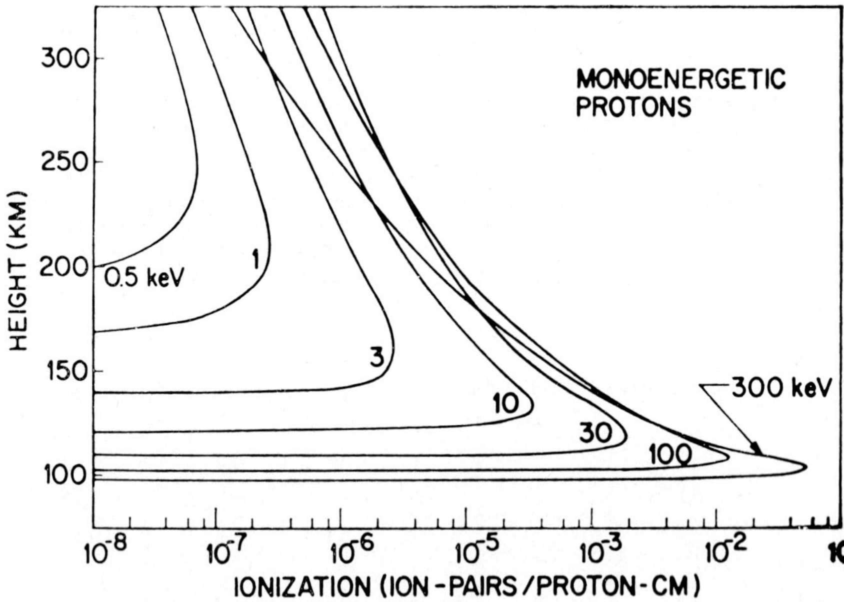

Fig. 4.38. Height profiles for ionization produced by monoenergetic protons with an isotropic pitch angle distribution by Eather (1970). (Courtesy *Annales de Géophysique*.)

4.3.1.3. *Secondary Electron Energy Distribution*

The energy distribution of the secondary electrons produced by proton and atomic hydrogen ionization is not as easily estimated as is the case for electron ionization. References to current knowledge are given by McNeal and Birely (1973). Unfortunately the distribution depends strongly on proton energy so that the secondary electron production rate must be integrated over the number of secondaries produced at each energy as the protons are decelerated, as well as over the initial proton energy distribution. The measured cross sections do not cover the important energy range below 20 keV. The only serious attempt to compute the secondary electron flux was made by Eather (1967a) on the basis of theoretical curves for ionization of Ne by protons calculated on the basis of the Born approximation by Bates *et al.* (1957). Eather gave a plot of the quantity,

$$\frac{1}{\eta_i}\int_E^\infty \eta_s(E)\,\mathrm{d}E, \tag{4.3.1j}$$

which is the appropriate term for use in Equation (4.2.1p) in evaluating the secondary electron flux. This quantity depends on the primary proton energy distribution for which a particular plausible form was assumed. Unfortunately an extrapolation of the Born approximation results below 100 keV is required and ionization by neutral H is neglected so that the calculated secondary electron flux may not be very accurate.

4.3.2. Excitation of Hydrogen Emission Lines and Their Doppler Profiles

4.3.2.1. *Hydrogen Line Emission*

In the course of penetrating into the atmosphere, protons undergo electron capture, neutral excitation and electron loss processes as indicated by Equations (4.3.1c), (f) and (d). For the capture and neutral collision processes there is a probability of the H atom being left in an excited state so that lines of Lyman, Balmer or other H series may be radiated. The hydrogen energy levels (terms) concerned may be designated by the quantum numbers n and l of which only n has a significant effect on the energy but l affects the transition probabilities and must be considered. To simplify the equations we will index individual terms by a single index j.

In the condition of ionization 'equilibrium', the number of excitations to the jth term per atmosphere-cm of path traversed by a proton will be

$$J(j) = N_0\,\sigma_{\text{eff}}(j) + \sum_i J(i)\left[\frac{A_{ij}}{\sum_k A_{jk}}\right], \tag{4.3.2a}$$

where

$$\sigma_{\text{eff}}(j) = \alpha\sigma_{10}(j) + (1-\alpha)\sigma_a(j).$$

N_0 is the Loschmidt number (the number of target molecules per cm^2 per atmos-cm);

α is the ionized fraction of the beam (Equation (4.3.1h)) while $\sigma_{10}(j)$ and $\sigma_a(j)$ are the partial cross sections for excitation of the jth H term (Equations (4.3.1c) and (f)). α and the σ's are functions of proton energy which can be expressed in terms of the residual range, R. The righthand term of Equation (4.3.2a) represents cascading for higher energy terms distinguished by the index i; the factor in square brackets is the branching ratio from these terms into the term j. The index k extends over all terms with energy lower than that of i.

Equation (4.3.2a) can easily be solved approximately by neglecting the cascade terms if the partial cross sections are known. In principle the first order solutions, for the $J(j)$'s can then be substituted back into the cascade terms to give successively better approximations. For a given hydrogen emission line between energy levels distinguished by the principal prime quantum numbers n,n' the production rate becomes,

$$J(n,n') = \sum_{\mathrm{j}} \left[\frac{\sum_{\mathrm{k}} A_{\mathrm{jk}}}{\sum_{\mathrm{l}} A_{\mathrm{jl}}} \right] J(j) = \sum_{\mathrm{j}} c(n,n',j)\, J(j), \tag{4.3.2b}$$

where j and k are summed over all terms with the quantum numbers n and n', respectively, while l is summed over all terms with energy lower than that corresponding to n. $c(n,n',j)$ then gives the contribution through each term to the total line intensity. If Equation (4.3.2b) is substituted into Equation (4.3.2a), then the direct excitation term becomes

$$J_{\mathrm{d}}(n,n') = N_0\, \alpha_{\mathrm{eff}}(n,n'), \tag{4.3.2c}$$

where

$$\sigma_{\mathrm{eff}}(n,n') = \sum_{\mathrm{j}} c(n,n',j)\sigma_{\mathrm{eff}}(j). \tag{4.3.2d}$$

The cross sections for the production of a given emission such as $\sigma_{10}(\mathrm{H}\beta)$ are suitable for calculating the direct term but must be resolved according to Equation (4.3.2d) to evaluate the cascade terms. In some cases as for σ_{10}(3s, 3p, 3d) the partial cross sections have been measured (see McNeal and Birely, 1973) but in others some assumptions must be made. Fortunately the cascade contributions are only about 10%. A more serious uncertainty lies in the values of $\sigma_a(nl)$ which have to be estimated (see eg. Eather, 1966a) for the Hβ case ($n = 4$) by an extrapolation and scaling of theoretical results for H–H collisions. Experimental values of $\sigma_a(\mathrm{L}\alpha)$ and $\sigma_a(\mathrm{H}\alpha)$ and their components are available.

The total yield of H-line quanta per proton can finally be evaluated by integrating $J(n,n')$ from R_0, the initial range, to zero and thus

$$Q(n,n') = \int_0^{R_0} J(n,n')\, \mathrm{d}R. \tag{4.3.2e}$$

Besides the uncertainties in the cross sections, there are other approximations implicit in Equation (4.3.2e). A single average cross section is used for N_2 and O_2 with dissociation of O being taken into account by using an atmospheric depth in atmosphere-cm reduced to O_2 and N_2 (Chamberlain, 1961, p. 575). Thus, changes in atmospheric composition are only partially taken into account. Moreover some protons are reflected magnetically without reaching zero energy. Finally Equation (4.3.2b) involves the assumption that radiation is the only loss process for the excited states of H. As originally pointed out by Omholt (1959), excited H could be collisionally deactivated before radiating. The most recent discussion of this effect is by McNeal and Birely (1973) who conclude that quenching may be complete for the metastable H(2s) term but that the higher levels should not be affected except possibly in the case of polar glow aurora. The latter occurs at heights in the region of 50 km where the time between collisions is much shorter.

Curves calculated by Eather (1966a, 1967b) $J(\mathrm{H}\alpha)$ and $J(\mathrm{H}\beta)$ are shown in Figure 4.39. Integrated values of $Q(\mathrm{H}\alpha)$ and $Q(\mathrm{H}\beta)$ for a range of initial energies are given in Table 4.20 which is based on data given by Eather (1967b) except as noted. Also plotted on Figure 4.39 is a curve for $J(\mathrm{H}\alpha)$ derived from the recent data of McNeal and Birely (1973) but with the cascade component neglected. This suggests that the neutral impact contribution may have been overestimated and that the Hα (and presumably Hβ) yields in Table 4.20 may be a little high.

4.3.2.2. *Ionization*

The integral

$$Q(\text{ionization}, R_0) = \int_0^{R_0} J(\text{ionization})\, \mathrm{d}R \tag{4.3.2f}$$

gives the total ionization per incident proton of initial range R_0. This integral can be evaluated from the expression given in Section 4.3.1.2 or directly from the ionization cross sections. Another useful quantity is the yield of 0,0 N_2^+ 1N band (λ3914) photons. The production of λ3914 in primary proton or hydrogen atom collisions should be given by the expression

$$J_p(3914) = N_0[\alpha\{\sigma_{10}(3914) + \sigma_{ip}(3914)\} + (1-\alpha)\sigma_{ia}(3914)]. \tag{4.3.2g}$$

This would not include the component produced by secondary electrons. Likewise the primary produced ionization should be given by

$$J_p(\text{ionization}) = N_0\{\alpha(\sigma_{10} + \sigma_{ip}) + (1-\alpha)\sigma_{ia}\}. \tag{4.3.2h}$$

The ratio, $\beta = Q(\text{ioniz. } R_0)/Q(3914, R_0)$, with the Q's defined analogously to Equation (4.3.2f) should give the fraction of ionizations to λ3914 photons for protons of initial range R_0. Values of β obtained range from an estimate of 7.5 obtained by Eather (1967b) for typical auroral protons from proton cross sections alone to about 20 which is obtained from Equations (4.3.2g) and (h) with recent cross sections

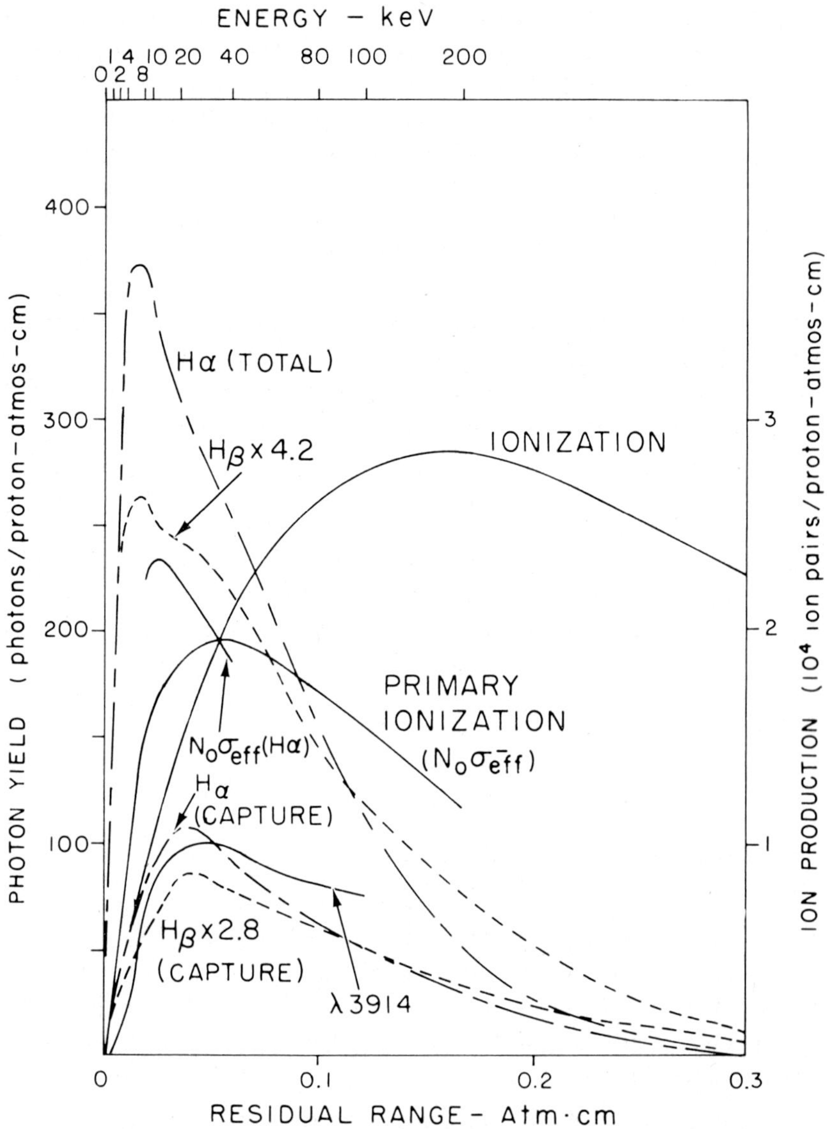

Fig. 4.39. Quantum and ionization yields per unit residual range for protons. λ3914 is for the N_2^+(0, 0) 1N band from the data of McNeal and Birely (1973) as are the primary ionization ($N_0\bar{\sigma}_{eff}$) and Hα production ($N_0\bar{\sigma}_{eff}$(Hα)) curves. The remaining curves are those used in the work of Eather (1966a, 1967b).

from Figure 4.37 (see also McNeal and Birely, 1973.) The corresponding ratio for electron excitation is about 18 for air at 110 km from the branching ratios of Figure 4.20 and Equation (4.2.1f). Thus it appears likely that β can be taken to have a similar value for proton and electron excitation.

TABLE 4.20

Initial proton (keV)	Lowest height of penetration (km)	Total emission photons/proton (Hα)	(Hβ)	Ion pairs/ Hβ photon	$I(3914)^a / I(H\beta)$	$I(4709)^a / I(H\beta)$
1	~156	<0.4	<0.1	?		
10	122	8.2	1.4	95	3.8	0.48
100	105	34.5	7.0	311	12.5	1.6

[a] Revised; based on $Q(\text{ion})/Q(3914) = 25$ for air.

4.3.2.3. *Hydrogen Line Profiles*

As was clearly demonstrated by Meinel (1951) the H-Balmer lines have a Doppler broadened profile when observed perpendicular to the terrestrial magnetic field direction (magnetic horizon observation) and show in addition a Doppler shift when viewed parallel to the field direction that is in the magnetic zenith. This is due to the motion of the protons along their trajectories as illustrated in Figure 4.36. A set of observed Hβ profiles is shown in Figure 4.40.

Originally it was hoped that analysis of the Doppler profile might give detailed

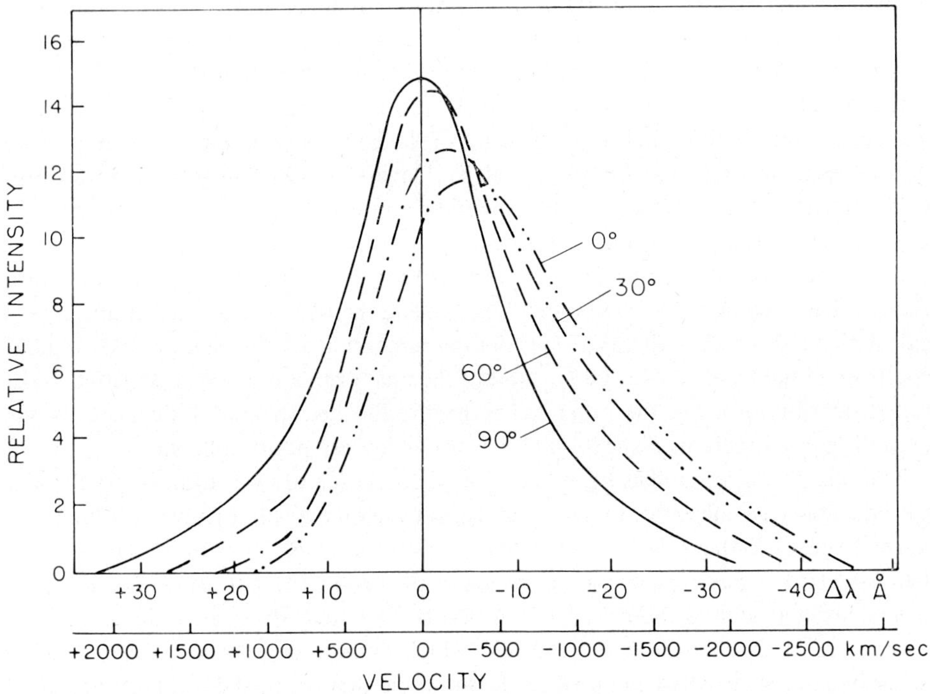

Fig. 4.40. Profiles of Hβ emission line in proton aurora as a function of magnetic zenith angle as measured by Vaisberg (1967).

information on the energy and pitch angle distributions of the incoming protons and consequently the theory of the Hα and Hβ profiles has been extensively developed. This theory has been presented in detail by Chamberlain (1961), Eather (1967b), and Omholt (1971).

It was quickly found by Chamberlain and Omholt that a range of both energies and pitch angles is necessary to generate magnetic zenith and horizon profiles similar to those observed. In order to obtain the best agreement one must also assume that the angular distribution is itself a function of energy. Vaisberg (1967) concluded that lower energy protons had a pitch angle distribution with more particles at higher pitch angles.

The observed Doppler profiles in the zenith, as may be seen from Figure 4.40 have an appreciable fraction of the total emission to the red side of the unshifted wavelength. Eather (1966b) demonstrated that this is a real effect and is due to the upward motion of protons which have been reflected in the converging terrestrial field. He showed that an effect of the magnitude observed could arise only if there was a component of low energy protons peaked at pitch angles of 70 to 90°.

In recent years the possibility of direct measurements of auroral proton energy and angular spectra has reduced interest in the problem of deriving these spectra from the optical profiles. The problem of predicting the optical profile from the observed fluxes remains of significance for a full understanding of processes involved. The analytical method of Chamberlain and Omholt discussed above neglected several important effects. These include the variation in atmospheric composition with height, the effects of spatial variations in the magnetic field and the effect of beam spreading after the first charge capture as illustrated in Figure 4.36. This last process will change the particle pitch angle because of the latitude variation of dip angle. The beam spreading itself was first analyzed by Davidson (1965) using a Monte Carlo technique and taking account of atmospheric composition. Further calculations have been carried out by Johnstone (1972).

A computation of the line profiles with all the effects discussed above taken into account has been done by Ponomarev (1972) using the Monte Carlo technique. These calculations show that there is considerable smearing of the energy and angular spectrum of the beam above 200 km so that the final profile is not very sensitive to the initial distributions. On the other hand the profiles are strongly influenced by the excitation cross sections $\sigma_{10}(H\beta)$ and $\sigma_a(H\beta)$ which are poorly known.

The observed line profiles have been found to show little variation in conformity with the above calculations. In a few rare cases rather narrow lines have been reported as reviewed by Eather (1967b). A further clear-cut case was observed by Harang and Pettersen (1967) who reported a narrowing of the profile (near the magnetic horizon) from a normal value of about 19 Å to 5 Å or less. This suggests a dominant component of protons with energies below 500 eV which should be stopped above 200 km according to the curves of Figure 4.38. It is notable that the λ6584 [N II] line appeared during the 30 min or so when the narrow Hα line was strong. It is not clear under what conditions the necessary large fluxes of soft protons are precipitated.

4.3.2.4. *Height Profiles of Balmer Lines*

Some references to height measurements of these lines in proton aurora were given in Section 3.4. In principle the height distribution may easily be determined from the function $J(n,n')$ of Equation (4.3.2b) after a suitable transformation of the independent variable from residual range to height with due allowance for pitch angle distribution. Such calculations have been done for Hα by Eather and Burrows (1966).

Because of the extended nature of proton aurora, approximating a uniform sheet, the height distribution cannot be measured from the ground but should be easily determined by rocket photometry. It should then be possible to derive some information as to the energy spectrum of the incident protons. This was attempted by Miller and Shepherd (1969) who encountered difficulties in reconciling the energy spectrum inferred from the height profile with the normal Doppler profile. This problem does not appear to have arisen in a more recent study by Egeland *et al.* (1973). It is valuable to bring theory and observation into harmony on the question of height profiles although it does not appear to be a practical method for measuring proton energy spectra unless a convenient routine method can be devised for measuring the Balmer line height profiles.

4.3.3. Spectrum of Proton Aurora

4.3.3.1. *Theory*

In contrast to the case of electron aurora little systematic work has been done in predicting the intensities of features excited by proton precipitation. In principle this might be done in a similar way if the cross sections for excitation of molecular and atomic states by proton and hydrogen atom impact were known together with the energy distribution of secondary electrons produced in ionizing collisions.

Direct excitation by H^+ and H. The direct excitation processes could be calculated from equations analogous to Equation (4.3.2g) with N_0 replaced by the number of target particles per cm^2 per atmos-cm and using the appropriate values of cross sections for the excited state considered. Unfortunately very few such cross sections are known. In many cases the cross sections $\sigma_p(X)$ would be expected to be small since a bare proton has a low probability of exciting a state for which a spin change is required. This applies to most of the strong features. However there is no such restriction on excitation by neutral H (Bates, 1960). In the case of simultaneous ionization and excitation the proton excitation cross sections $\sigma_{ip}(e)$ and $\sigma_{10}(e)$ can be large.

The most tractable case of direct excitation, that of the N_2^+ 1N bands, has already been considered in Section 4.3.2.2. In this instance the cross sections for the excitation of the $v' = 0$ level have been reliably determined so that now the intensity of these bands should be accurately calculable from the incident proton flux. (Unfortunately there are still uncertainties in calculating $I(H\beta)$ although $I(H\alpha)$ should be accessible.)

The only other state for which the direct excitation rate should be obtainable is the $C^3\Pi$ state of N_2 which is discussed by McNeal and Birely (1973); this involves mostly excitation by neutral H. For other transitions the cross section data, as reviewed by McNeal and Birely, is either absent or unreliable. Eather (1967a) estimated an upper limit to $I_\alpha(5577)$ by assuming the cross section to be equal to that for electrons of the same velocity. This gave $I_d(5577)/I(H\beta) \leq 3$.

Excitation by secondary electrons. Excitation and ionization by secondary electrons produced from primary protons might be expected to be of importance and indeed was originally thought to be dominant (e.g., Omholt, 1959). For ionization processes, Eather (1967a) showed that the H^+–Ne secondary electron function of Bates *et al.* (1957) led to a 7% contribution to ionization by secondaries for the particular primary energy distribution assumed ($\sim e^{-E/20}$ for $E < 100$ keV and $\sim e^{-E/100}$ for $E > 100$ keV). This calculation may be done using Equations (4.2.1p) and (4.2.2a) and yields a very similar result when the loss functions of Figure 4.21 are employed. It is quite sensitive to the shape of the cross section for secondary electron production which is poorly known. For 1 MeV protons the calculated fraction of secondary ionization is about 50% and it is believed to be about 2/3 at higher energies.

Secondary electron excitation might be expected, by analogy with electron aurora, to be much more important in the case of the optically forbidden excitations. In the case of $O(^1S)$ we compute an approximate value of 1.5% for the ratio $I_s(5577)/I_p(\text{ionization})$ or $I_s(5577)/I_p(3914) \sim 0.37$ assuming $I_p(\text{ionization})/I_p(3914) = 25$ for air.

A similar calculation for the excitation of the $B^3\Pi$ state of N_2 gave a ratio $I(B^3\Pi)/I_p(\text{ionization})$ of 9.3% or $I(1PN_2)/I(3914)$ of about 2.3.

Excitation by dissociative recombination. A third contribution to the excitation of the $O(^1S)$ and $O(^1D)$ states in particular, is dissociative recombination. This can in principle be calculated according to the method of Section 5.1 from a knowledge of the production rates of O_2^+, O^+, N_2^+ and N^+ in the case of the H–H^+ beam. Cross sections for the production of these ions have been measured and are reviewed by McNeal and Birely (1973). As yet the ion chemistry has not been computed in detail since the preliminary study of Eather (1967a). It would appear that the results of a detailed calculation would be similar to the solution for electron aurora since the branching ratios O^+/O_2^+ and N^+/N_2^+ are similar as is the ratio $I(3914)/I(\text{ionization})$. In this case we might expect the contribution to $I(5577)$ from dissociative recombination to be perhaps 30 to 100% of that due to secondary electron excitation depending on the NO concentration.

4.3.3.2. *Observed Spectrum and Comparison with Theory*

The spectrum of proton aurora has not yet been determined throughout the full wavelength range. A difficulty arises in ensuring that a given luminosity is excited by protons alone. In the red and IR regions the nightglow OH bands become comparable with or stronger than the features of a typical proton aurora so that the analysis of the spectrum is difficult.

A photographic spectrum of what is presumably a proton aurora was published by Vallance Jones (1971). Photometric measurements of intensity ratios of features in proton aurora have been reported by Eather (1968a) and Vaisberg (1962).

Qualitatively the proton auroral spectrum resembles that of electron aurora apart from the presence of the Balmer lines. There is also an apparent increase in the ratio of $I(1,3;\ 1\mathrm{N}\ N_2^+)/I(0,2;\ 1\mathrm{N}\ N_2^+)$ at 10 Å resolution. Thus Ahmed (1962) showed that this apparent ratio increases from about 0.3 to 0.75 as $I(\mathrm{H}\beta)/I(0,2\ N_2^+)$ rises from 0 to 1.

TABLE 4.21

Intensity ratios in proton and electron aurora

Emission	Intensity predicted				Intensity[c] observed	Intensity[e] electron aurora
	Direct	Secondary	Indirect	Total		
Hβ	100[a]			100	33	–
Hα	400			400	–	
N_2^+ 1N						
N_2^+(1N 0, 0)	500[b]	30	–	530	330–600	(500)
N_2^+(1N 0, 2)	46	3		49	30–43	
[O I] λ5577	<300[f]	~200[f]	(100?)	<600	300–400	600
[O I] λ6300					80–100	
N_2(1P)	?	1150[f]	–	1150	(2400)	4200
N_2(1P; 6, 3)					53[d]	95
N_2^+(M; 1, 0)					180–240[d]	970
N_2^+(M)					760–1020	4100

[a] The table has been adjusted to give a predicted value of 100 R.
[b] Value derived from Table 4.20 for 20 keV protons.
[c] The observed intensities are normalized to the theoretical for λ3914. Based on Eather (1968a) except as noted.
[d] Based on values relative to $I(5577)$ measured by Vaisberg (1962).
[e] Scaled down from IBC3 aurora.
[f] See Section 4.3.3.1.

The observed intensities normalized to a value of about 500 R for λ3914 are set out in Table 4.21 together with the theoretical estimates discussed in the preceding section. On the whole the agreement is as good as can be expected. In the table scaled values of the intensity of electron aurora are given. Compared to these λ5577 and N_2 1P seem perhaps 50% weaker in proton aurora while N_2^+ M may be 3 or 4 times weaker. This is in accordance with the original laboratory finding of Fan (1958) that the N_2^+ M is less easily excited by a proton beam than by electrons.

Vibrational band ratios. The question of possible changes in the relative population rates of the N_2^+ 1N bands for proton or H excitation is an interesting one. The laboratory measurements of Moore and Doering (1969b) show clearly that for proton energies less than about 5 keV, a marked enhancement of the ratio sets in gradually, reaching a factor of 2 at about 1 keV.

This effect was sought but not found in proton aurora by Harang and Gérard (1970) in a careful measurement of the $\Delta v = -1$ sequence. This may be because a dominant part of the emission is excited by protons with energy above 5 keV. There is also a possibility that the effect may not arise for neutral H excitation which might conceivably dominate below 4 keV (although the latest cross sections of McNeal and Birely (1973) indicate that proton excitation remains more important at the lowest energies).

The apparent enhancement of the 0,2 to 1,3 band ratio in low resolution spectra noted above is ambiguous because of the presence of atomic lines underlying the 1,3 band. Multiplet 1 of O^+ (see Figure 4.9) and 5 of N^+ (see Figure 4.11) have components between 4630 and 4650 Å and so can contribute to the 1,3 band maximum which lies in the same region (see Figure 4.4). Harang and Gérard have shown that multiplet 36 of O^+ (4185–90 Å) is strongly enhanced in proton aurora and consequently it is plausible that the observed changes in $I(1,3)/I(0,2\ N_2^+)$ may be due to this effect.

Clark and Belon (1959) observed a strong apparent enhancement of the $I(1,3)/I(0,2)$ ratio in the great type-d aurora of February 11, 1958. As pointed out by Harang and Gérard there is again an uncertainty as to the relative contributions to the effect of enhanced atomic lines (pronounced in this type of aurora), of increased N_2 vibrational temperature or of the enhancement of the cross section for $v' = 1$.

Rotational temperatures in proton aurora. Here again abnormal enhancements in rotational populations have been found in the laboratory by Moore and Doering (1969a) for proton energies below 5 keV. There is no real evidence for or against such effects in proton aurora. The high rotational temperatures reported by Clark and Belon (1959) and discussed in Section 4.2.5.5 could in part be due to this effect although the alternative explanation of a high atmospheric temperature seems possible.

4.3.4. Polar cap glow

As noted in Section 3.2.5 this is excited by the 1 to 100 MeV protons responsible for polar cap absorptions. The earlier spectral measurements and their interpretation were discussed by Sandford (1963). Later work has been reviewed by Sandford (1967) and Alcaydé (1968). The theoretical approach is analogous to that for ordinary proton aurora. Brown (1964a, b) has also discussed the problem. The observed value of 0.2 for the ratio $I(5577)/I(3914)$ is much lower than normal aurora but the theoretical models predicted even lower ratios as well as lower absolute values for $I(3914)$. Clearly the ratio $I(5577)/I(3914)$ presents theoretical difficulties and the conflict cannot be taken too seriously. The absolute value of $I(3914)$ should be accurately calculable provided the proton (and electron) fluxes are known over the entire energy range. It is clear that much interesting work remains to be done on this topic.

CHAPTER 5

AURORA AND THE IONOSPHERE

Auroral particles produce important effects on the atmosphere besides the excitation of the auroral spectrum. These include heating effects and the production of ionization. Auroral ionization is responsible for enhanced electrical and thermal conductivity and the absorption and reflection of radio waves. The reflection effect known as radio aurora is certainly related to structure induced in auroral ionization by various instabilities or to structure in the precipitation pattern which is likely due to instabilities in the magnetospheric plasma. Electric fields and currents appear in association with aurora as a result of magnetospheric processes. The magnetic perturbations and probably the infrasonic emissions from aurora are generated by the current systems. These topics will be discussed in the following sections.

5.1. Auroral Ionization and Its Effects

As a result of the influx of auroral particles, ions are produced at a rate η_i ions per unit volume. In the simplest approximation these ions are lost at a rate, L given by

$$L = \alpha n(\mathrm{M}^+)\, n_e,$$

where α is the effective recombination coefficient, $n(\mathrm{M}^+)$ is the ion concentration and n_e is the electron concentration. Here, it is assumed that all ions present recombine at the same rate. If the ratio of negative ions to electrons is λ, then $n_e = n(\mathrm{M}^+)/(1+\lambda)$ since $n_e + n(\mathrm{M}^-) = n(\mathrm{M}^+)$ for charge neutrality.

Thus, the equation governing the electron concentration, n_e, (if transport is neglected) is

$$\frac{dn_e}{dt} = \frac{\eta_i}{1+\lambda} - \alpha n_e^2. \tag{5.1a}$$

In the steady state

$$n_{e\infty} = \sqrt{\frac{\eta_i}{(1+\lambda)\alpha}}. \tag{5.1b}$$

The time dependent solutions for the growth of electron concentration from an initial value of zero, if ion production is turned on at $t = 0$, is

$$n_e = n_{e\infty} \frac{(1 - e^{-2\alpha n_{e\infty} t})}{(1 + e^{-2\alpha n_{e\infty} t})}. \tag{5.1c}$$

The time for n_e to equal $n_{e\infty}/2$ is about $0.55/n_{e\infty}\alpha$. Likewise the decay, if η_i is reduced to zero, is given by

$$n_e = \frac{n_{e0}}{1+\alpha n_{e0}t} \tag{5.1d}$$

for which the time constant for a decay to $n_{e0}/2$ is given by the expression,

$$t_{\frac{1}{2}} = 1/\alpha n_{e0}\,. \tag{5.1e}$$

We can use these equations to estimate the order of magnitude and other characteristics of auroral ionization before continuing with a more detailed discussion in the following sections. It is convenient to relate the ionization rate η_i to the emission of a conveniently measured auroral band. The 4278 Å 0,1 band of the N_2^+ 1N system is very suitable. Since it is a fully allowed transition the band emission is proportional to the excitation rate. Its intensity is close to 1/3 that of the 0,0 band for which according to Figure 4.20 the production rate is $1/14\,\eta(N_2 \to \text{ions})$. Thus, from Equation (4.2.1f) and (g), taking for 110 km, $n(O_2) = 0.4 \times 10^{12}$, $n(N_2) = 0.191 \times 10^{13}$ and $n(O) = 0.19 \times 10^{12}$ cm^{-3}, one gets

$$\eta_i \simeq 70\eta(4278). \tag{5.1f}$$

The dissociative recombination rates of NO^+ and O_2^+, the dominant ions at auroral heights, are $\sim 2 \times 10^{-7}$ cm^3 s^{-1} and consequently we can adopt this value for α. In an IBC3 aurora the total observed intensity of the 4278 Å band is about 30 kR (see Section 4.2.4). If we assume this to be distributed over 20 km (a reasonable half-intensity height range for normal aurora), $\eta(4278)$ at its peak would be about 1.5×10^4 quanta cm^{-3} s^{-1}) and the ionization rate $\simeq 1 \times 10^6$ ions cm^{-3} s^{-1}. Introducing this into Equation (5.1b) (taking $\lambda = 0$) we get

$$n_{e\infty\text{max}} = \sqrt{1 \times 10^6/2 \times 10^{-7}} \simeq 2.2 \times 10^6 \text{ e cm}^{-3}.$$

From Equation (5.1c) the time constant for the build up of ionization would be about 1 s at this height.

More generally on this simple approach

$$n_{e\infty\text{max}} \simeq 1.3 \times 10^4 \sqrt{I(4278)}, \tag{5.1g}$$

where I is given in rayleighs. If the vertical half width exceeds 20 km the value would be proportionately smaller. (More exact calculations (see Figure 5.6) suggest the coefficient should be closer to 0.7×10^4.)

These relations give the order of magnitude of the peak electron concentrations to be expected in aurora. To obtain more refined results we must consider ion production, ion reactions and transport effects in more detail.

The ionization produced by auroral particles with different energy spectra is illustrated in Figure 4.19.

The rates of ionization of individual ions are given by Equations (4.21f) to (l).

Most of the ionization produced by normal auroral forms, with distinct lower borders in the 100 to 110 km range, is due to electrons with average energies of the order of 10 keV. Such particles produce most of their ionization below 160 km. Some of the rarer auroral forms such as type-a, -d or -f or the weaker LAR-arcs are probably due to electrons with average energies of 1 keV or less and produce appreciable ionization up to 200 km or higher.

5.1.1. The Ion and Electron Continuity Equation in Aurora

The general problem of accounting for electron and ion concentrations in the E- and F-regions is exceedingly complex and one which is the subject of much effort at the present time. The basic equation to be solved is the equation of continuity for the jth ion,

$$\frac{\mathrm{d}n_j}{\mathrm{d}t} = \eta_j - L_j - \mathrm{div}(n_j\mathbf{v}_j), \tag{5.1h}$$

where n_j is the ion concentration, η_j is the ion production rate, L_j the loss rate and $\mathbf{v}_j$ the average velocity. The last term represents the losses due to transport effects. These arise from (i) diffusion, (ii) neutral wind drag, (iii) vertical neutral atmospheric motions, and (iv) electric field drifts.

In the solution of the Equation (5.1h) for the normal E and F regions the transport terms are very important. Stubbe (1970) and Strobel and McElroy (1970) have published studies of the F region problem.

An accurate theoretical treatment of the ion and electron transport terms is quite involved. A basic discussion has been given by Schunk and Walker (1970b, 1971). The order of magnitude of some of the effects can be estimated from the following simplified treatment.

For each ion, from conservation of momentum, there is a steady state equation* of the form

$$-\nabla p_j + n_j e_j\left[\mathbf{E} + \mathbf{u}\times\mathbf{B} + \frac{m_j}{e}\mathbf{G}\right] + n_j e_j \mathbf{v}_j \times \mathbf{B} = + n_j m_j \nu_j \mathbf{v}_j, \tag{5.1i}$$

where p_j is the ion pressure, $\mathbf{u}$ is the centre of mass velocity of the atmosphere, i.e., very nearly the neutral wind velocity, $\mathbf{v}_j$ is the ion drift velocity relative to the center of mass velocity. $\mathbf{G}$ is the gravitational acceleration and ν_j is the ion collision frequency†. Since the concentration of doubly charged ions is negligible we write e instead of e_j. The term on the right hand side represents the drag term due to ion collisions with neutral particles.

* Electrical equations are given in SI unit form.

† In this case the frequency of collisions in which the total momentum of the particle is transferred to the atmospheric particles. Stubbe (1968) has discussed the exact definition of the various collision frequencies appropriate to the ionisphere.

A similar equation can be written for electrons for which the gravitational term is negligible, so that

$$- \nabla p_e - n_e e[\mathbf{E} + \mathbf{u} \times \mathbf{B}] - n_e e \mathbf{v}_e \times \mathbf{B} = n_e m_e \nu_e \mathbf{v}_e \tag{5.1j}$$

ν_e is the electron collision frequency*.

Let us consider first the motion of particles parallel to the magnetic field. In this case the mobility of the electrons is much higher than that of the ions and the parallel component of Equation (5.1j) determines the parallel electric field. Let $\mathbf{E}' = \mathbf{E} + \mathbf{u} \times \mathbf{B}$ be the effective electric field seen in a coordinate system moving with the neutral wind velocity. The parallel component of Equation (5.1j) is

$$- \nabla p_e \cdot \mathbf{b} - n_e e \mathbf{E}' \cdot \mathbf{b} = n_e m_e \nu_e \mathbf{v}_e \cdot \mathbf{b}. \tag{5.1k}$$

(The $\mathbf{v}_e \times \mathbf{B}$ term has zero parallel component.) Thus

$$\mathbf{E}' \cdot \mathbf{b} = - \frac{1}{n_e e} \nabla p_e \cdot \mathbf{b} + \frac{\mathbf{j} \cdot \mathbf{b}}{\sigma_{\parallel}}, \tag{5.1l}$$

where $\mathbf{b}$ is unit vector in the field direction and

$$\mathbf{j} \simeq \mathbf{j}_e = - n_e \mathbf{v}_e e \tag{5.1m}$$

is the total parallel current.

$$\sigma_{\parallel} = n_e e^2 / m_e \nu_e \tag{5.1n}$$

is a good approximation to the electrical conductivity parallel to $\mathbf{B}$.

If we likewise take the parallel component of Equation (5.1i) and substitute $\mathbf{E}' \cdot \mathbf{b}$ from Equation (5.1l) we obtain

$$- \left(\nabla p_j + \frac{n_j}{n_e} \nabla p_e \right) \cdot \mathbf{b} + n_j e_j \frac{\mathbf{j} \cdot \mathbf{b}}{\sigma_{\parallel}} + n_j m_j \mathbf{G} \cdot \mathbf{b} = n_j m_j \nu_j \mathbf{v}_j \cdot \mathbf{b}. \tag{5.1o}$$

Finally for the parallel ion flux

$$\begin{aligned} \phi_{j\parallel} = n_j \mathbf{v}_j \cdot \mathbf{b} = - \frac{1}{m_j \nu_j} \Bigg(k \left(\nabla n_j T_i + \frac{n_j}{n_e} \nabla n_e T_e \right) \cdot \mathbf{b} \\ + n_j e_j \frac{\mathbf{j} \cdot \mathbf{b}}{\sigma_{\parallel}} + n_j m_j \mathbf{G} \cdot \mathbf{b} \Bigg), \end{aligned} \tag{5.1p}$$

where T_i is the ion temperature, assumed the same for all ions. We assume the partial pressure of each species to be given by a term of the form, nkT. This equation determines the parallel component of $n_j \mathbf{v}_j$ in Equation (5.1h).

The perpendicular component is also needed to evaluate all the contributions to the transport term in Equation (5.1h). For convenience let us put

$$\mathbf{E}_j = \left[\mathbf{E} + \mathbf{u} \times \mathbf{B} + \frac{m_j \mathbf{G}}{e} - \frac{1}{n_j e_j} \nabla p_j \right] \tag{5.1q}$$

* See footnote †, p. 180.

where $\mathbf{E}_j$ becomes an effective field term for the jth ion. Then (5.1i) becomes

$$\frac{e_j}{m_j}\mathbf{E}_j + \mathbf{v}_j \times \mathbf{\Omega}_j = \nu_j \mathbf{v}_j, \tag{5.1r}$$

where

$$\mathbf{\Omega}_j = e_j \mathbf{B}/m_j \tag{5.1s}$$

$\mathbf{\Omega}_j$ is the ion gyro frequency vector. If we take the z-axis in the field direction and solve for the x- and y-components of $\mathbf{v}_j$ we obtain the result

$$\mathbf{v}_{j\perp} = \frac{e_j}{m_j}\left(\frac{\nu_j}{\Omega_j^2 + \nu_j^2}\right)\mathbf{E}_{j\perp} - \frac{e_j}{m_j}\left(\frac{\Omega_j}{\Omega_j^2 + \nu_j^2}\right)\mathbf{b} \times \mathbf{E}_{j\perp}, \tag{5.1t}$$

where the x- and y-components of $\mathbf{v}_j$ and $\mathbf{E}_j$ have been recombined into a vector equation. The factors controlling the perpendicular fields are not well known. Experimental evidence cited in Section 3.6.3 indicates fields of the order of 100 mV m^{-1} near auroral forms; within the forms the evidence is conflicting. Most measurements indicate a significant reduction.

Equation (5.1t) leads to the Pederson and Hall conductivity terms for ions although the field $\mathbf{E}_j$ is here an effective value given by Equation (5.1q).

As will be shown in Section 5.1.3 the transport term in Equation (5.1h) is usually small in comparison to the production and loss terms *within* auroral forms over the height range of auroral emissions. Consequently the solution of Equation (5.1h) can be considered without the transport term.

5.1.2. Ion Chemistry in the Auroral Atmosphere

The equations of Section 4.2.1 should give the production rates of the different ions produced directly from electron impact on N_2, O_2 and O. Besides the ground state ions of N_2^+, O_2^+, O^+ and N^+, metastable $O^+(^2D)$, $O^+(^2P)$ and $O_2^+(a^4\Pi)$ are important because their lifetimes are sufficiently long for them to undergo reactions which are different from those of the ground state ions. Very recently Zipf (1973a) has suggested that the large observed UV flux in the N_2 Birge-Hopfield region (850 to 1000 Å) (see Section 4.2.4.4) should be very effective in producing ionization of O, O_2 and NO. The additional ion production terms could have pronounced effects on the ion chemistry and their evaluation will be an interesting problem.

The principal reactions of the ions are summarized in Figure 5.1. Reaction rates are indicated in units of 10^{-11} cm^3 s^{-1}. (Further data are presented in Table 4.6.) It will be seen that N_2^+ and O^+ are rapidly converted to O_2^+ and NO^+ by reactions mostly with O and O_2. $O^+(^2D, ^2P)$ and $O_2^+(a^4\Pi)$ react in addition with N_2 to give N_2^+. O_2^+ may itself be converted to NO^+ in reactions with the minor constituents N and NO.

The appearance of N and NO in the reaction scheme requires that the reactions of

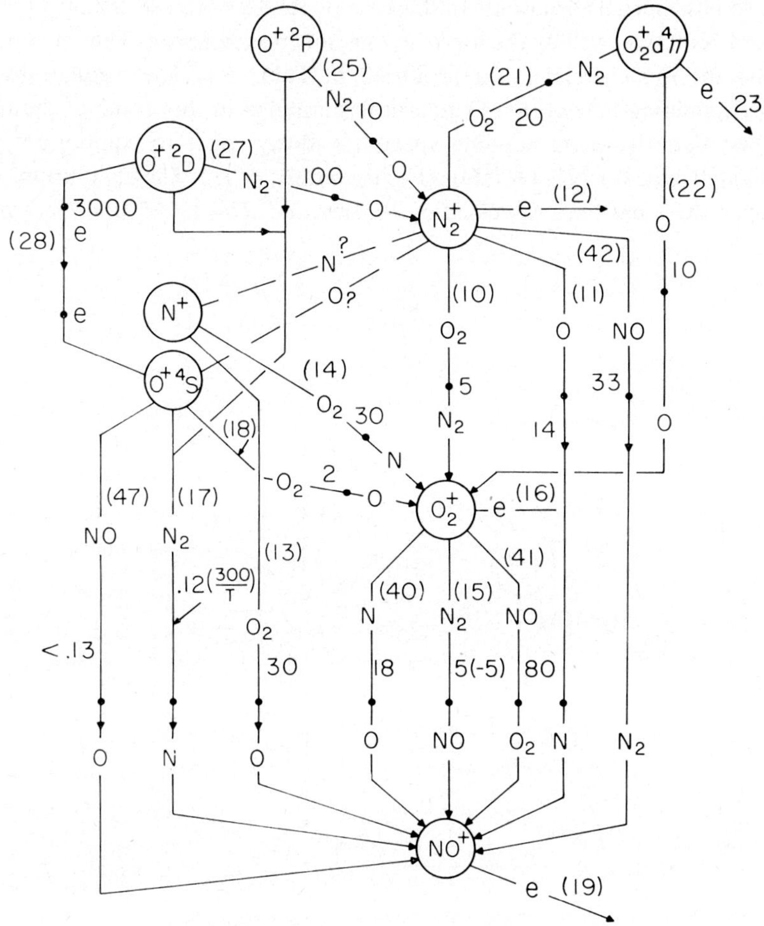

Fig. 5.1. Reaction scheme for ions in auroral atmosphere. Reactions are indicated by lines joining initial and final ions with the other reactant and product species indicated. The number in brackets is the reaction coefficient subscript to be used in this book while the unbracketed numbers are values of the rate constant in units of 10^{-11} molecular cgs units. Additional data including temperature coefficients are given in Table 4.6.

these species as shown in Figure 5.2 be considered. The yields of $N(^2D)$ and $N(^4S)$ are not well known but energetically they can be produced in the dissociative recombination of N_2^+, NO^+ and in the reaction of N_2^+ with O. In addition they can be produced directly in the dissociation of N_2 by electron impact. The cross sections for this process leading to $N(^4S)$ and $N(^2D)$ are not known. The overall cross section $\sigma(N_2 \xrightarrow{e} N+N \text{ or } N+N^+)$ has been measured by Winters (1966); it has a peak value of 2.1×10^{-16} cm^2 compared to the ionization cross section of 2.5×10^{-16}. This suggests the dissociation cross section is of the same order of magnitude as that for the production of N_2^+. Consequently we take $\eta(N) = \eta(N_2^+)$ and assume $\eta(N^2D) = (2/3)\eta(N)$. Strobel (1971) has discussed the problem of N and NO production in relation to NO diurnal variations. In this case photoelectrons are assumed to produce N(^{4}S) and N(^{2}D) as well as the ion reactions mentioned above. The set of reactions governing the ion chemistry is reproduced in Table 5.1. These equations can be solved by numerical integration. The task is simplified in that some of the reactions are so fast that the corresponding species is always near to equilibrium. This is approximately true for N_2^+, $O^+(^2D)$, $O^+(^2P)$ and $O_2^+(a^4\Pi)$. The equilibrium solution for some of these has been discussed by Walker and Rees (1968) and Gérard (1970).

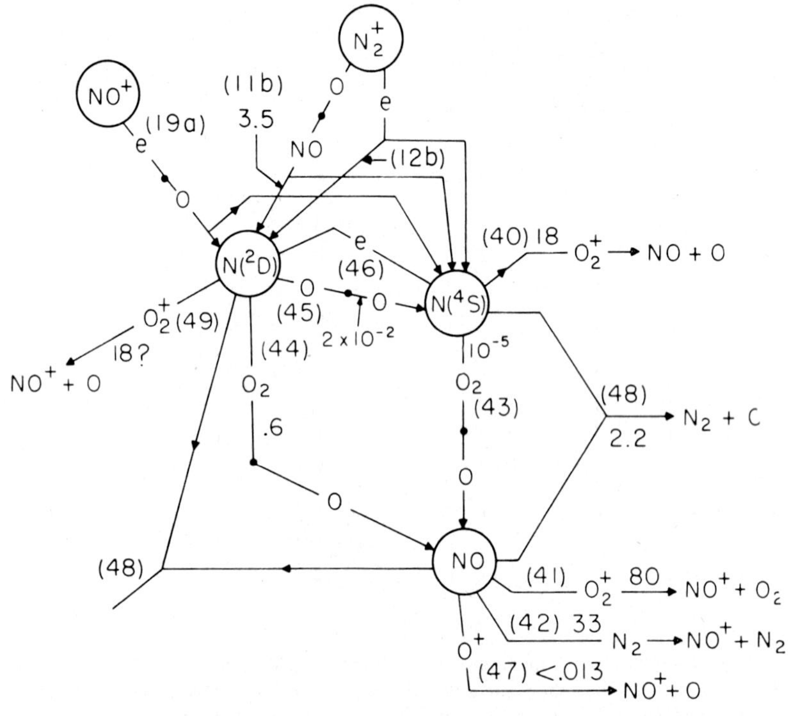

Fig. 5.2. Reactions of N(^{2}D), N(^{4}S) and NO following conventions of Figure 5.1.

For N^+ Equation (5) in Table 5.1 is linear and can be integrated analytically while the same is true for $O^+(^4S)$ (Equation (3) of Table 5.1) if the small terms involving NO and $O^+(^2D)$ are neglected.

The growth of ion concentrations obtained by a numerical solution of the equations of Table 5.1 at different heights is illustrated in Figure 5.3. These curves exhibit some interesting features. At 110 km $n(O_2^+)$ reaches a maximum value about 6 s after the beginning of the excitation and almost equals $n(NO^+)$. After 15 s NO^+ becomes dominant. At 130 km the $n(O_2^+)$ maximum is not reached until 20 s and O_2^+ is never dominant. At 150 km the growth curves do not exhibit maxima and NO^+ is always dominant. It appears that the slower buildup of NO is responsible for the suppression

Table 5.1*

Differential equations for production and loss of ionic and neutral species in the auroral atmosphere

1 $\frac{d[N_2^+]}{dt} = \eta(N_2^+) - \{k_{10}[O_2]+k_{11}[O]+k_{12}[e]+k_{42}[NO]\}[N_2^+] \quad k_{27}[N_2][O^{+\prime}]+k_{25}[N_2][O^{+\prime\prime}]+k_{21}[N_2][O_2^{+\prime}]$

2 $\frac{d[O_2^+]}{dt} = \eta(O_2^+)+k_{10}[O_2][N_2^+]-\{k_{15}[N_2]+k_{16}[e]+k_{40}[N]+k_{41}[NO]+k_{49}[N']\}[O_2^+]+k_{18}[O_2][O^+]+k_{14}[O_2][N^+]+\{k_{22}[O]+k_{23}\}[O_2^{+\prime}]$

3 $\frac{d[O^+]}{dt} = \eta(O^+) \quad -\{k_{17}[N_2]+k_{18}[O_2]+k_{47}[NO]\}[O^+]+\{A_6+k_{28}[e]\}[O^{+\prime}]$

4 $\frac{d[NO^+]}{dt} = \quad +\{k_{11}[O]+k_{42}[NO]\}[N_2^+]+\{k_{15}[N_2]+k_{40}[N]+k_{41}[NO]\}[O_2^+]+\{k_{17}[N_2]+k_{47}[NO]\}[O^+]-k_{19}[e][NO^+]+k_{13}[O_2][N^+]$

5 $\frac{d[N^+]}{dt} = \eta(N^+) \quad -(k_{13}+k_{14})[O_2][N^+]$

6 $\frac{d[O^{+\prime}]}{dt} = \eta(O^{+\prime}) \quad -\{A_6+k_{28}[e]+(k_{17}+k_{27})[N_2]\}[O^{+\prime}]$

7 $\frac{d[O^{+\prime\prime}]}{dt} = \eta(O^{+\prime\prime}) \quad -\{A_7+ \quad +k_{25}[N_2]\}[O^{+\prime\prime}]$

8 $\frac{d[N]}{dt} = \eta(N) \quad +\{k_{11}[O]+k_{12}[e]\}[N_2^+]-k_{40}[N][O_2^+]+k_{17}[N_2][O^+]+k_{19}[e][NO^+]+\{k_{45}[O]+k_{46}[e]\}[N']-k_{48}[N][NO]+k_{14}[O_2][N^+]$

9 $\frac{d[N']}{dt} = \eta(N') +\{k_{11b}[O]+k_{12b}[e]\}[N_2^+]-k_{49}[N'][O_2^+] \quad +k_{19a}[e][NO^+]-\{k_{45}[O]+k_{48}[NO]+k_{46}[e]\}[N']$

10 $\frac{d[NO]}{dt} = \quad -k_{42}[NO][N_2^+] \quad -k_{41}[NO][O_2^+]+k_{15}[N_2][O_2^+] \quad +\{k_{43}[O_2]-k_{48}[NO]\}[N]+\{k_{44}[O_2]+k_{48}[NO]\}[N']$

11 $\frac{d[O_2^{+\prime}]}{dt} = \eta(O_2^{+\prime}) \quad -\{k_{22}[O]+k_{21}[N_2]+k_{23}[e]\}[O_2^{+\prime}]$

*$O^{+\prime}$, $O^{+\prime\prime}$, N' and $O_2^{+\prime}$ refer respectively to $O^+(^2D)$, $O^+(^2P)$, $N(^2D)$ and $O_2^+\ a^4\pi$. A_6 and A_7 are the sums of the downward transition probabilities for $O^+(^2D)$ and $O^+(^2P)$ respectively. The subscripts on the rate constants refer to the reaction numbers of Figs. 5.1, 5.2 and Table 4.6.

of O_2^+ at the lower heights. These results are quite dependent on atmospheric composition and on the adopted ion reaction rate constants so that they should be regarded as illustrating the possibilities rather than as quantitative measures. Since bright aurora may move fairly rapidly it is clear that equilibrium values of electron and ion densities may never be attained. Recently Jones and Rees (1973) have published curves for ion growth in aurora calculated on the basis of a similar model. Their results are comparable with those of Figure 5.3 although there are differences reflecting different initial conditions.

The recent suggestion by Zipf (1973a) that EUV emissions will produce significant additional ionization of O, O_2 and NO would also lead to important changes in the ion chemistry.

Figure 5.4 shows the growth of electron concentration with height for various times. It is clear that times of the order of 15 to 100 s are required to approach equilibrium values as the height increases from 110 km.

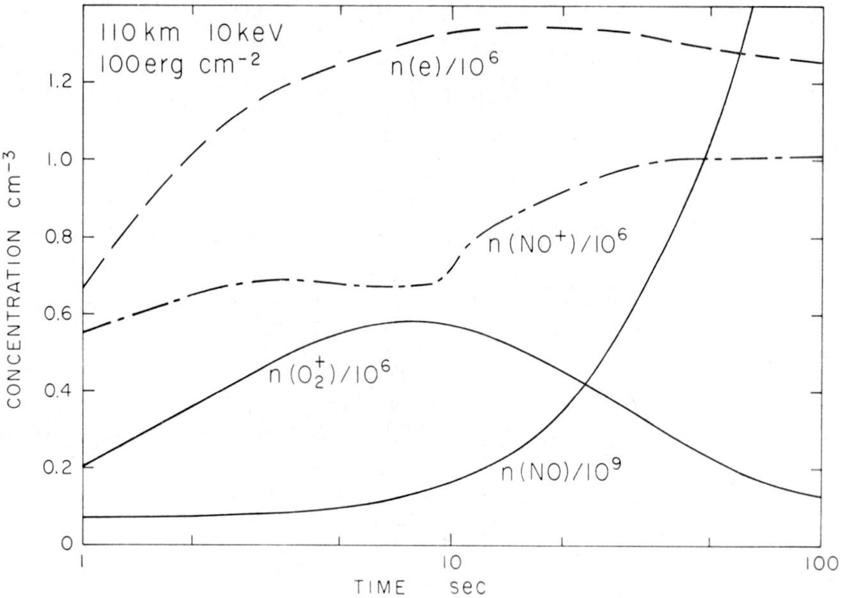
110 km 10 keV
100 erg cm⁻²
n(e)/10⁶
n(NO⁺)/10⁶
n(O₂⁺)/10⁶
n(NO)/10⁹
CONCENTRATION cm⁻³
TIME sec
0
0.2
0.4
0.6
0.8
1.0
1.2
1
10
100

(a)

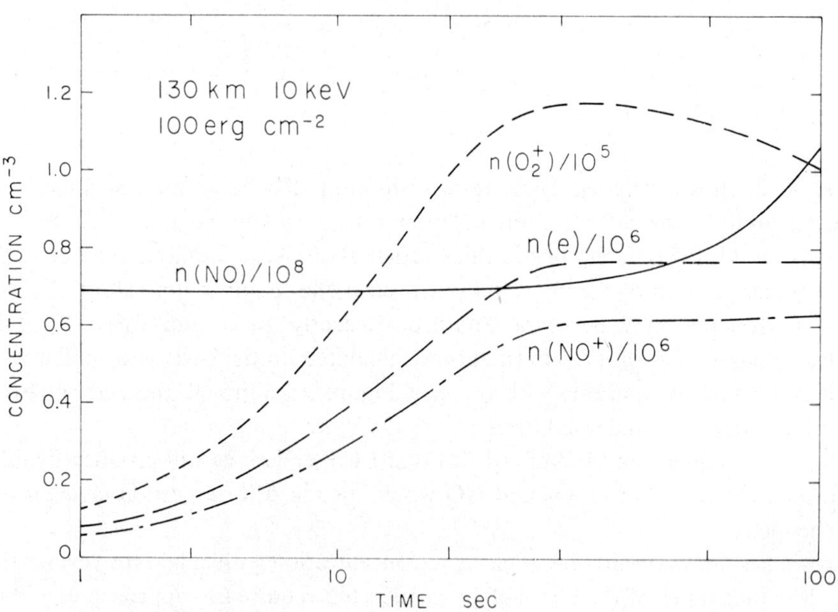
130 km 10 keV
100 erg cm⁻²
n(O₂⁺)/10⁵
n(e)/10⁶
n(NO)/10⁸
n(NO⁺)/10⁶
CONCENTRATION cm⁻³
TIME sec
0
0.2
0.4
0.6
0.8
1.0
1.2
1
10
100

(b)

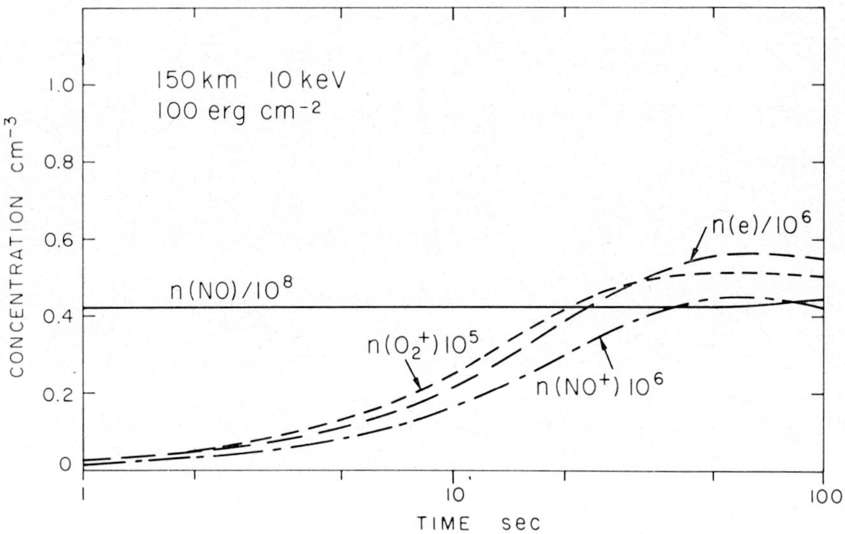

(c)

Fig. 5.3. Growth of ion concentrations for input flux of 100 erg cm² of 10 keV electrons at (a) 110 km (b) 130 km (c) 150 km. The initial values assumed were $n(O_2^+) = 2,0.91$ and 0.8×10^2, $n(NO^+) = 20$, 35 and 6.6×10^2 $n(NO) = 7.2$, 6.9 and 4.2×10^7 cm^{-3} at the three heights respectively.

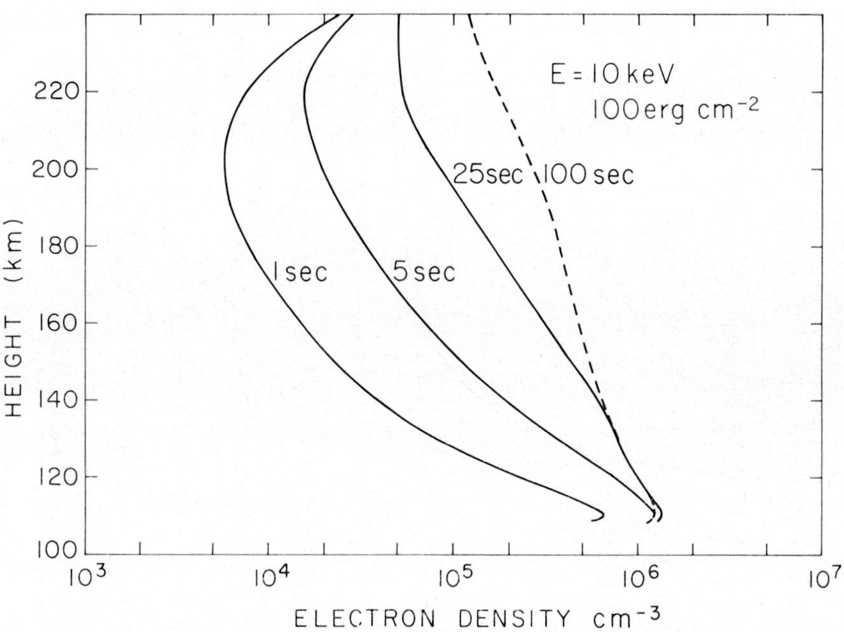

Fig. 5.4. Growth of electron concentration as a function of height.

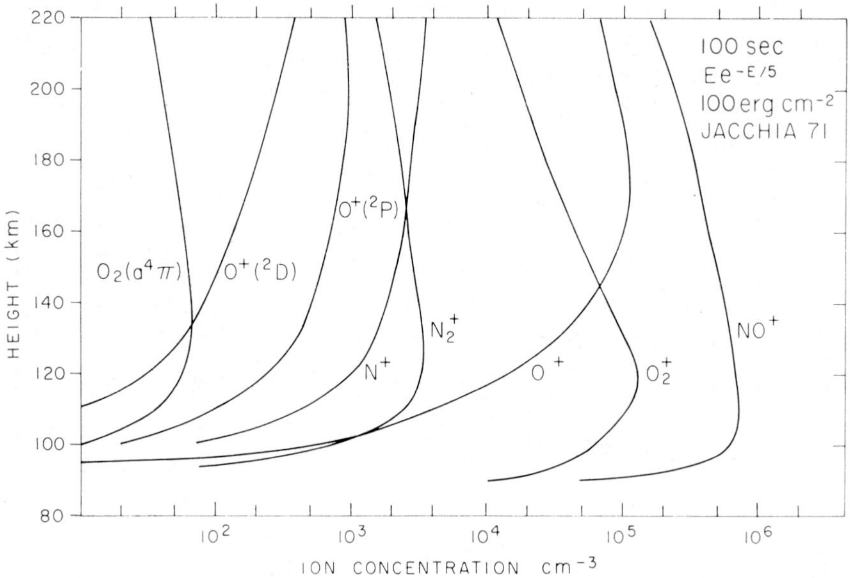

(a)

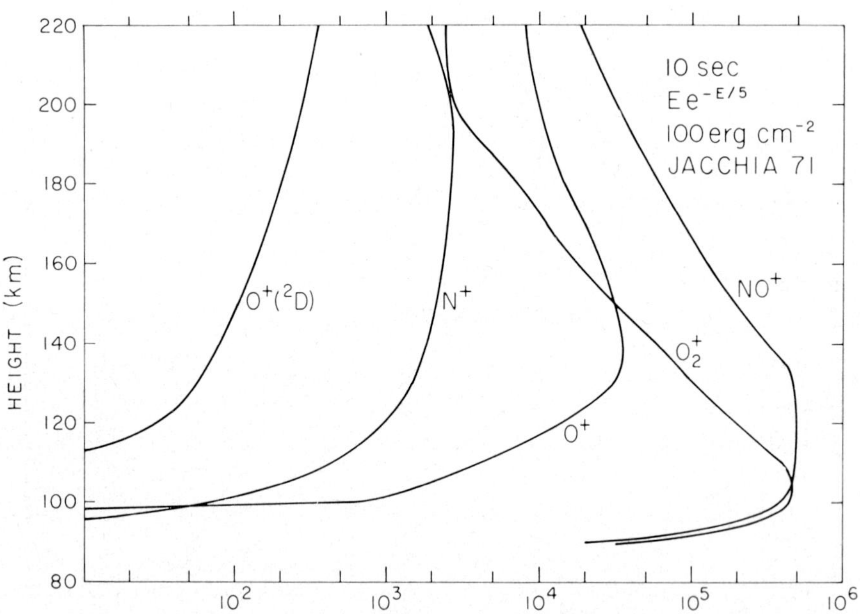

(b)

Fig. 5.5. Height profiles of calculated ion concentrations after (a) 100 s (b) 10 s for 100 erg cm^2 electron flux with distribution $\sim E \exp(E/5)$.

Figure 5.5a shows the calculated ion concentrations after 100 s as a function of height while Figure 5.5b shows the calculated situation after 10 s. The electron spectrum in this case is of the form $E \exp(-E/5)$.

5.1.3. Effects of Transport on Auroral Ionization

The contributions to the transport term will now be considered to see in what degree the approximate solutions of Section 5.1.2 might be perturbed.

(i) *Diffusion*

Diffusion parallel to the field lines is important in determining *F* region ion densities over times of the order of an hour especially at heights above 300 km. It should not be important for forms lasting a few minutes below 250 km. In order to verify this the $\nabla\phi_{j\|}$ terms obtained using Equation (5.1p) were calculated for the case of monenergetic 31.6 keV electrons and were compared with the production and loss terms η and L of Equation (5.1h). The comparison was made at a time of 200 s after the beginning of excitation. In all cases the $\nabla\phi_{j\|}$ terms were negligible compared to the production term, η_j. This indicates that the solution would be little altered by including the transport term.

Diffusion perpendicular to the field lines is considered together with the electric field effects below.

(ii) *Neutral Wind Drag*

Parallel component. $\mathbf{u} \times \mathbf{B}$ has a zero component parallel to $\mathbf{B}$ and consequently in Equation (5.1q) the neutral wind does not affect the solution in Equation (5.1p). Since $\mathbf{v}_j$ is the velocity relative to the center of mass, the ion velocity relative to the Earth has added to it the parallel component of the neutral wind. Such winds in the E and F regions may have horizontal velocities of the order of 200 m s^{-1} (Rees, 1973) and consequently vertical components as high as 35 m s^{-1} in auroral regions. As a rough guide to the value of $\nabla\phi \simeq v\partial n/\partial z$, we find, for example from Figure 5.5, that at 200 km $\partial n(\mathrm{NO}^+)/\partial z \simeq 0.1$ cm^{-4} so that $\nabla\phi \simeq 3.5 \times 10^2$ cm^{-3} s^{-1}. The value of $\nabla\phi$ due to diffusion for this case is -0.7 cm^{-3} s^{-1} so that the wind term dominates the diffusion term and is of the same order as the production term which is about 3×10^2 cm^{-3} s^{-1}. Clearly towards 200 km this transport term is important.

Perpendicular components. At lower heights where $v_j \gg \Omega_j$ the ions are carried by the wind and large values of $\nabla\phi$ could arise in narrow arc elements. Such effects have not been extensively considered in the past.

(iii) *Vertical Atmospheric Motions*

These are relatively slow and should not affect auroral ionization directly,

(iv) *Electric Field Drifts*

Parallel components. As discussed in Section 5.1.1 a parallel electric field gives rise to an electron current given by Equation (5.11). The **j** term in Equation (5.1p) thus corresponds to the externally applied E field. Schunk and Walker (1971) find that a parallel current density of greater than 1 amp m^{-2} would be required to affect ion densities within arcs. This is much larger than the values reported for example in Figure 3.17. In other words it does not appear that parallel electric fields of sufficient intensity to produce significant transport can exist without giving rise to impossibly large currents. This assumes that the parallel conductivity is given by Equation (5.1n). Nevertheless as noted in Equation (3.6.3b) there is some evidence that such parallel fields can be supported. This would imply anomalously low parallel conductivities caused perhaps by some kind of plasma instability.

Perpendicular components. Perpendicular electric fields of the order of 50 mV m^{-1} have been measured, at least outside auroral forms. The evidence is conflicting as to whether this value is maintained within the form. Schunk and Walker (1971) have analyzed the contribution of the ion pressure gradient and electric field terms to a model of an auroral arc. They conclude that the contribution of $\nabla\phi$ is negligible inside the form but significant outside. The effect could be of importance in very thin auroral form elements although the situation is complicated by the rapid motions of such fine structure.

5.1.4. Observations of Electron Concentrations in Aurora

The observation of electron and ion densities in aurora has presented difficulties because of the lack of resolution of the indirect methods such as ionospheric sounders, as well as the spatial and temporal limitations of direct rocket measurements. The use of incoherent scatter systems promises to provide more complete data but this study is just beginning (e.g., Baron, 1973).

Ionospheric sounder data, as discussed in Section 2.4.3, gave some of the first reliable information as to electron densities in auroras. This has recently been reviewed by Omholt (1971) on the basis of Equations (5.1b) and (5.1f). In this work electron densities of 10^5 to 10^6 were measured as $I(4278)$ varied from about 0.5 kR to 30 kR. These are of the same order as those estimated in Section 5.1 although agreement is not perfect; Omholt's (1955) data suggest a lower value of α than the same analysis applied to Knecht's (1956) results. This could be a photometric calibration problem or a result of the approximations in the analysis.

Figure 5.6 shows points from the model described in Section 5.1.2 plotted against $I(4278)$ as well as the two lines which summarize the experimental radio-sonde data from Omholt (1971). It may be seen that (i) the agreement is quite good and (ii) variations in the electron energy spectrum do have some effect in producing a scatter of points.

More detailed observations on auroral electron densities are given by rocket measurements. Reddy *et al.* (1969) have measured height profiles of n_e under quiet and disturbed conditions at Ft. Churchill. Some typical profiles are shown in Figure 5.7. The result by Bryant *et al.* (1970) in a post-break-up glow (~10 kR λ5577) together with a measurement of I(4278) and the corresponding point is plotted on Figure 5.6. It fits well with the theoretical points. The results of Reddy *et al.* (1969) did not include auroral intensity measurements and so cannot be related directly to

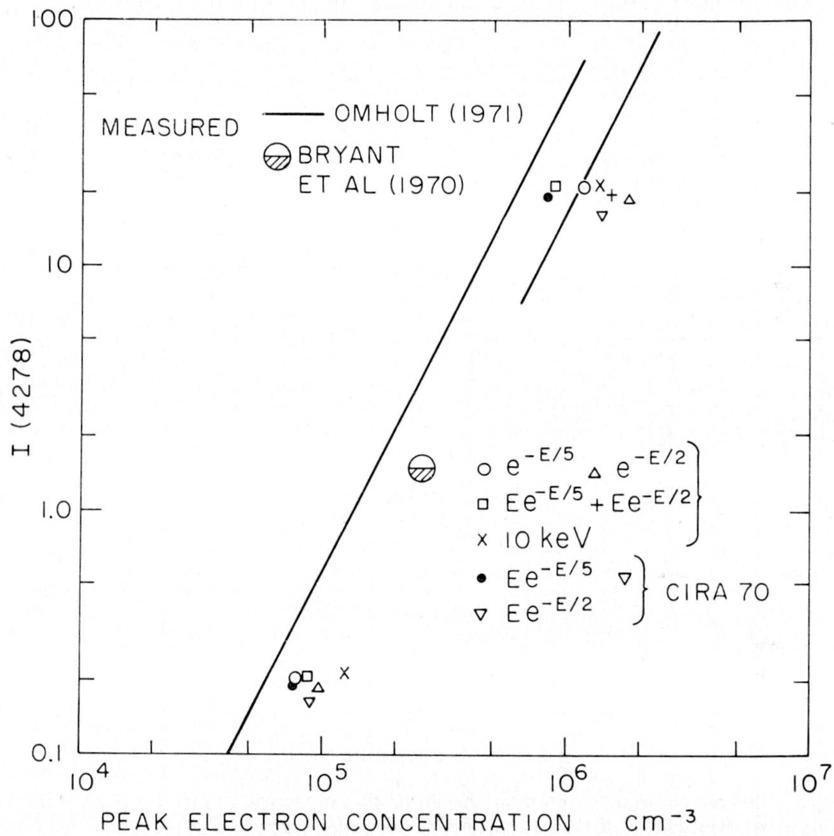

Fig. 5.6. I(4278) (kR) vs theoretical and measured values of peak electron concentration. The theoretical points are given for various initial energy distributions. The solid lines summarized the observed points cited by Omholt (1971).

visible aurora. Their work gave a careful comparison with ionosonde records. They concluded that small scale irregularities (≪1 km) were responsible for the non-blanketing E layer reflection during substorms and that the peak electron density corresponded to the upper frequency of the blanketing (i.e., totally reflecting) layer (fbES).

A further important observation is that of McNamara (1969) who succeeded in passing an ejected Langmuir probe through a rapidly moving active band during the break-up phase of a strong substorm. The auroral form appeared with a rise time of 10 to 20 s in the field of the fixed photometer monitoring the 100 km point on the downward leg of the probe trajectory. The measured $n_e(h)$ profile is shown in Figure 5.7. The narrow height range of the high values of n_e may be due in part to the horizontal component of the probe velocity; it is however not very different from the calculated profiles of Figure 5.4. The intensity estimate of $I(5577) \simeq 60$ kR which would correspond to $I(4278) \simeq 10$ to 20 kR also fits as well on Figure 5.6.

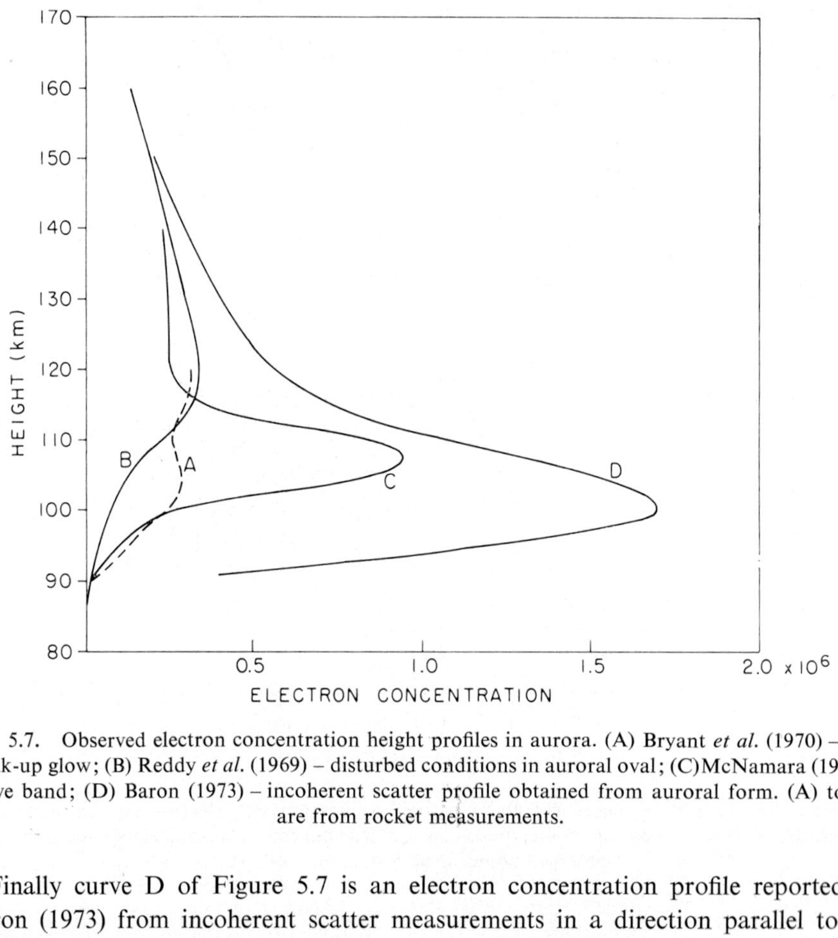

Fig. 5.7. Observed electron concentration height profiles in aurora. (A) Bryant *et al.* (1970) – post break-up glow; (B) Reddy *et al.* (1969) – disturbed conditions in auroral oval; (C)McNamara (1969) – active band; (D) Baron (1973) – incoherent scatter profile obtained from auroral form. (A) to (C) are from rocket measurements.

Finally curve D of Figure 5.7 is an electron concentration profile reported by Baron (1973) from incoherent scatter measurements in a direction parallel to the magnetic field. The auroral conditions were not well defined but from the time variation it would appear to have been an active or at least rapidly moving form since the ionization varied rapidly over 20 s periods. This curve also resembles the theoretical ones of Figure 5.4.

At greater heights very striking effects in auroral ionization, apparently due to neutral winds have been demonstrated by Bates *et al.* (1973).

5.1.5. Comparison of Theoretical and Observed Ion Concentrations in Aurora

Measurements of ion concentrations in aurora have been made quite frequently although the optical details of the auroral forms are often lacking. Swider and Narcisi (1970) have given references leading back to the earlier work. The observations are somewhat variable in detail but in general show ratios of $n(NO^+)/n(O_2^+)$ as high as 10 in the range 100 to 140 km. Figure 5.8 shows ion concentrations measured

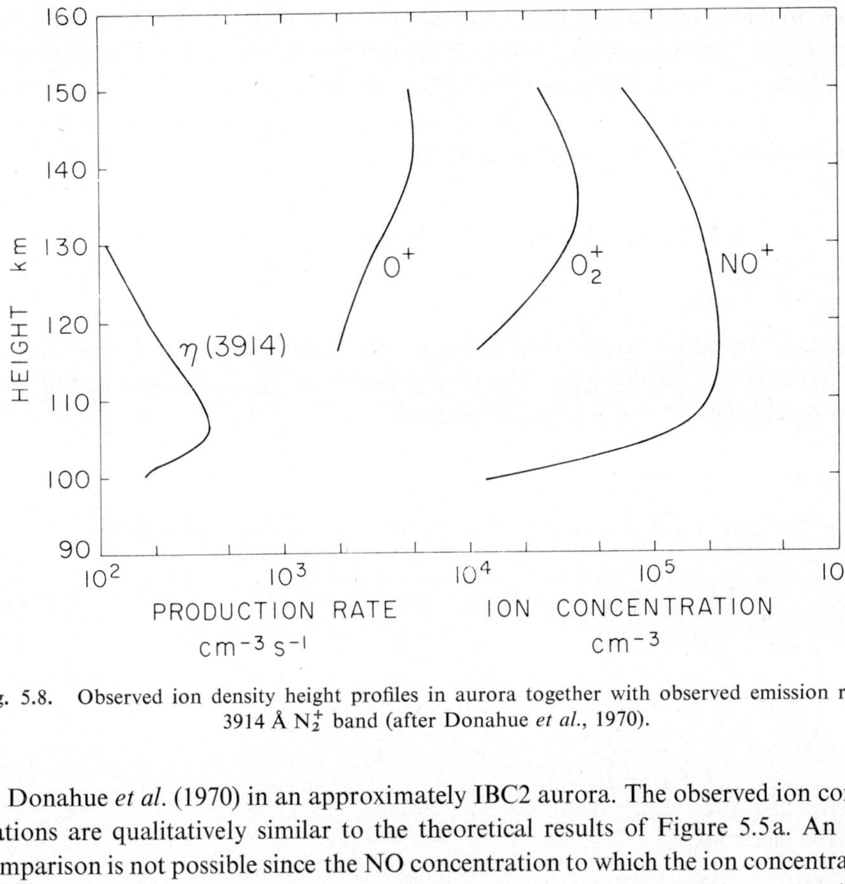

Fig. 5.8. Observed ion density height profiles in aurora together with observed emission rate of 3914 Å N_2^+ band (after Donahue *et al.*, 1970).

by Donahue *et al.* (1970) in an approximately IBC2 aurora. The observed ion concentrations are qualitatively similar to the theoretical results of Figure 5.5a. An exact comparison is not possible since the NO concentration to which the ion concentrations are very sensitive is unknown. As shown by Figure 5.3 the NO concentration is sensitive to the past history of excitation. Swider and Narcisi (1970) have carried out some model calculations which also show how values of $n(NO^+)/n(O_2^+)$ are affected by n(NO). As noted in Section 4.2.4.7 Zipf *et al.* (1970) have measured n(NO) and $n(O_2)$ to be 3.8×10^{10} and 2.3×10^{10} at 120 km inside an IBC2 arc. This exceeds the 100 s value of n(NO) on Figure 5.3b by a factor of about 500. There may be difficulties in forming so much NO on energy grounds. The data of Table 4.5 (cf., also Dalgarno *et al.*, 1965) indicate that an IBC3 aurora requires of the order of 100 erg cm^{-2} for excitation. For each two NO molecules produced by the mechanisms considered above, one molecule of N_2 must be dissociated requiring about 5 eV per molecule of NO. Thus the peak production of NO would not exceed about 12×10^{12} cm^{-2} s^{-1} or about 12×10^{6} cm^{-3} s^{-1} assuming a vertical extent of 10 km. Even in the absence of losses a time of about 3000 s or $\sim$1 h would be needed to reach the measured concentrations. Further measurements (Zipf, 1974) have given NO concentrations to be typically 2 to 3% of those of O_2 in intense substorms.

This argument, however, is not completely sound since NO could be produced with a smaller energy input from N_2 and O or O_2 through a catalytic cycle involving some species produced by the auroral excitation. No such cycle has yet been suggested. Alternatively a process deriving energy from auroral electric fields could be involved.

In any case it is clear that values of n(NO) considerably less than 10^{10} cm^{-3} would have a strong effect on ion composition in aurora and consequently on the dissociative recombination process for producing $O(^1S)$.

5.2. Electrical Conductivity, Electric Fields and Currents

5.2.1. Conductivities

Equation (5.1r) which relates the ion velocity to the effective electric field may be used to derive the conductivity. The same equation is applicable to electrons with e_j and m_j changed to $-e$ and m_e, respectively.

(a) *Parallel Conductivity*

Taking the component (Equation (5.1r)) parallel to the field one obtains

$$\frac{e_j}{m_j} E_{j\parallel} = \nu_j v_{j\parallel}, \tag{5.2a}$$

ν_j is to be understood as the total collision frequency summed over all particles. For ions it is usually sufficient to include only neutral particles in the summation so that

$$\nu_j = \sum_n \nu_{jn} + \Big(\sum_i^{i\neq j} \nu_{ji}\Big) \tag{5.2b}$$

while for electrons

$$\nu_e = \sum_n \nu_{en} + \sum_i \nu_{ei}. \tag{5.2c}$$

Consequently the total ion parallel current is given by

$$\begin{aligned} j_{i\parallel} &= \sum_j n_j e_j v_{j\parallel} \\ &= \sum_j \frac{n_j e_j^2}{m_j \nu_j} E_{j\parallel}. \end{aligned} \tag{5.2d}$$

The electron current is

$$j_{e\|} = -n_e e v_{e\|} = +\frac{n_e e^2}{m_e \nu_e} E_{e\|} . \tag{5.2e}$$

The effective electric field is given by Equation (5.1q). If we consider only that component of the current driven by the apparent electric field $\mathbf{E}' = \mathbf{E} + \mathbf{u} \times \mathbf{B}$ then the conductivity is given by

$$\sigma_\| = (j_{i\|} + j_{e\|})/E'_\| = \sum_j n_j e_j^2/m_j \nu_j + n_e e^2/m_e \nu_e . \tag{5.2f}$$

Since $m_j > 10^4\, m_e$ while $\nu_e < 10^2\, \nu_i$ (see Table 5.2) $m_j \nu_j > 10^2\, m_e \nu_e$ and consequently the electron term dominates.

The collision frequencies appearing in these expressions have a somewhat complicated theoretical basis. Exact expressions are given by Banks (1966a, b) and

TABLE 5.2

Approximate collision frequencies for calculated conditions corresponding to Figure 5.5a (~ IBC3 aurora)

h (km)	n_n (cm^{-3})	n_e (cm^{-3})	T_e (K)	ν_{en} (s^{-1})	ν_{ei} (s^{-1})	ν_{in} (s^{-1})	Ω_e (s^{-1})	Ω_i (s^{-1})
100	0.11(14)	0.67(6)	211	8.0(4)	1.2(4)	7.4(3)	2(6)	3.5(1)
120	0.55(12)	0.91(6)	530	6.3(3)	4.0(3)	3.7(2)		
140	0.86(11)	0.70(6)	1106	1.4(3)	1.0(3)	5.7(1)		
160	0.31(11)	0.53(6)	1483	6.0(2)	5.0(2)	2.1(1)		
180	0.14(11)	0.41(6)	1848	3.0(2)	2.8(2)	9.4(0)		
200	0.77(10)	0.29(6)	2154	1.8(2)	1.6(2)	5.2(0)		

Stubbe (1968) as modified by Rees and Walker (1968). The order of magnitude of the frequencies is given by the following equations:

$$\begin{aligned} \nu_{en} &\simeq 5 \times 10^{-16}\, n_n T_e^{1/2} \\ \nu_{ei} &\simeq 5.4 \times 10^{-5}\, n_i / T_e^{3/2} \\ \nu_{in} &\simeq 2.6 \times 10^{-15}\, n_n / \mu^{1/2} \end{aligned} \tag{5.2g}$$

where μ is the reduced mass in amu of the ion-neutral pairs. The concentrations here are per m^3. Some calculated values using these equations are given in Table 5.2 for the ion densities of Figure 5.5a. These indicate that electron-ion collisions must be taken into account throughout the auroral region.

(b) *Perpendicular Conductivity*

The perpendicular conductivity can be derived from the solution of the perpendicular component of Equation (5.1r) as given in Equation (5.1t). In this case, the ion current is

$$\begin{aligned} \mathbf{j}_{i\perp} &= \sum n_j e_j \mathbf{v}_{j\perp} \\ &= \sum_j \frac{n_j e_j}{B} \left\{ \left(\frac{\nu_j/\Omega_j}{1+\left(\frac{\nu_j}{\Omega_j}\right)^2} \right) \mathbf{E}_{j\perp} - \left(\frac{1}{1+\left(\frac{\nu_j}{\Omega_j}\right)^2} \right) \mathbf{b} \times \mathbf{E}_{j\perp} \right\} \end{aligned} \tag{5.2h}$$

where Equation (5.1s) has been used.

The electron current can be derived by changing the subscripts, noting that $e_e = -e$ and that $|\Omega_e| = -\Omega_e$.

Consequently

$$\mathbf{j}_{e\perp} = \frac{n_e e}{B} \left\{ \left(\frac{\nu_e/|\Omega_e|}{1+\left(\frac{\nu_e}{\Omega_e}\right)^2} \right) \mathbf{E}_{e\perp} + \left(\frac{1}{1+\left(\frac{\nu_e}{\Omega_e}\right)^2} \right) \mathbf{b} \times \mathbf{E}_{e\perp} \right\}. \tag{5.2i}$$

Both Equations (5.2h) and (i) have components parallel to $\mathbf{E}_\perp$ and in the $\mathbf{b} \times \mathbf{E}_\perp$ direction. These are the Pederson and Hall currents, respectively. If again the currents are limited to the components driven by the apparent electric field,

$$\mathbf{E}'_\perp = \mathbf{E}_\perp + \mathbf{u} \times \mathbf{B},$$

the conductivities take the form

$$\sigma_P = \frac{e}{B} \left[\frac{n_e(\nu_e/|\Omega_e|)}{1+(\nu_e/\Omega_e)^2} + \sum n_j \left\{ \frac{(\nu_j/\Omega_j)}{1+(\nu_j/\Omega_j)^2} \right\} \right]. \tag{5.2j}$$

(Concentrations of doubly charged ions are trivial so $e_j = e$), and

$$\sigma_H = \frac{e}{B} \left[\frac{n_e}{1+(\nu_e/\Omega_e)^2} - \sum n_j \left\{ \frac{1}{1+(\nu_j/\Omega_j)^2} \right\} \right]. \tag{5.2k}$$

The physical meaning of these equations is of interest. In the absence of collisions and of electric fields the motion of electrons and ions in the plane perpendicular to $\mathbf{B}$ is circles of radius,

$$r_L = \frac{m|v_\perp|}{|B|e}, \tag{5.2l}$$

traversed at an angular frequency Ω_e or Ω_j, as shown in Figure 5.9a. If an electric field is applied as shown in Figure 5.9b the ion acquires a larger velocity and a smaller curvature in the down-field segment of its orbit and a higher curvature in the up-field segment. The result motion may be described as the original circular one, plus a transverse component

$$\mathbf{v}_d = \frac{\mathbf{E} \times \mathbf{B}}{B^2}. \tag{5.2m}$$

Electrons and ions drift with the same velocity (in the same direction).

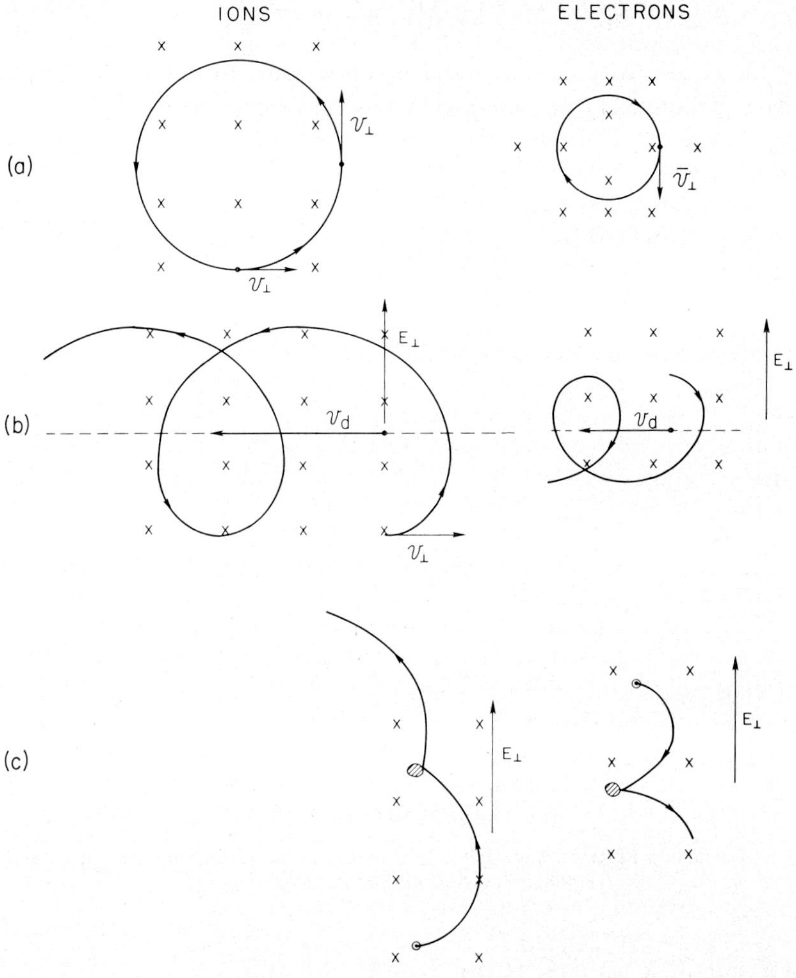

Fig. 5.9. Motions of ions and electrons in (a) magnetic field (b) crossed electric and magnetic field without collisions and (c) with collisions.

In Equation (5.2j) it is clear that if $\nu \ll \Omega$ then the corresponding terms of the Pederson conductivity become negligible while the two terms in Equation (5.2k) become equal and opposite corresponding to the identical drift of ions and electrons. (The currents derived by applying 5.2 m are the same as those from Equations (5.2h) and (5.2i) when $\nu \ll \Omega$.)

When collisions can occur progress in the direction of the field can take place as shown in Figure 5.9c. This gives rise to the first terms in Equations (5.2h) and (i) and to the Pederson conductivity. If $\nu > \Omega$ then it is clear that the motion parallel to the electric field becomes dominant. In the limit $\nu \gg \Omega$, σ_P becomes

$$\frac{e}{B}\left[n_e\left(\frac{|\Omega_e|}{\nu_e}\right)+\sum n_j\left(\frac{\Omega_j}{\nu_j}\right)\right]=\frac{n_e e^2}{m_e \nu_e}+\sum\frac{n_j e^2}{m_j \nu_j}$$

so that the conductivity decreases and becomes equal to the parallel conductance given by Equation (5.2f), i.e., B ceases to have any influence.

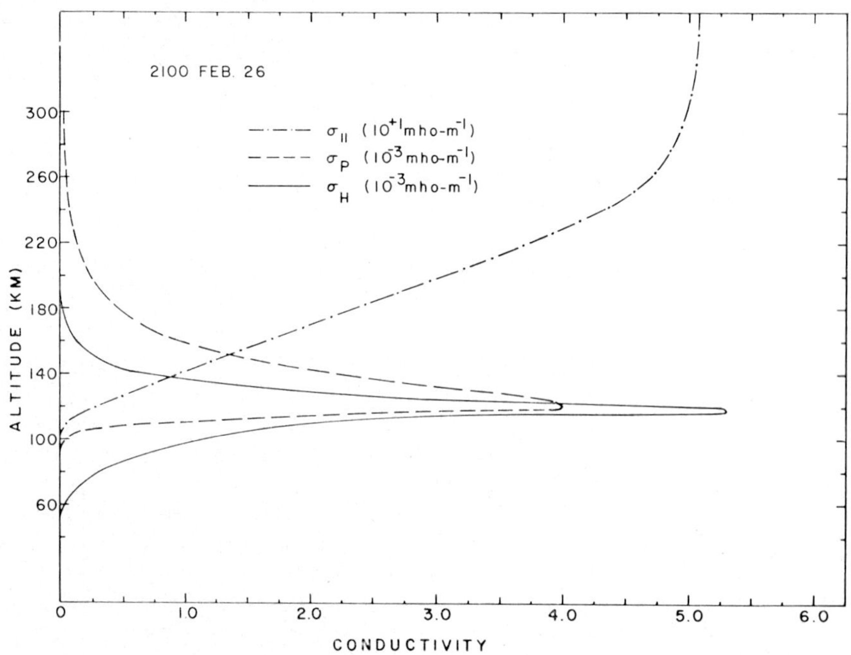

Fig. 5.10. Calculated Pederson and Hall conductivities in an auroral arc after Rees and Walker (1968). (Courtesy *Annales de Géophysique.*)

In the case of auroral ionization Table 5.2 shows that $\nu_e \ll \Omega_e$ and so for electrons the Hall conductivity is dominant, i.e., the electrons drift in the $\mathbf{E}\times\mathbf{B}$ direction. For ions $\nu_j > \Omega_j$ below perhaps 140 km so that both Pederson and Hall conductivity are important. Since the resultant Hall conductivity is the difference between the ion and electron terms, σ_H has a maximum value in the region usually somewhere below the height where $\nu_j \simeq \Omega_j$; the exact height dependence involves also the variation of $n_e = \Sigma n_j$ as determined by Equation (5.2k). These effects are illustrated in Figure 5.10, taken from the work of Rees and Walker (1968) who calculated the electron and ion densities in an auroral arc from observed $I(1N_2^+)$ emission profiles. The peak value of n_e used was 2.6×10^6 at 150 km which should according to Figure 5.6, correspond to $I(N_2^+, 4278) \simeq 100$ kR. (In fact the observed intensity was about ten times smaller so the conductances of Figure 5.10 should be taken as corresponding to an aurora in the IBC3 to 4 range.) McDiarmid and McNamara (1969) have calculated σ_H and σ_P for an aurora which would correspond to IBC1–2 according to Figure 5.6.

5.2.2. Currents and Electric Fields in the Auroral Ionosphere

(a) *Parallel Currents*

As noted in Section 3.6.2 parallel sheet currents have been measured in association with auroral forms. Their magnitude is of the same order as that of the precipitating electron flux. Thus Park and Cloutier (1971) were able to fit their observations as shown in Figure 3.17 with a parallel current of 0.16 amp m^{-1}. The current which would arise from an IBC3 aurora 10 km wide according to Table 4.5 would be about 0.18 amp m^{-1}. Park and Cloutier found that there was a similar downward sheet current just equatorwards of the arc.

The electric field which would be required to drive the return current may be estimated from Equation (5.1l). This equation indicates that the field has two components: (i) the ∇p_e term and (ii) the $j_\parallel/\sigma_\parallel$ term.

The first represents the field required to prevent the electrons establishing a different scale height from the heavy ions; (this would violate charge neutrality.) In the simple case of a single dominant ion with $d(T_i + T_e)/dz = 0$ and $\bar{v}_{\parallel e} = \bar{v}_{\parallel j} = 0$, one may deduce from Equation (5.1p) that

$$\frac{dn_i}{dz} = \frac{dn_e}{dz} = -\frac{n_e m_i g}{k(T_i + T_e)} \tag{5.2n}$$

and hence from Equation (5.1l) if $dT_e/dz = 0$ that

$$E'_\parallel = -\frac{1}{n_e e}\nabla p_e \cdot \mathbf{b} = \frac{m_i g}{e}\left(\frac{T_e}{T_i + T_e}\right). \tag{5.2o}$$

B was taken as vertical so that $G_\parallel = g$. From this, one obtains a field of the order of 1.5 $\mu V\ m^{-1}$.

The $j_\parallel/\sigma_\parallel$ term corresponds to the field required to drive the parallel current. $\sigma_\parallel$ may be obtained from Equation (5.1n) or (5.2f) by substituting the collision frequency $\nu_e = \sum \nu_{ei} + \sum \nu_{en}$. From Table 5.2 for strong aurora it appears that ν_e is of the order of $2\nu_{ei}$. At greater heights electron-ion collisions will dominate. Consequently we can estimate the magnitude $\sigma_\parallel$ by substituting $\nu_e = \nu_{ei}$ from Equation (5.2g) so that

$$\sigma_\parallel \simeq \frac{e^2 T_e^{3/2}}{m_e\, 5.4} \times 10^5. \tag{5.2p}$$

This expression is independent of n_e. For $T_e = 3500\,K$ it gives 1.1×10^2 mho m^{-1}. This may be compared with the exact value of Figure 5.10 of $\sim 0.5\times10^2$ mho m^{-1} above 300 km. At lower heights $\sigma_{\parallel}$ decreases because of the electron temperature factor and the growing importance of ν_{en}. At greater heights $\sigma_{\parallel}$ tends to remain constant.

For the current of 0.2 amp m^{-1} quoted at the beginning of this section, if we assume an arc width of 10 km we obtain a current density of 0.2×10^{-4} amp m^{-2} which combined with a conductance of 10^2 mho m^{-1} gives a field of the order of 0.2 μV m^{-1}. A similar result follows from the currents for the IBC3 aurora example of Table 4.5. At lower heights and for more intense electron beams as in active IBC4 rays the calculated field might rise to perhaps 20 μV m^{-1}. This result has led to the belief that fields of the order of 20 mV m^{-1} as reported in Section 3.6.3b should not occur.

There is a possibility that the parallel conductivity could drop to much lower values than those of Equation (5.2f) if the parallel current becomes large enough to generate unstable plasma waves. This was first suggested by Swift (1965) for the case of ion-acoustic waves. Ossakow (1968) argued that the instability would not be excited outside the plasmasphere. More recently Kindel and Kennel (1971) have shown that ion-cyclotron waves can be excited at lower current thresholds and that this instability can arise for current densities of the order of 10^{-6} to 10^{-5} amp m^{-2}, i.e., within the range of auroral parallel currents. Kindel and Kennel believed that ion-neutral collisions would prevent excitation below the *F* region maximum so that this does not explain the directly observed fields.

Boström (1971) has discussed the possibilities of the production of parallel field due to space charges which may form when the electron drift velocity exceeds the thermal velocity. Hultqvist (1971) has discussed the possible production of a parallel field by a thermoelectric effect between a hot plasma sheet and a cold ionosphere.

Other types of transient parallel electric fields are possible and are discussed in Section 6.2.3.

(b) *Perpendicular Currents*

The existence of large horizontal currents in association with auroral displays has long been recognized from magnetometer studies. The deduction of the nature of the real current system from ground and rocket magnetometer observations is difficult as has been discussed in Section 3.6.2. Even when a fairly definite line or sheet current can be reasonably inferred to be present it may still be hazardous to attempt to reconstruct the electric field because the conductivity profiles such as those of Figure 5.10 are dependent on the electron density profile. If we accept the theoretical conclusion that appreciable vertical electric fields do not exist below 200 km then the field lines, at least below this height, should be equipotentials so that the perpendicular currents flowing should be dependent on the height integrated conductivities (since $\mathbf{E}_{\perp}$ should be independent of height). Clearly the relative magnitudes of $\int \sigma_P\,dh$ and $\int \sigma_H\,dh$ and hence the relative directions of $\mathbf{E}_{\perp}$ and $\mathbf{j}$ depend on $n_e(h)$.

It is consequently dangerous to attempt to infer electric fields from magnetic observations although in the past attempts have been made to do this in the absence of other information. Thus Taylor and Hones (1965) derived ionospheric and magnetospheric electric fields from the equivalent current system for S_D. Bullen (1968) deduced information on the behavior of electric fields in aurora from magnetograms and from conductivities calculated from simultaneous ionosonde data. Föppl *et al.* (1968) note that the currents predicted from observed values of $n_e(h)$ and **E** (from Ba clouds) would have produced magnetic variations closely resembling those observed. Wescott *et al.* (1969) from similar measurements concluded that the currents inferred from ground observations during negative bays were the Hall currents corresponding to southward electric fields. The corresponding Pederson currents could be the horizontal segment joining the parallel sheet currents illustrated in Figure 4.17. The latter currents would produce little effect at the ground.

In the localized region of enhanced ionization, such as might correspond to an auroral band, polarization effects could occur as envisaged in the theory of the auroral electrojet advanced by Coronti and Kennel (1972a); (see also Mozer, 1971).

Although detailed knowledge of the configuration of electric currents and fields associated with auroral forms is incomplete, the observed currents and fields are generally consistent with calculated conductances. A typical horizontal field perturbation during the break-up phase of a strong substorm is 500 γ (1 $\gamma = 1$ nanotesla). The corresponding horizontal line current is 2.5×10^5 A for a 100 km height. If we assume the current flows along a form 10 km wide the integrated current density $\int j \, dh = 25$ A m^{-1}. Such disturbances are generally associated with bright forms so let us take the electron density from Figure 5.6 for IBC3 ($I(4278) \simeq 30$ kR) namely 1.4×10^6 cm^{-3}. According to the calculations of Rees and Walker (1968), $\int \sigma \, dh$ for Hall conductivity was about 125 mho for a form with $n_e(\text{max}) = 2.6 \times 10^6$ so $\int \sigma \, dh$ would scale to about 70 mho. The driving field would then be about 350 mV m^{-1}. This is several times higher than the fields reported outside auroral forms and more than an order of magnitude larger than those inside if the reduction discussed in Section 3.6.3 is real. Although it is possible that electric field measurements have not been successfully made within arc elements carrying the strongest currents it seems likely that the major part of the magnetic effect is due to horizontal current sheets several degrees of latitude wide. The perturbation discussed above could be produced by a sheet current with a height integrated density of about 1 A m^{-1} which could be driven within the arc discussed by a field of ~ 14 mV m^{-1}. If the field outside were 50 mV m^{-1} the same current could be driven if the electron concentration were reduced by a factor of 3.5; this would correspond to a reduction in auroral excitation by a factor of 10. It thus appears necessary to assume that a large part of the current is carried outside discrete auroral forms in a broad region where the ionization is enhanced by background precipitation or decaying after the passage of brighter aurora. This is in agreement with the conclusions of Rostoker (1972) that the electrojet may be several degrees wide in latitude and does not necessarily flow within bright visual arcs.

5.3. Heating Effects in Aurora

The atmosphere may be heated by aurora in various ways such as

(i) direct heating from energy of auroral particle fluxes,

(ii) joule heating as a consequence of ionospheric currents, and

(iii) conduction downwards in the electron gas from the magnetosphere.

The atmosphere, particularly during aurora, is not in thermal equilibrium and consequently in general cannot be assigned a definite temperature. In particular the exchange of energy between electrons, ions and neutral particles may be so comparatively slow that the average kinetic energy possessed by each of these may be different. On the other hand the interchange of energy among like particles may be fast enough to give each an approximately Maxwellian energy distribution characterized by a temperature. Thus we can consider the separate heating and cooling effects for electrons, ions and neutrals and ascribe a temperature to each. Likewise the vibrational temperature of neutral molecules could differ from their translational temperature.

5.3.1. Heating of the Neutral Atmosphere by Auroral Particle Fluxes

Auroral particle fluxes should provide an appreciable heat input to the atmosphere. The energy of the incident electrons of, for example, an IBC3 aurora is of the order of 100 erg cm^{-2} s^{-1}. For particles of normal auroral energies, for example 10 keV monoenergetic, the energy is absorbed over about 10 km near 110 km and hence the energy input rate per unit volume would be about 10^{-4} erg cm^{-3} s^{-1}. The heat capacity per unit volume is of the order of $n(\mathrm{M})\,7k/2 = 2.45 \times 10^{-4}$ erg cm^{-3} deg^{-1} and consequently, if the effect were simply one of heating the maximum temperature rise would be about 0.4 K s^{-1}. At greater heights for softer fluxes the effects would be larger for the same total energy input. As shown recently by Hays *et al.* (1973) the most important effect is to produce an upward convection which may be very important. Discussion of this is postponed until Section 5.3.11 since other modes of heating must also be included.

5.3.2. Heating of Electrons by Auroral Particles

The ambient electrons of the ionosphere can be heated very effectively in collisions with low energy secondary electrons. The theory has already been developed in Section 4.2.1. As a secondary electron is slowed through the interval dE, the energy diverted to heating the ambient electrons is

$$\frac{(\mathrm{d}E/\mathrm{d}x)_{\mathrm{ee}}\,\mathrm{d}E}{(\mathrm{d}E/\mathrm{d}x)_{\mathrm{en}} + (\mathrm{d}E/\mathrm{d}x)_{\mathrm{ee}}} = \varepsilon(E)\,\mathrm{d}E$$

where the energy losses per unit path are as defined in Section 4.2.1, and consequently the total heating from a secondary of initial energy E_s is given by

$$\int_0^{E_s} \varepsilon(E)\,\mathrm{d}E$$

so that the total heating rate per unit volume is

$$\int_0^{\infty} \eta(E_s) \int_0^{E_s} \varepsilon(E)\,\mathrm{d}E\,\mathrm{d}E_s\,. \tag{5.3a}$$

It is assumed that the secondaries lose their energy locally.

5.3.3. Heating of Ions by Auroral Particles

Ions are not heated appreciably by the secondary electrons but rather by the hot electron gas. The rate is given by the expression for electron cooling by ionic collisions from Table 5.3.

5.3.4. Heating of Ions and Electrons by Electric Fields

As discussed in Section 5.2.1 electrons and ions are accelerated by an electric field and in the absence of collisions acquire a drift velocity $(\mathbf{E}\times\mathbf{B})/B^2$ as a result of $\mathbf{E}_\perp$ and a parallel velocity if there is a parallel component to the electric field. Since this is an ordered motion it does not correspond to a rise in temperature. When collisions occur the drift kinetic energy of the accelerated particle may be considered as being converted to thermal energy. The energy so converted from this simple viewpoint would be

$$\tfrac{1}{2}n_1 m_1 v_1^2 \nu_{12}\,,$$

where n_1 particles of mass m_1 and drift velocity $\mathbf{v}_1$ have their energy thermalized in collisions with massive stationary particles of species 2 at a collision rate ν_{12}. The exact expression given by Walker and Rees (1968) is

$$Q_{12} = n_1 \mu_{12} |\mathbf{v}_1 - \mathbf{v}_2|^2 \nu_{12}\,, \tag{5.3b}$$

where μ_{12} is the reduced mass of the colliding particles and $\mathbf{v}_2$ is the drift velocity of the thermalizing particles. This collision rate is similar but not quite the same as those involved in the electrical conductivity equations.

The heating of ions by the field can be evaluated from Equation (5.3b) by introducing the drift velocity $\mathbf{v}_{j\perp}$ (which is relative to CM velocity) from Equation (5.1t) so that $\mathbf{v}_{j\perp} = \mathbf{v}_1 - \mathbf{v}_2$ hence

$$Q_j = \sum_n \mu_{jn} \nu_{jn} |\mathbf{v}_{j\perp}|^2 n_j\,. \tag{5.3c}$$

Since $v_{j\perp}$ has orthogonal components it follows that

TABLE 5.3

Values of cooling rates of electron gas in $\mathrm{eV\,cm^{-3}\,s^{-1}}$

Process	Collision partner	Cooling rate	Reference
Elastic collisions with neutrals	N_2	$n_e n(N_2)\ 1.8\times10^{-19}\ T_e(T_e-T_n)(1-1.21\times10^{-4}\ T_e)$	Dalgarno *et al.* (1963)
	O_2	$n_e n(O_2)\ 1.2\times10^{-19}\ T_e^{1/2}(T_e-T_n)(1+3.6\times10^{-2}\ T_e^{1/2})$	Dalgarno *et al.* (1963)
	O	$n_e n(O)\ 5.3\times10^{-19}\ T_e^{1/2}(T_e-T_n)(1+5.7\times10^{-4}\ T_e)$	Dalgarno *et al.* (1963)
Elastic collisions with ions	All	$n_e\, 8\times10^{-16}\,T_e^{-3/2}(T_e-T_i)\left\{\frac{n(O^+)}{16}+\frac{n(O_2{}^+)}{32}+\frac{n(NO^+)}{30}\right\}$	Dalgarno *et al* (1967)
Rotational excitation	N_2	$n_e\, n(N_2)\ 2.9\times10^{-14}(T_e-T_n)\,T_e^{-1/2}$	Dalgarno and Henry (1965)
	O_2	$n_e n(O_2)\ 6.9\times10^{-14}(T_e-T_n)\,T_e^{-1/2}$	Hanson (1963)
Vibrational excitation	N_2	$n_e n(N_2)\ 6.5\times10^{-10}\ T_e^{-1/2}(T_e/2000)^P \left\{\frac{\exp\left(\frac{-3400}{T_e}\right)-\exp\left(\frac{-3400}{T_N}\right)}{\{1+\exp(-3400/T_N)\}}\right\}$ $P=1.4\ T_e<2000^\circ$ $P=6.0\ T_e>2000^\circ$	Empirical fit to data computed by Rees *et al.* (1967)
	O_2	$n_e n(O_2)\ 8.4(-10)\ T_e^{-1/2}(T_e/1000)^{1.735}\left\{\frac{\exp\left(\frac{-2280}{T_e}\right)-\exp\left(\frac{-2280}{T_N}\right)}{\{1+\exp(-2280/T_N)\}}\right\}$	Empirical fit to data of Lane and Dalgarno (1969)
Excitation of $O(^1D)$	O	$n_e n(O)\ 1.1\times10^{-10}\ T_e^{1/2}\exp\left(\frac{-2.27\times10^4}{T_e}\right)\times$ $\times\left\{0.406+0.357\times10^{-4}T_e-(.333+1.83\times10^{-4}\ T_e)\exp\left(\frac{-1.37\times10^{-4}}{T_e}\right)\right.$ $\left.-(0.456+0.174\times10^{-4}T_e)\exp\left(\frac{-2.97\times10^{-4}}{T_e}\right)\right\}$	Empirical fit to data computed by Rees *et al.* (1967)
Excitation of O^3P_1		$n_e n(O)\ 3.4\times10^{-12}(T_e-T_n)\,T_n^{-1}(1-7\times10^{-5}\ T_e)$	Dalgarno and Degges (1968)

$$|v_{j\perp}|^2 = \frac{E_\perp^2 e_j^2}{m_j^2}\,\frac{1}{(\Omega_j^2+\nu_j^2)} = \frac{E_\perp^2}{B^2}\,\frac{1}{1+\left(\frac{\nu_j}{\Omega_j}\right)^2}\,. \tag{5.3d}$$

As before ν_j is the sum over all neutral collisions. Substituting Equation (5.3d) into (5.3c) and summing over all ions gives a total ion heating rate of

$$Q_i = \frac{E_\perp^2}{B^2}\sum_j n_j \frac{\sum\limits_n \mu_{jn}\nu_{jn}}{1+\left(\frac{\nu_j}{\Omega_j}\right)^2}\,. \tag{5.3e}$$

The heating of electrons follows similarly except that collisions with ions as well as neutrals must be considered so that the ion velocity enters into Equation (5.3b). The electron drift velocity may be obtained from Equation (5.2i) since $\mathbf{j}_{e\perp} = -n_e e\mathbf{v}_{e\perp}$. Moreover in the region of interest $\Omega_e \gg \nu_e$ (see Table 5.3) and hence to a good approximation

$$\mathbf{v}_{e\perp} = -\frac{(\mathbf{b}\times\mathbf{E}_\perp)}{B}\,. \tag{5.3f}$$

When this is substituted into Equation (5.3b) one obtains

$$Q_{en} = \frac{E_\perp^2}{B^2}\,n_e m_e \nu_{en} \tag{5.3g}$$

since $\mu_{en} \simeq m_e$.

For collisions with ions $\mathbf{v}_1 - \mathbf{v}_2 = \mathbf{v}_{e\perp} - \mathbf{v}_{j\perp}$ so that from Equation (5.1t)

$$\mathbf{v}_{e\perp} - \mathbf{v}_{j\perp} = -\frac{1}{B}\left\{\left(\frac{\nu_{jn}/\Omega_j}{1+\left(\frac{\nu_j}{\Omega_j}\right)^2}\right)\mathbf{E}_\perp + \left[1 - \frac{1}{1+\left(\frac{\nu_j}{\Omega_j}\right)^2}\right]\mathbf{b}\times\mathbf{E}_\perp\right\},$$

where the non-electrostatic terms of Equation (5.1q) are neglected, so that

$$|\mathbf{v}_{e\perp} - \mathbf{v}_{j\perp}|^2 = \frac{E_\perp^2}{B^2}\,\frac{1}{1+\left(\frac{\Omega_j}{\nu_j}\right)^2}$$

and consequently

$$Q_{ei} = \frac{E_\perp^2}{B^2}\,n_e m_e \sum_j \frac{\nu_{ej}}{1+\left(\frac{\Omega_j}{\nu_j}\right)^2} \tag{5.3h}$$

where the summation is over the various species of ions.

5.3.5. Heating of neutral particles by electric fields

The process described in Section 5.3.4 for ions results in a sharing of the randomized kinetic energy between ions and neutrals in the thermalizing collisions since the ionic and neutral masses are about the same and consequently Equation (5.3e) can be taken also as giving the neutral heating rate. Ions also are cooled by neutral collisions and consequently there is also the heating term given by Equation (5.3k) as soon as $T_i > T_n$.

5.3.6. Transport effects in the electron and ion gas

The conduction of heat by the electron gas becomes important above 250 km. In addition to purely conductive effects, there are additional heat flow and cooling terms when there is a parallel electric current. Such currents might be expected to balance the inflow of auroral particles although as discussed in Section 3.6.2 the return current appears to be displaced a few kilometers in latitude from the particle flux, at least in some observed cases.

In the absence of parallel currents the net heat gain as a result of conduction in the electron gas is $-\nabla \cdot \mathbf{q}$ where

$$\mathbf{q} = -K^e \nabla \cdot T_e ; \tag{5.3i}$$

K^e is electron thermal conductance for zero current. Rees *et al.* (1971) used the expression,

$$K^e = \frac{7.7 \times 10^5 \, T_e^{5/2}}{1 + 3.22 \times 10^4 (T_e^2 / n_e) \sum_k n_{nk} \mathbf{Q}_k} \tag{5.3j}$$

where $\mathbf{Q}_k$ is the average momentum transfer cross section for the kth neutral species of concentration n_{nk}. This expression is due to Banks (1966c). K^e is in units of eV cm^{-1} s^{-1} deg^{-1} when cgs units are used for the arguments. In the presence of a current, extra terms arise. The first is the addition of a term $-\beta \mathbf{j}$ to Equation (5.3i). β is the thermoelectric coefficient and arises because the faster electrons have a smaller cross section for collisions with ions and hence carry more than their share of the current so that heat flow results. The second effect is a cooling due to the adiabatic expansion associated with an upward motion of the electron gas. These effects are discussed by Schunk and Walker (1970a) and applied to an auroral case by Rees *et al.* (1971).

According to Walker and Rees (1968) conduction in the ion gas should not be important.

5.3.7. Cooling of the electron gas

The final electron temperature is determined by the balance between heating, cooling and transport.

The most important modes of cooling are as follows:

(i) Elastic collisions with neutrals.
(ii) Elastic collisions with ions.
(iii) Rotational excitation of N_2 and O_2.
(iv) Vibrational excitation of N_2 and O_2.
(v) Excitation of $O(^1D)$ level from $O(^3P)$.
(vi) Excitation of the fine structure levels of $O(^3P)$ i.e. $^3P_1$ and $^3P_0$.

Expressions for the contributions to electron cooling rates corresponding to these processes have been given by various authors and these are presented in Table 5.3.

5.3.8. Cooling of ion gas

Ions are cooled by collisions with neutrals. Walker and Rees (1968) used the expression $R(n)\,(T_i - T_n)$ for this cooling rate where

$$R(n) = 3k \sum_j n_j \sum_n \frac{\mu_{jn} \nu_{jn}}{m_n}. \tag{5.3k}$$

5.3.9. Electron temperatures in aurora

Theory

The steady state temperature of electrons in aurora is determined by solving the energy equation. In the absence of electric currents (or electron fluxes) the equation is

$$-\nabla \cdot \mathbf{q} + Q_e - L_e = 0, \tag{5.3l}$$

where Q_e is the heating rate and L_e is the cooling rate. The solution for T_e determines the electron temperature. Solutions of Equation (5.3l) have been given by Walker and Rees (1968) for electron impact heating and by Rees and Walker (1968) for the case where heating by a perpendicular field is included. An example of their results is shown in Figure 5.11. This shows the electron temperatures calculated for the same aurora discussed in Section 5.2.1 in relation to Figure 5.10. As noted this was probably in the IBC3–4 range. T_e is raised to values in the 3000 to 4000 K range from 250 to 350 km. Walker and Rees showed that thermal conductance does not affect the results significantly below 250 km. The effect of electric field heating is less than 500 K for electrons above 250 km and negligible below.

Rees *et al.* (1971) have extended Equation (5.3l) for the case where the electrons have a drift velocity $\mathbf{v}_e$. In this case one must consider the energy balance for a volume element moving with the drift velocity, so that in place of Equation (5.31) the electron energy equation becomes

$$-\tfrac{5}{2} n_e k T_e \nabla \cdot \mathbf{v}_e - \nabla \cdot \mathbf{q} + Q_J + Q_e - L_e = \tfrac{3}{5} k \frac{D(n_e T_e)}{Dt} \tag{5.3m}$$

The first term represents the energy loss due to the adiabatic expansion. Q_J is the Joule heating term while $D/Dt = d/dt + \mathbf{v}_e \cdot \nabla$. Rees *et al.* have solved Equation (5.3m) for the steady state assuming that there is a return parallel current equal and opposite to the primary electron flux determined from Section 4.2.1(v). This is sufficiently accurate to give the order of magnitude of the effect. The results show that the cooling

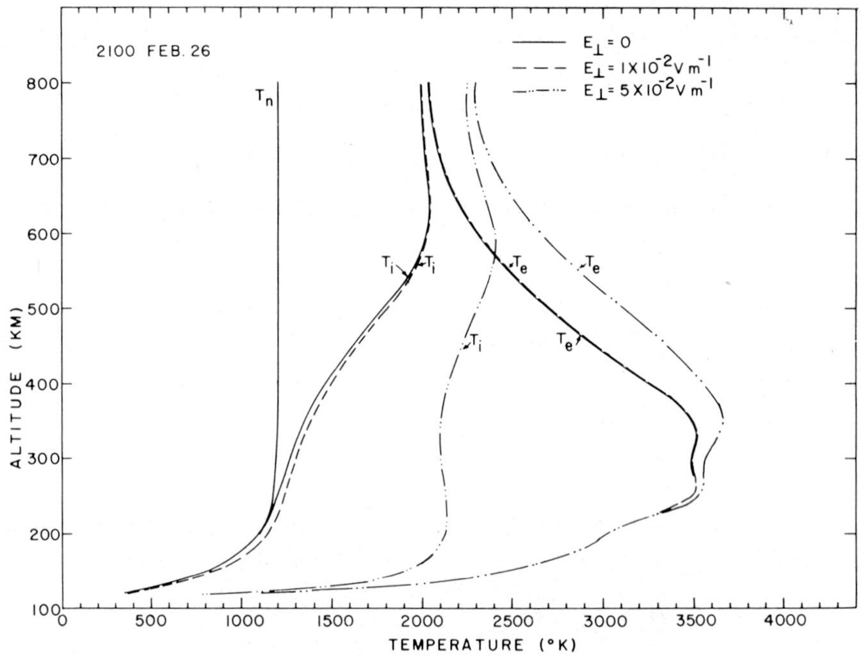

Fig. 5.11. Electron and ion temperatures in aurora with and without perpendicular electric fields after Rees and Walker (1968). (Courtesy *Annales de Géophysique.*)

effect of the parallel current can produce a significant reduction of T_e in the 300 to 1000 km region.

Comparison with Observations

Electron temperature measurements during auroral displays have been reported by Holt and Lerfald (1967) and McNamara (1969) and in the auroral oval at night by Brace *et al.* (1969), Berthelier and Sturges (1967), and Walker *et al.* (1968). Other references may be found in the review of Pfister (1967). The measurement of McNamara is one of the best defined and corresponds to the electron concentration measurements illustrated in Figure 5.7. It is reproduced in Figure 5.12. As noted in Section 5.1.4, this aurora had an estimated intensity of 60 kR and apparently a peak electron concentration of $\sim 10^6$ cm^{-3} at 108 km. This seems comparable with the case discussed by Rees *et al.* (1971) for which the intensity was ~ 40 kR and

$n_e = 1.4 \times 10^6$ cm^{-3} at 115 km. The theoretical electron temperatures for this case are plotted on Figure 5.12 together with results calculated by means of using the equations described above. The results of Holt and Lerfald are less certain in their interpretation but suggested electron temperatures higher than McNamara's in the 100 to 140 km

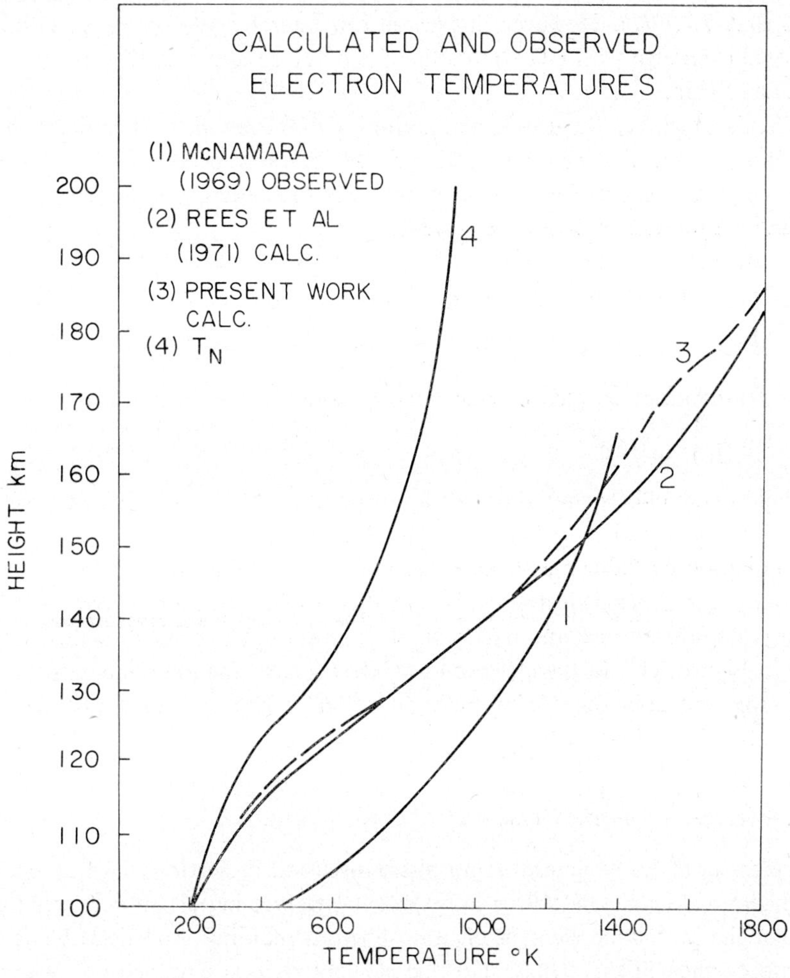

Fig. 5.12. Calculated and observed electron temperatures in aurora. T_N is the neutral gas temperature.

range. Berthelier and Sturges obtained even higher temperatures. Walker *et al.* (1968) also found calculated values to be up to 500 K too low in the 100 to 130 km region.

The discrepancies between calculated values of T_e and those measured by probes are also found in the absence of aurora and at midlatitudes (Brace *et al.*, 1969). Temperatures measured by incoherent scatter are close to those of the neutral temperature up to 120 km while this was found only occasionally in the direct Langmuir probe measurements.

Walker (1968) considered the problem of the need for an additional mechanism for heating electrons in the 100 to 130 km region and explored the possibility that N_2 had an enhanced *vibrational* temperature, T_V, as a result of $O(^1D)$ quenching and the N–NO reaction. This appeared attractive (except that there is evidence from auroral spectra that $T_V \simeq T_N$). However, it was shown later by Walker *et al.* (1969) that vibrational exchange with CO_2 limits the rise of T_V to about 20 K min^{-1} for bright aurora at 120 km.

It must be concluded the problem of accounting for even the fairly modest enhancements of T_e such as that of Figure 5.12 has not been solved. Smith (1972) has shown that probe measurements for electron temperatures less than 1000 K are very sensitive to contamination and may not be reliable.

5.3.10. Ion temperatures in aurora

Theory

The ion temperature should be determined by solving an equation of the form

$$Q_i - L_i = 0 \tag{5.3n}$$

since conduction effects should not be important. L_i, the ion cooling rate, would be obtained from Equation (5.3k) while the heating rate would be derived from the fourth equation of Table 5.3 (heat transfer from hot electrons) and Equation (5.3e) (heating by a transverse electric field). Equation (5.3n) has been solved by Rees and Walker (1968) for auroral and background conditions. The results for the former are shown in Figure 5.11. In the absence of an electric field the ion temperature remains close to the neutral below 200 km but a 50 mV m^{-1} field leads to a very steep rise in temperature above 120 km to over 2000 K at 200 km.

5.3.11. Effects of neutral particle heating in aurora

The energy input to the neutral atmosphere discussed in Sections 5.3.1 and 5.3.5 has been shown by Hays *et al.* (1973) to be able to produce important convective effects. If we imagine a volume element of atmosphere containing n particles being heated at a rate Q, then in the steady state the element rises at a velocity W which must increase upwards so that the element can expand as the pressure decreases. The expansion leads to external work being done which reduces the temperature rise. In the steady state it is easy to show from classical thermodynamics that

$$W = \frac{Q}{nC_p \dfrac{dT}{dz} + nmg},$$

where m and C_p are the average mass and heat capacity per particle. This equation was solved by Hays *et al.* who found upward winds of the order of 10 cm s^{-1} at 125 km, and 100 cm s^{-1} at 170 km increasing to a maximum of several m s^{-1} over 200 km for heating by a 50 mV m^{-1} perpendicular field. The heating by particle fluxes produces winds of comparable magnitudes, with softer particles being particularly effective (for a given energy flux).

These vertical winds have superposed on them the diffusive velocities of the neutral

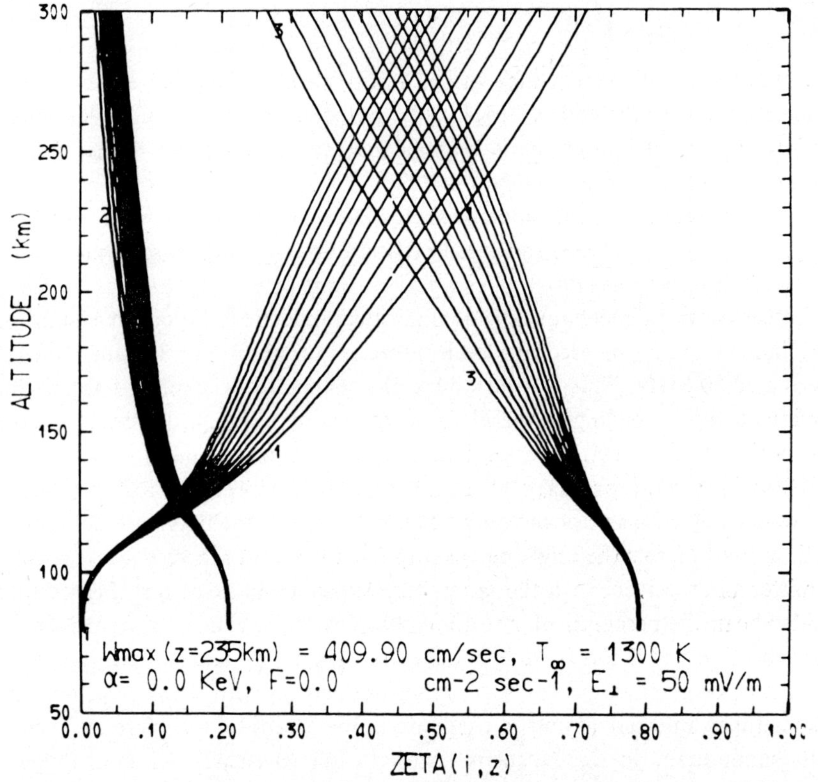

Fig. 5.13. Changes in atmospheric composition due to neutral wind produced by auroral heating caused by a 50 mV m^{-1} perpendicular field. Curves 1, 2, and 3 give the initial composition for O O_2 and N_2 while the remaining curves give the composition at successive 1 h intervals, (Hays *et al.*, 1973). (Courtesy *Planetary and Space Science.*)

components which arise from the balance between partial pressure gradients, gravitational forces and drag forces due to collisions with other species. The equations for the concentration changes of the major neutral constituents have been formulated and solved by Hays *et al.* One of their results is reproduced in Figure 5.13. In this case the heating is due to a 50 mV m^{-1} perpendicular field. After several hours significant changes are predicted in the composition especially above 200 km.

5.4. Reflection of Radio Waves from the Auroral Ionosphere

5.4.1. Reflection mechanisms

The application of ionospheric sounders to the determination of electron concentrations in aurora has been mentioned in Section 2.4.3. These measurements depend on the total reflection of vertically incident radio waves. Total or critical reflection occurs to a first approximation when the electron concentration reaches a value

$$n_{ec} = m_e \varepsilon_0 \omega^2 / e^2, \tag{5.4a}$$

where ω is the angular frequency of the radio wave. Evaluating the constants one obtains $n(\text{cm}^{-3}) = 1.24 \times 10^4 f^2(\text{MHz})$. As noted by Reddy *et al.* (1969) the interpretation of auroral ionograms is complicated by reflections at higher frequencies which are due to structured ionization.

The higher frequency reflections *generally called radio aurora* were discovered by Harang and Stoffregen (1938) and extensively studied by using pulsed radar techniques after 1945. Chamberlain (1961) gives a full account of the development of this field. The simplest possible mechanism is again critical reflection. Equation (5.4a) however would indicate that an electron concentration of about 3×10^7 cm^{-3} would be necessary at 50 MHz. This is well above the values to be expected for commonly occurring aurora according to the ideas of Equation (5.2) (e.g., Figure 5.6), although it is conceivable that such values might exist in limited volumes.

Another important property of radio aurora is aspect sensitivity. This is the tendency for reflections to occur only when a normal to the magnetic field line at the reflection point bisects the angle between the incident and reflected beams. When the transmitter and receiver are at the same place (so-called backscatter) this requires that the radar beam be perpendicular to the field lines at the reflection point. The height of the reflection point can be determined from the range and direction of the echo. These can be determined precisely at higher frequencies with the narrow field systems and sometimes inferred at lower frequencies by means of interference techniques. Results summarized in the excellent review by Hultqvist and Egeland (1964) show that at all frequencies (i.e., at least 40 to 500 MHz) the reflections come from within a few kilometers of 110 km. The deviation from the specular reflection condition can be expressed in terms of the aspect angle, α, the value by which the angle between the line from the transmitter to a point at the scattering height, and the field direction differs from 90°. Contours of constant aspect angle for a given radar generally show a region of minimum or zero values poleward of the station. Since the field is close to vertical in auroral zone regions the elevation angle of the line of sight to reflection points of small or zero aspect angle is generally small. Comparison of observed reflections vs aspect angle, as summarized by McDiarmid (1972) show that the scattered intensity falls off at a rate of up to 10 dB/°. Some of the Canadian measurements indicate a lower value of 1 to 2 dB/° at least in certain cases.

The need to understand reflections from lower electron densities and to account for high aspect sensitivity led to the development of theories which ascribe the reflections to multiple weak scatterings from fluctuations in electron density. Booker (1956) developed the theory of the scattering in terms of reflections from a random distribution of ellipsoidal scattering regions elongated in the direction of the field lines and this theory was further developed by Moorcroft (1961a, b) who allowed also for the possibility of critical reflection in the cores of the ellipsoids. As shown by Moorcroft (1966) this type of model fails to account for observations of the variation of scattered intensity with frequency. In fact the random models predict a scattering cross section monotonically decreasing with frequency whereas observations show that this is often not fulfilled.

A more fundamental interpretation of the radio reflection is an approach which also arises from Booker's (1956) paper. This will be outlined after a brief consideration of the scattering properties of a single electron.

The scattering cross section of a free electron can be defined as the ratio of the total power scattered to the incident intensity and represents the effective area of the electron in taking energy from the incident beam. It is easily calculated by considering the oscillatory motion of a free electron in the electric field of the incident radiation and calculating the power re-radiated. As shown, for example by Stone (1963), the cross section is given by the relation

$$\sigma = \frac{8\pi}{3}\left(\frac{\mu_0 e^2}{4\pi m_e}\right)^2 = 6.7 \times 10^{-25}\ \mathrm{cm}^2. \tag{5.4b}$$

When scattering is produced by a large number of electrons the phase of the scattered waves must be taken into account in a manner analogous to the scattering of X-rays by a crystal lattice. As a consequence the differential cross section (per unit volume per unit solid angle in the direction of the receiver) is

$$\frac{\mathrm{d}\sigma}{\mathrm{d}\Omega} = \left(\frac{\mu_0 e^2}{4\pi m_e}\right)^2 S(\mathbf{k}), \tag{5.4c}$$

where $S(\mathbf{k})$ is the three-dimensional spectral power density of the electron density fluctuations evaluated at a wavenumber $\mathbf{k}$ where

$$\mathbf{k} = \mathbf{k}_1 - \mathbf{k}_2$$

in which $\mathbf{k}_1$ and $\mathbf{k}_2$ are the vector wavenumbers of the incident and scattered waves (i.e., $2\pi/\lambda \times$ unit vector in the propagation direction). For backscatter $|k| = 4\pi/\lambda$ and consequently one can say that the cross section for a radar wavelength λ is determined by the component of spatial electron density power distribution with a wavelength $\lambda/2$ measured in the direction of the radar beam. More generally this becomes $\lambda/[2 \sin(\theta/2)]$ when the scattered wavenumber vector makes an angle θ with the incident wavenumber vector. Thus, in the so-called 'forward-scatter' measurements with two or more stations mentioned in Section 2.4.2 where $\theta \ll 180°$ the effective wavelength of irregularities in $(\Delta n_e)^2$ is much longer.

The significance of the interpretation of the radio reflections in terms of spatial frequencies of $(\Delta n_e)^2$ is that such wave structures may have a real physical existence within the ionized regions associated with aurora. (It may be noted in passing that the ellipsoidal structures of the earlier treatments were suggestive of the ray structure of some auroral forms.) According to recent ideas fine structure in the electron density distribution can arise from instabilities which will be discussed in the following section.

5.4.2. Fine structure of auroral ionization

Some structure is undoubtedly impressed on auroral ionization by the spatial and temporal structure of the incident particle beam while other contributions can arise during the lifetime of the ionization produced.

As discussed in Section 3.3.2, rayed auroral forms frequently or always consist of *curls* in a narrow band of precipitated particles. These curls are formed in less than 1 s and have velocities of up to 20 km s^{-1} parallel to the forms in which they appear. As will be discussed in Section 6.3.2.1 they are probably due to instabilities in the sheet of particles responsible for the auroral form. However the rapid motions associated with such fine structure would probably smooth out much of the irregularity in the ionization since the time constant for the ion recombination is of the order of 1 s or more.

The detailed relation between visual aurora and radio aurora reflections has been the subject of much debate. Unwin and Baggaley (1972) tend to consider the association to be poor although Forsyth (1968) interpreted the evidence more optimistically. Very recently Balsley *et al.* (1973) have reported results with a narrow field radar showing a very definite association with auroral forms in cases which could be well defined. This will be discussed further below. It is possible as has been argued by Sofko and Kavadas (1969) that much of the structure causing radar reflections is due to the incoming particles. This is difficult to test since the optical structure can only be seen directly in the magnetic zenith by means of TV techniques while most radars have had wide azimuthal beam widths. It has been generally felt however that some further source of fluctuations was required.

Two processes have been suggested as being responsible for the production of structure in pre-existing auroral ionization.

(i) *Two Stream Instability – Ion Acoustic Waves*

This process, well known in laboratory plasmas, was first applied by Farley (1963) in the ionosphere as an explanation of radio reflections from the equatorial electrojet. It can occur when the difference between the drift velocities of electrons and ions reaches a critical value. An ion-acoustic wave is a longitudinal compressional traveling wave involving changes in ion density. In order to maintain charge neutrality the electrons are constrained to follow the ions and a small longitudinal **E** oscillation develops to achieve this coupling. The instability is analogous to the generation of

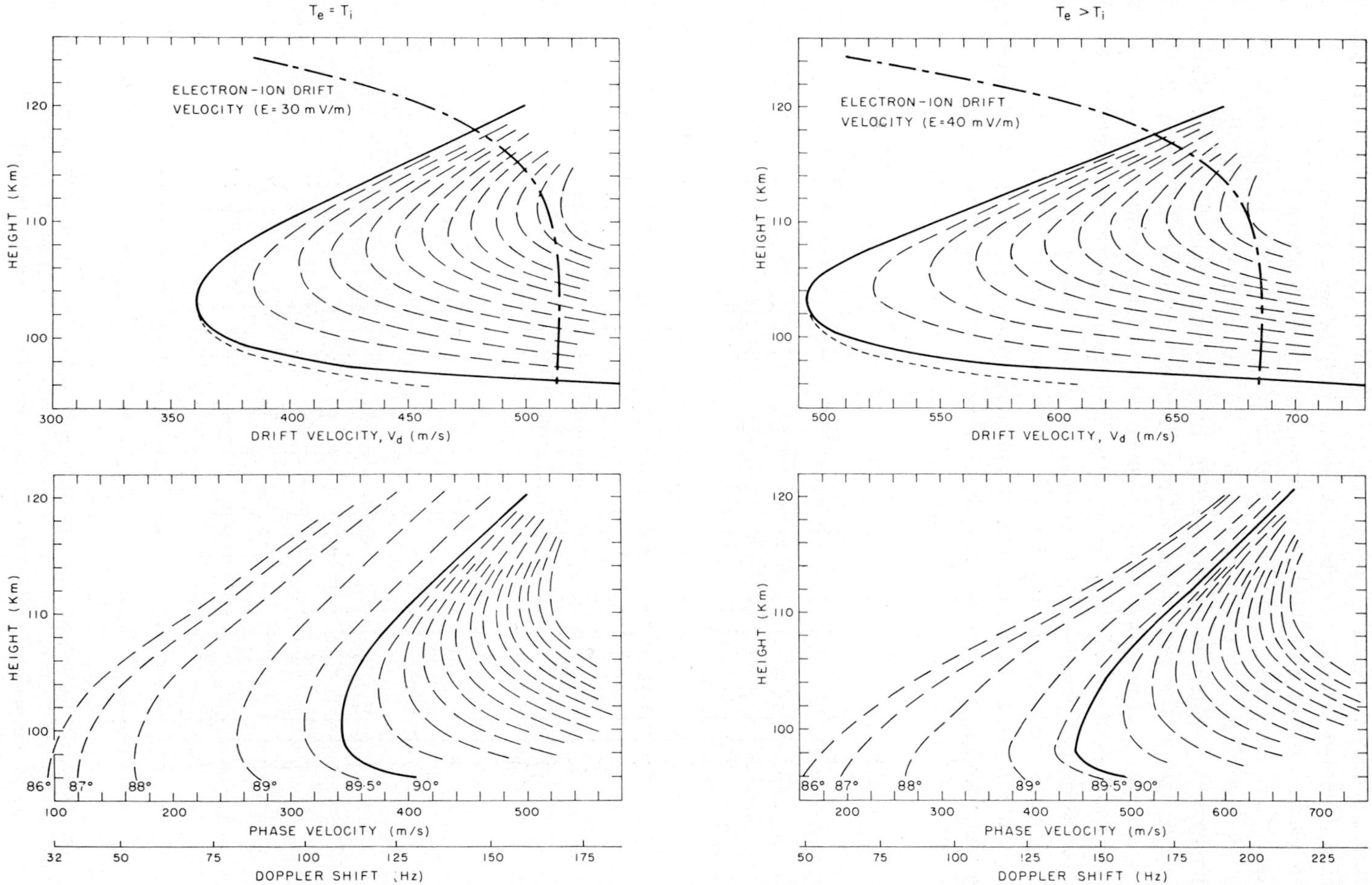

Fig. 5.14. Differential electron-ion drift velocities required for the excitation of ion-acoustic waves. The solid line represents the critical values while the light dashed lines give the values for growth rates increasing in steps of 20 s^{-1}. The short-long dash lines give calculated electron-ion drift velocities (McDiarmid, 1970). (Courtesy *Canadian Journal of Physics*; reproduced by permission of the National Research Council of Canada.)

electromagnetic waves in a traveling wave tube in which the wave grows at the expense of the kinetic energy of electrons moving with the wave but bunched so they are decelerated by the wave field (like a linear accelerator in reverse).

These waves are subject to strong Landau damping except when their propagation vector, **k**, is perpendicular to the magnetic field, **B**. The component of the differential drift velocity $\mathbf{v}_{ei} = \mathbf{v}_e - \mathbf{v}_i$ in the propagation direction must reach the critical value for the excitation of waves in that direction. Figure 5.14, from McDiarmid (1970) shows values of $\mathbf{v}_{ei}$ required at auroral heights to give various growth rates at an ion-acoustic wavelength for reflections at 48 MHz. The abscissa in the upper curves may be taken as the component of $\mathbf{v}_{ei}$ in the direction of the radar beam.

The ion-acoustic waves excited as described conform naturally to the description of the scattering process expressed by Equation (5.4c) since the presence of the plane wave with wavelength $\lambda/2$ propagating in the direction of the radar beam provides a component of $S(\mathbf{k})$ determining the back scattering at the radar wavelength λ. Aspect

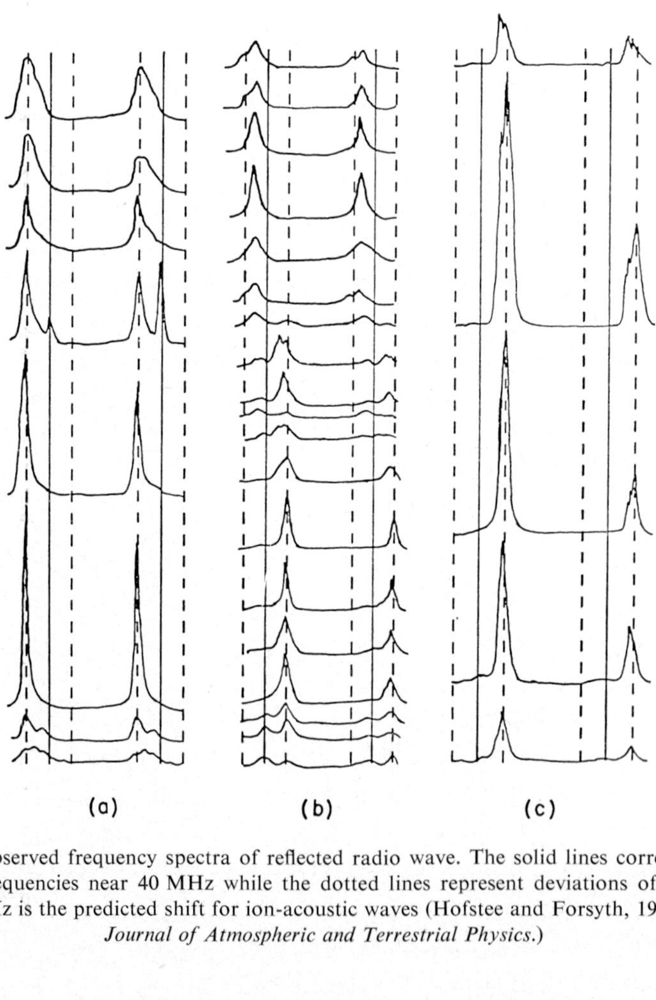

Fig. 5.15. Observed frequency spectra of reflected radio wave. The solid lines correspond to two transmitted frequencies near 40 MHz while the dotted lines represent deviations of ±100 Hz. In this case 100 Hz is the predicted shift for ion-acoustic waves (Hofstee and Forsyth, 1972). (Courtesy *Journal of Atmospheric and Terrestrial Physics.*)

sensitivity arises because ion-acoustic waves do not propagate except in a direction nearly perpendicular to **B**. The predicted degree of aspect sensitivity may be higher than is observed (cf., McDiarmid, 1972; and Hofstee and Forsyth, 1972). As suggested by the latter authors the direction of **B** may be perturbed by auroral currents. Moorcroft (1972) has argued that aspect angle deviations up to 10° may thus be possible.

Reflections from ion-acoustic waves should show a Doppler shift corresponding to the phase velocity of the waves. The occurrence of reflections having the expected Doppler shift for reflections of this sort has been demonstrated by Hofstee and Forsyth (1972) from whose paper Figure 5.15 is reproduced. However the extensive analyses of Hofstee and Forsyth (1972) and Balsley and Ecklund (1972) show that only a minority of echos have the expected frequency spectrum. They have concluded that some other process must be important in producing the structure from which the reflections arise.

(ii) *Drift Gradient Instability*

Another process which should be capable of structuring existing ionization is the $\mathbf{E} \times \mathbf{B}$ or drift gradient instability. This has received wide attention in laboratory plasmas as well as in ionospheric applications. References to earlier work may be found in the review of Unwin and Baggaley (1972) and Hooper and Walker (1971). Knox (1972) has applied the treatment of Whitehead (1968) to radio aurora. The physical nature of the instability may be understood from Figure 5.16. Suppose CD in (a) represents the boundary between a band of ionization and a less ionized background region and that **E** and **B** are as shown. Then there will be a drift of ions

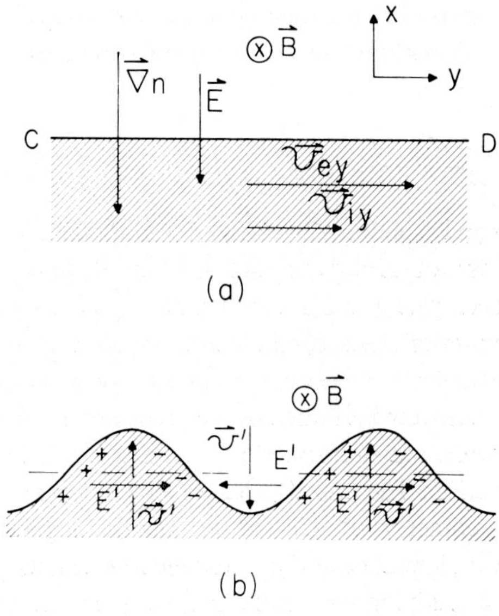

Fig. 5.16. Mechanism of drift-gradient instability.

and electrons in y direction as shown, with $v_{ey} > v_{iy}$. If a perturbation disturbs the interface as in (b), the differential electron-ion drifts will produce polarization and associated perturbation fields E' as shown. These will produce the drift velocities v' which tend to cause the perturbation to grow. If the **E** field is reversed the perturbation will be stabilized. Diffusion tends to destroy instabilities of short wavelength while recombination damps out those at longer wavelength.

The analysis of Hooper and Walker applied to the same aurora as discussed by Rees and Walker in Section 5.3.9 indicated that the instability could grow for wavelengths between 40 m and 20 km. This falls short of the range needed for VHF radar reflections (<1 m). Knox concluded that instabilities of $\simeq 1$ m could easily be generated with the high plasma density gradients implied by the fine structure discussed in Section 3.3.2. Hallinan and Davis (1970) also suggest that values of **E** up to 1000 mV m^{-1} may occur in association with auroral fine structure and consequently the possibilities of transient instabilities of short wavelength would be enhanced.

(iii) *Ion-Cyclotron Instability*

Very recently D'Angelo (1973) has suggested that the ion-cyclotron wave instability excited by parallel electron currents as discussed in Section 5.2.2 could be responsible for the third class of echos noted below in Section 5.4.3 (ii). Such waves would propagate almost normally to the magnetic field and issue radially from any current filament in which the current density exceeds the instability threshold. D'Angelo argues that the velocity spectrum of short-wavelength electron density variations would match the observed flat Doppler distribution. It is not yet clear whether the required relationship between strong parallel currents and this type of echo can be demonstrated. The association of the evening diffuse echoes with the eastern electrojet (see Section 5.4.3(v)) is disturbing since this region is not expected to be associated with strong parallel currents.

5.4.3. Comparison of Theory and Observations of Radio Aurora

Many points in this comparison have been noted or implied in the above in discussing the theory. The reviews of Hultqvist and Egeland (1964), Lange-Hesse (1967), McNamara (1971), Czechowsky and Lange-Hesse (1971), and Unwin and Baggaley (1972) give good accounts of the systematization of data in terms of temporal and latitude variations, and behavior with magnetic activity. It is clear that in a general way the occurrence of radio aurora indicates the presence of auroral ionization having structure associated with magnetospheric processes or being structured in the ionosphere after production. The actual reflections seen depend on the characteristics of the radar and the geometry relative to the Earth's magnetic field. Hitherto the emphasis has been on understanding the phenomenon and as yet it has not proved very useful as a tool for studying the ionization. Some critical points to be explained by the theoretical interpretations will now be considered.

(i) *Aspect Sensitivity*

Since auroral primaries are guided by the magnetic field and since ionization diffuses most easily along the field direction, aspect sensitivity is to be expected even in the case of irregularities due to the precipitation pattern. The two stream instability mechanism implies a very high degree of aspect sensitivity but as noted above this may be relaxed in the presence of currents. In any case it appears that this mechanism is of minor importance. The aspect sensitivity to be expected in other cases is not so well defined.

(ii) *Frequency Spectrum of Reflected Waves*

Observations of the frequency distribution of the reflected wave provide a critical test of the nature of the structuring mechanism. On this basis Balsley and Ecklund (1972) have shown that only about 10% of echos satisfy the stringent requirements of the two stream instability process. Another 20% show some tendency towards a peak in the frequency spectrum and could possibly indicate drifts associated with the drift gradient instability. The majority of echos shows a very broad power spectrum (the last are of the diffuse type).

(iii) *Polarization of Reflected Waves*

The polarization state of the reflected waves has been studied by Sofko and Kavadas (1971) and by Harang and Tröim (1970) as well as in earlier work reviewed by Hultqvist and Egelund (1964). Very recent work with a new fast system by Sofko (1974) appears to indicate that strong echoes which probably correspond to single scattering remain highly polarized and show only a small change in polarization direction consistent with Faraday rotation by the component of the terrestrial field parallel to the beam. Weaker echoes, presumably due to multiple reflections, show larger polarization changes consistent with larger Faraday rotations along path segments parallel to the field as well as depolarization effects analogous to those of optical multiple scattering. This study resolves the problems encountered in the earlier work.

(iv) *Association with Visual Aurora*

The two stream instability mechanism does not require a close association with very bright aurora whereas drift gradient and critical reflections would be likely to be associated with rather intense precipitation. The study of one event by Balsley *et al.* (1973) with an azimuthal field of 1° provided evidence for an association of narrow reflecting regions with the equatorward edge of visual bands before midnight and the poleward edge after. Moreover the Doppler shifts in the frequency spectrum of the returned pulses were consistent with the line of sight component of an assumed constant drift velocity along the edge of the band. This is consistent with drift of auroral structures noted in 3.7 I(4), II(4) and IV(4). A similar pattern was observed earlier by Lyon and Kavadas (1958) for the apparent motions of radar echos. The pattern observed by Balsley *et al.* would be consistent with a drift gradient instability produced by poleward **E** before midnight and equatorward after. This appears to be consistent with the fields noted in Section 3.6.3(a) and with the fields needed to produce the optically observed drift patterns. The recent paper of Davis (1971) inferring general magnetospheric convection patterns and electric fields is of interest in this connection. As noted in the next paragraph premidnight diffuse echoes may be associated with a region equatorwards of visual forms. Greenwald *et al.* (1973) however found some associations between premidnight discrete echoes and surges or spirals on the poleward edge of the oval.

(v) *Association with Electric Currents*

According to the evidence summarized by Unwin and Baggaley (1972) there is a maximum chance of radio reflections in association with the auroral electrojets. Such an association is consistent with the instability theories which require electric fields and ionization. A detailed comparison of the visual forms, magnetograms and radio echos has been made by McDiarmid and McNamara (1969). The required electric field and currents were calculated for a model auroral arc on the basis of the ion-acoustic wave mechanism. A westward **E** of 30 mV m^{-1} was sufficient to excite the instability. The current flowing contained a large N-ward Hall current component similar to the configuration of Figure 3.17. McDiarmid (1970) has carried out a similar analysis for the event studied by Hofstee and Forsyth (1969) as described in Section 5.4.2(i) and Figure 5.16. In this case also a westward **E** was required

Czechowsky and Lange-Hesse (1971) have presented evidence for the connection between radio reflections and the electrojet as well as with visual break-up aurora.

Very recently Greenwald *et al.* (1973) and Ecklund *et al.* (1973) have associated the diffuse type of evening echos with the eastward electrojet just equatorward of the region of visual auroral activity. Phase velocities up to 1000 m s^{-1} were observed. There are still difficulties in identifying the structuring process.

5.5. Radio Absorption

The absorption of energy from radio waves in the ionosphere is a consequence of the loss of energy from oscillating free electrons in collisions with neutral particles. As described in Section 2.6.2 measurements of the absorption of cosmic noise provide a convenient way of monitoring such absorptions and consequently of obtaining an indication of the electron concentrations at heights where absorption is effective. It turns out that this phenomenon provides a good measure of the higher energy component of auroral electron fluxes.

5.5.1. Theory of Absorption

After traversing a distance, s, the amplitude of a wave is attenuated according to a relation of the form

$$E = E_0 e^{-ks}, \tag{5.5a}$$

where k is given, to the approximation adopted earlier in this chapter, by the simple relation,

$$k = \frac{e^2}{2m_e c \varepsilon_0} \frac{n_e \nu_e}{(\omega^2 + \nu_e^2)}, \tag{5.5b}$$

provided that the wave frequency, ω, is much larger than the plasma frequency. In the presence of a magnetic field ω must be replaced by $\omega \pm \Omega_e$ but this, as may be seen from Table 5.2, has only a small effect at riometer frequencies around 30 MHz. The power attenuation in dB cm^{-1} for a concentration expressed per cm^3 becomes

$$\frac{dA}{ds} = 0.46 \frac{n_e \nu_e}{\omega^2 + \nu_e^2} \tag{5.5c}$$

so the total attenuation becomes

$$A = 0.46 \int \frac{n_e \nu_e \, ds}{\omega^2 + \nu_e^2}. \tag{5.5d}$$

If the electron density is only important at heights where $\nu_e \ll \omega$ then

$$A \simeq \frac{0.46}{\omega^2} \int n_e \nu_e \, ds. \tag{5.5e}$$

Because of the factor ν_e in the integrand, the electron densities at the lowest heights are weighted most heavily and consequently it is the incident particles which penetrate below 100 km (i.e. having $E > 20$ keV) which are especially important in producing absorption.

Equations (5.5d) and (e) are approximations in that the simple derivation assumes that the electron collision frequency is independent of energy. More accurate expressions have been derived by Sen and Wyller (1960).

5.5.2. Application to Aurora

The calculation of the riometer absorption to be expected from a given input flux of particles reduces to that of finding the electron concentration profile together with a knowledge of the atmospheric parameters necessary to determine the collision frequencies. Hargreaves (1969b) gives ıeferences to model calculations which show that auroral electron fluxes are sufficient to account for observed absorption levels. As a further illustration taken from this review, Figure 5.17 shows rocket measured electron concentration profiles (Jespersen *et al.*, 1968) under auroral conditions and the corresponding calculated absorption profiles and integrated absorptions. The latter agree well with the simultaneously observed values.

Considerable work has also been done in matching the variation of A with ω in

Equation (5.5d) to give the height of maximum absorption. Clearly where $\nu_e \gg \omega$ the integrand becomes independent of ω whereas in the opposite case it is proportion to ω^{-2}. This provides a means of estimating the height of maximum absorption. The results however give values which appear too low.

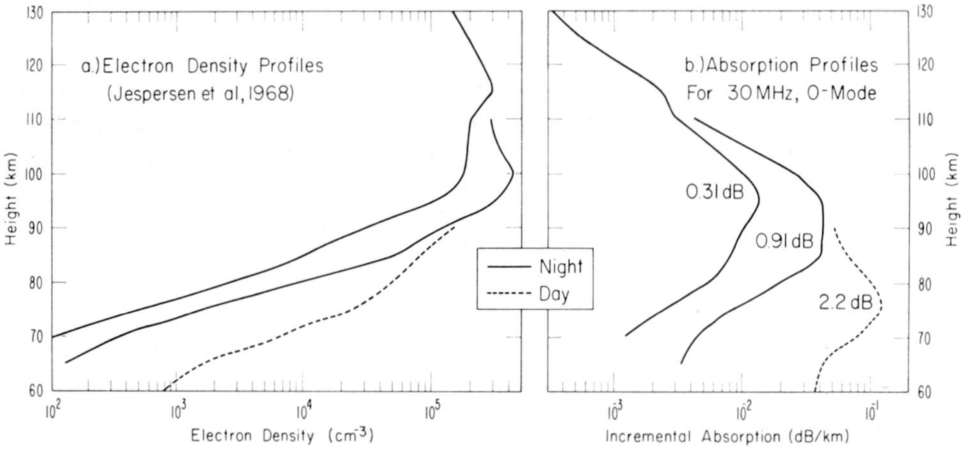

Fig. 5.17. Measured electron densities and computed absorption during auroral events (Hargreaves, 1969b). (Courtesy *Proceedings IEEE.*)

The correlation between riometer measurements and auroral intensity has been extensively studied in Norway. Omholt (1971) has reviewed this work. To the approximations involved in their derivation Equations (5.1b) and (5.1f) imply that

$$n_{e\infty} = c_3(h)\,[c_1(h)\,\eta(4278)]^{\frac{1}{2}}, \tag{5.5f}$$

where $c_3(h)$ and $c_1(h)$ are the (height variable) proportionality constants in these equations respectively. Moreover Equation (5.5c) implies that

$$\frac{\mathrm{d}A}{\mathrm{d}s} \simeq \frac{\mathrm{d}A}{\mathrm{d}h} = c_2(h)n_e\,;$$

therefore if $n_e = n_{e\infty}$

$$A = \int_0^\infty [c_1^{\frac{1}{2}}(h) c_2(h)\, c_3(h)]\,\sqrt{\eta(4278)}\,\mathrm{d}h$$

$$= \frac{1}{c_4}\int_0^\infty \sqrt{\eta(4278)}\,\mathrm{d}h \simeq \frac{\Delta h^{\frac{1}{2}}}{c_4}\sqrt{I(4278)}, \tag{5.5g}$$

where $1/c_4$ is the average value of the product in square brackets weighted by the $\sqrt{\eta}(4278)$ height profile. For the last equality of Equation (5.5g) to be valid the ion production must be assumed confined to a narrow height range Δh within which its value is constant. Since $c_2(h)$, in particular, is strongly height sensitive, it is clear that the relation between absorption and auroral brightness is very sensitive to the height of the emission which is in turn controlled by the energy spectrum of the incident particles.

The approximation that $n_e = n_{e\infty}$ implies that a deviation from Equation (5.5g) will occur for fluctuating intensities.

5.5.3. Results of Riometer Studies

Two main applications are discernible in systematic observations of radio absorption.

(i) Studies of the relationship between optical brightness and absorption as an indication of particle energy according to Equation (5.5g).

(ii) Studies of riometer observations as an approximate index of the influx of higher energy auroral particles.

These topics are the subject of an extensive literature which is well covered in the reviews of Hargreaves (1969b), Hultqvist (1969), Jelly (1969), and Omholt (1971). Only the salient results will be summarized.

(i) *Optical-Riometer Studies*

Holt and Omholt (1962) and later Johansen (1965) studied the relationship between absorption and emission on the basis of Equation (5.5g). Further studies of this type with improved riometer spatial resolution have been continued by Ansari (1964), Eather and Jacka (1966a), Berkey (1968), and Gustafsson (1969). In general these observations indicate a hardening of the spectrum in the transition from quieter evening aurora to morning forms. A supposed correlation between pulsating aurora and a hardening of the spectrum appears not to hold according to the recent work of Brekke (1971, 1972).

Time lags between absorption and auroral emission have been studied by Holt and Omholt (1962), Eather and Jacka (1966a), Gustafsson (1964), and Reid and Phillips (1971). It appears that the longer values of the order of several minutes are due to differences between the fields of the optical and riometer systems since rather shorter time constants should be typical of the height profiles shown in Figure 5.17. For example for the 0.91 dB case at 90 km, the center of the absorption region, τ would be about 50 s for $n_e = 10^5\ \mathrm{cm}^{-3}$ and $\alpha = 2 \times 10^{-7}\ \mathrm{cm}^3\ \mathrm{s}^{-1}$ according to Equation (5.1e). As shown by Eather and Jacka (1966a) time constants between 7 and 60 s (for 90% of equilibrium absorption) would be expected for exponential energy spectra with E_0 from 5 to 20 keV and for 1 dB equilibrium absorption. The time constant is inversely proportional to the equilibrium absorption value. Their results gave values of c_4 appropriate to E_0 in the range 5 to 24 keV. Reid and Phillips measured time lags of ~ 2 s and showed these were consistent with expected values.

(ii) *Synoptic Studies*

The simplicity of riometer observations makes them well suited for synoptic studies of the higher energy component of auroral electrons. Unfortunately the same absorption can correspond to a wide range of auroral brightness (or energy flux) dependent upon the incident particle spectrum. The occurrence of absorption can of course be studied without relating it to auroral intensity. As a rough guide the mean value for $A/\sqrt{I(5577)}$ from Eather and Jacka's (1966a) data is about 0.5 dB $(\mathrm{kR})^{-\frac{1}{2}}$. The observed brightness can easily vary by a factor of 10 or more for a given value of absorption.

As with visual auroral observations plots have been made of auroral absorption vs magnetic time and latitude. There is a maximum in average absorption on the oval around 2000 h magnetic dipole time and another by day at 0900 h at around 67° geomagnetic latitudes on the zone of mantle aurora. The observations have been extended to take account of substorm time in the work of Hargreaves (1968) and Jelly (1968, 1970). The substorms appear to originate just before midnight with a daytime maximum at 0600 h growing after about 1 h. The absorption spreads poleward in the midnight sector in the same way as the optical substorm. Hargreaves (1969a, 1970, 1971) has demonstrated conjugate point and local correlation effects with a set of riometers in Canada and Antarctica.

5.6. X-Rays from Auroral Electrons

The higher energy parts of the auroral electron energy spectrum can generate detectable intensities of bremsstrahlung X-rays at balloon heights. These emissions have been extensively studied and the subject has been reviewed by Brown (1966) and Trefall (1971).

5.6.1. Theory of X-ray Production

The differential cross section, $\sigma(E, E_\lambda)$, for the production of an X-ray photon from an incident electron of kinetic energy E, is such that $\sigma(E, E_\lambda)\,\mathrm{d}E_\lambda$ is the effective cross section for the production of a photon in the energy range $\mathrm{d}E_\lambda$ at E_λ from a target nucleus. In general $\sigma(E, E_\lambda)$ is given (Evans, 1955) by an expression of the form

$$\sigma(E,E_\lambda) = \sigma_0 BZ^2(E + m_e c^2)/(E,E_\lambda)\,, \tag{5.6a}$$

where $\sigma_0 = 0.58 \times 10^{-27}$ cm^2, Z is the atomic number of the target nucleus, and B is a quantity which is of generally the order of 10. B is a function of E_λ, E and Z. For auroral electrons $E < m_e c^2 = 511$ keV so the cross section is not independent of E as is the case at higher energies.

The energy loss rate, $-\mathrm{d}E/\mathrm{d}x$ analogous to the quantities discussed in Section 4.2.1 is given by contributions of the form

$$-\left(\frac{\mathrm{d}E}{\mathrm{d}x}\right)_{\mathrm{rad}} = n\int_0^E \sigma(E,E_\lambda)E_\lambda\,\mathrm{d}E_\lambda \tag{5.6b}$$

analogous to Equations (4.2.1r) and (s).

In the non-relativistic case this reduces to

$$n\sigma_0 Z^2 m_e c^2 \int_0^1 B\,\mathrm{d}\left(\frac{E_\lambda}{E}\right) = n\sigma_0 Z^2 m_e c^2 \bar{B}, \tag{5.6c}$$

where $\bar{B}(\sim 16/3)$ is an average value of B so that the effective loss function analogous to Equation (4.2.1s) is

$$L_\lambda(E) \simeq 16\sigma_0 Z^2 m_e c^2/3. \tag{5.6d}$$

For air $Z = 7.2$ so $L_\lambda(E) \sim 10^{-19}$ eV cm^2. From the loss function data of Figure 4.21 $L(E)$ is $\sim 10^{-16}$ eV cm^2 for ionization processes at 30 keV so that only about $1/10^3$ of the electron energy is converted to X-rays. As shown by Chamberlain (1961) the ratio in general is about E(MeV)/100.

Values of $\sigma(E, E_\lambda)$ for the energy range appropriate to auroral particles have been calculated by Rees (1964).

When X-rays are produced by auroral electrons having an initial spectral distribution $\Phi_{\mathrm{po}}(E)$, each primary produces a spectrum of X-rays and this spectrum varies as the electron loses energy. The nature of the total spectrum can be determined as follows. The number of X-ray photons of energy E_λ emitted per electron in traversing a thickness dz measured in gm cm^2 is given by

$$\mathrm{d}^2 N(E_\lambda) = \sigma(E,E_\lambda)\,\mathrm{d}E_\lambda\quad(\mathrm{d}z/m),$$

where m is the average atomic mass since (dz/m) is the number of atoms cm^{-2} along the path; hence the total number of such photons along the whole path is

$$\mathrm{d}N(E_\lambda) = \frac{\mathrm{d}E_\lambda}{m}\int_0^{R_0}\sigma(E,E_\lambda)\,\mathrm{d}z = \frac{\mathrm{d}E_\lambda}{m}\int_0^{R_0}\sigma(R,E_\lambda)\,\mathrm{d}R,$$

where R_0 is the initial range of the electron in gm cm^{-2}. This can be integrated by means of a relation between energy and residual range $R = R_0 - z$. For an incident flux of electrons expressed as a function of initial range, the final X-ray distribution will be

$$\frac{\mathrm{d}N(E_\lambda)}{\mathrm{d}E_\lambda} = \frac{1}{m}\int_0^\infty \Phi(R_0)\int_0^{R_0}\sigma(R,E_\lambda)\,\mathrm{d}R\,\mathrm{d}R_0\,. \tag{5.6e}$$

This equation is based on the concept of a straight path for the primary electron and depends upon an empirical range-energy relation. Much of the work which has been done on the basis of Equation (5.6e) has also employed simplified *relativistic* expressions for the $\sigma(E, E_\lambda)$ and approximate range-energy relations. The pioneering work of Anderson and Enemark (1960) and others (for reference see Brown, 1966; Berger and Seltzer, 1972) was based on such approximations. The calculations of Rees (1964), Bailey *et al.* (1970) and Berger and Seltzer have employed a more accurate method. These will be described in the next section since the production and transmission of the bremsstrahlung are considered together.

5.6.2. X-RAY PROPAGATION

The X-rays are produced in the same general height range as auroral ionization, that is, mostly above 90 km. Most observations have been at balloon heights from 30 to 40 km. The X-ray spectrum is thus further modified in passing through the atmosphere. The main loss processes are photoelectric absorption and Compton scattering. Values of the extinction coefficients for these are given for example by Brown (1966). Photoelectric absorption becomes dominant and rises rapidly below 20 keV and consequently the X-ray spectrum observed from balloon heights rapidly cuts off in the 20 to 30 keV region. The transmission from the production region however is a complicated process because of the importance of scattering both Compton and coherent.

Rees (1964) formulated and carried out an approximate solution of the height profiles of X-ray production and absorption. In this treatment the angular distribution of the X-ray photons, the variations in the energy spectrum of primary electrons with height and the variation in the X-ray spectrum with height, were all considered. The electron part of the problem is the same as for auroral primaries as discussed in Section 4.2.1 except that a knowledge of the primary spectrum is crucial. Rees also calculated the height distribution of the ionization produced by auroral bremsstrahlung.

If only the X-ray spectrum well below the production region is of importance, then the height integrated production spectrum should be given by Equation (5.6e). The X-ray flux at lower heights can then be calculated taking into account the absorption and scattering processes. Exact solutions of this problem have been obtained for a number of particular cases by Pilkington and Anger (1971) who employed the Monte Carlo technique to follow the downward path of photons to balloon heights. In general it was found that, if the initial X-ray spectrum has the form

$$\frac{\mathrm{d}N}{\mathrm{d}E_\lambda} \simeq e^{-E_\lambda/E_0},$$

the effective value of E_0, E_0', at balloon heights decreases (i.e., the spectrum softens) although the effect is small for $E_0 < 20$ keV. (Photoelectric absorption, of course, produces a peak at about 40 keV so E_0' refers to the slope of the spectrum at higher energies.) E_0' also depends on the detector 'field of view.' Because of scattering, high count rates are obtained from sources even 75 km away from the balloon especially with omnidirectional detectors. A narrow field detector (vertical) has more spatial resolution but suffers from large shifts in E_0' for forms out of the main field. Pilkington and Anger assumed that the X-rays were produced isotropically at 90 km; they found that there was little change in going to a more realistic height distribution.

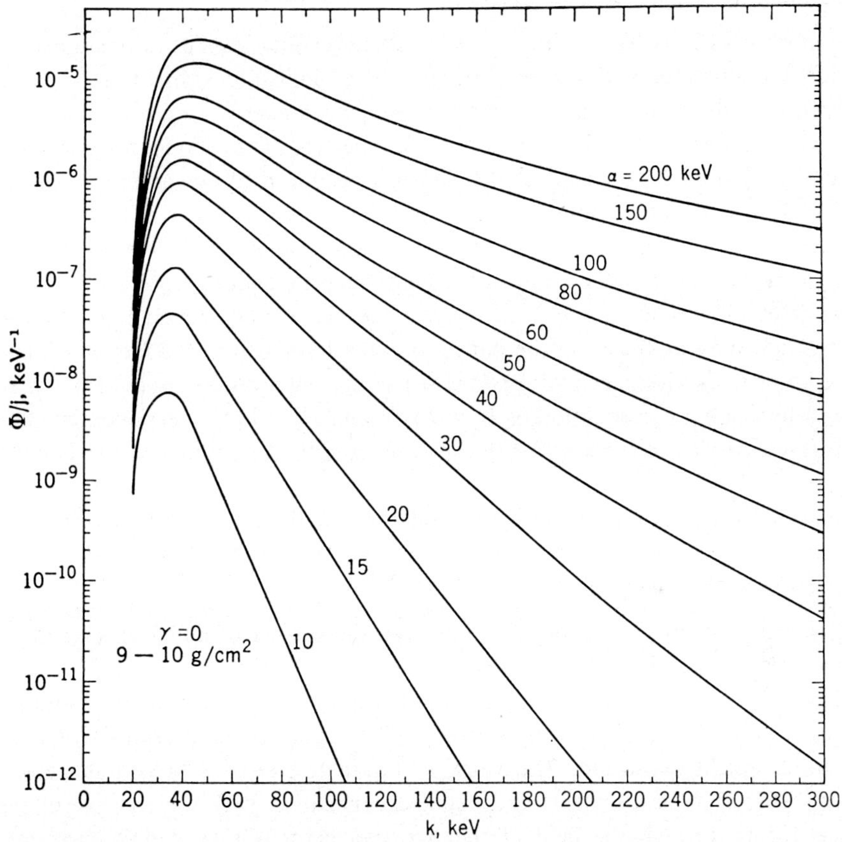

Fig. 5.18. Theoretical Monte Carlo X-ray fluxes at an atmospheric depth of 9–10 g cm^{-2} for incident electron energy spectra of form exp $(-E/\alpha)$ (Berger and Seltzer, 1972). (Courtesy *Journal of Atmospheric and Terrestrial Physics.*)

A more complete treatment of the X-ray spectrum problem has been carried out by Berger and Seltzer (1972) who have added X-ray photon production and transmission into the Monte-Carlo calculation of electron penetration of Berger *et al.* (1970). Figure 5.18 shows the resulting X-ray spectrum at an atmospheric depth of 9 to 10 gm cm^{-2} for a series of isotropic exponential incident electron spectra with various values of E_0.

5.6.3. Observations of Auroral X-Rays

Brown's (1966) and Trefall's (1971) reviews provide a guide to the extensive literature on X-ray observations. Modern balloon borne systems with NaI(Tl) scintillators make it possible to measure the energy spectrum of the photons. Four main features have been considered.

(a) *Relation to Electron Energy Spectrum*

Balloon observations have a much longer duration than rocket measurements and pıovide uninterrupted data near a particular geographical location. Thus in principle they can provide valuable information of the electron energy spectrum. It is obvious however that there is little information on the spectrum for energies less than 30 keV because of absorption of most of the photons produced by lower energy electrons. Nevertheless it is thought that by comparing the observed spectrum with theoretical ones for various initial electron energy distributions that good determinations can be made of the shape of the latter in the higher energy region.

Examples of careful attempts to recover the electron spectrum from the X-ray observations may be found in the papers of Maral (1970b) and Bailey *et al.* (1970). In the latter work riometer and forward scattering observations were also available; it was found that a combination of $E_0 = 9$ keV and $E_0 = 40$ keV exponential electron spectra were needed to account for all the observations of a daytime auroral precipitation event.

Somewhat more qualitative spectral estimates are discussed below under (d).

(b) *Relation to Visual Aurora*

As with radio-aurora this association has been the subject of much uncertainty. The reason is partly the difficulty of observing the same effective field optically as contributes to the X-rays at a balloon (i.e., ~ 100 km radius above the payload) and partly because a hard electron flux with $E_0 > 40$ keV can generate a substantial X-ray flux with little optical excitation. The work of Pilkington *et al.* (1968) and Clark and Anger (1967) showed a definite and detailed correlation with the passage of type-b auroral forms through the field of the X-ray detector. Other definite associations between visual and X-ray auroral pulsations have been found by Hoffmann and Green (1972), Scourfield *et al.* (1970), and Rosenberg *et al.* (1971). Rosenberg (1968) had earlier shown that visual and X-ray aurora are sometimes but not in general well correlated.

A more indirect association is implicit in the work of Bjordal *et al.* (1971) who show a definite relation between a well defined westward electrojet and auroral X-rays. Such an electrojet is almost certainly accompanied by intense aurora.

It is clear that, as with cosmic noise absorption, the degree of correlation with aurora depends strongly on the shape of the electron energy spectrum. Synoptic studies of substorm and time latitude variations should reveal some of the underlying properties of the particle acceleration process.

(c) *Time Variations of X-Ray Emission*

Many striking quasi-periodic phenomena appear in X-ray intensities. Figure 5.19 (taken from Parks *et al.*, 1968) summarizes the main types of variations and their characteristic times of appearance. Somewhat longer period fluctuations (> 50 s) have been reported by Ullaland *et al.* (1967), Brönstad and Trefall (1968), Evans (1963), and Barcus and Christensen (1965). It has been suggested that these slow variations

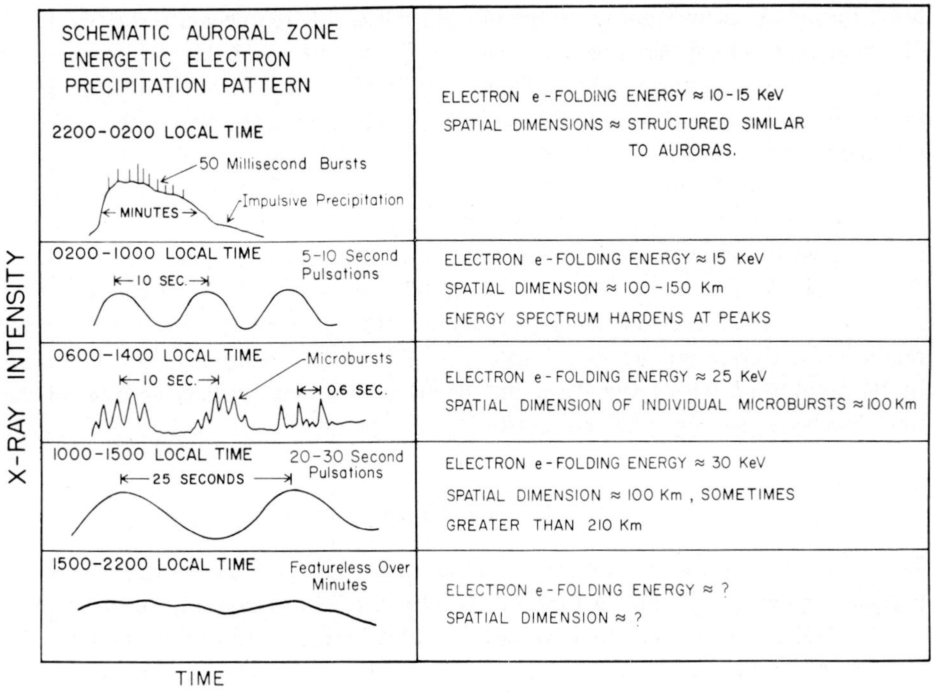

Fig. 5.19. Summary of types of X-ray pulsations and their times of occurrence (Parks *et al.*, 1968). (Courtesy *Journal of Geophysical Research*; copyright American Geophysical Union.)

are due to a modulation of a continuing precipitation process by a hydromagnetic wave (Section 6.3.3).

The 5 to 15 s period pulsations have been widely studied as detailed by Brown (1966). They appear to be related to auroral and magnetic field pulsations as discussed in the previous section.

The shorter period microbursts are an interesting phenomenon. The parent electron spectra have been studied and correlated with the X-ray microbursts and VLF chorus in a series of papers by Venkatesan *et al.* (1968), Oliven *et al.* (1968), and Oliven and Gurnett (1968). In this work there appeared to be some problems in accounting for the observations in different energy channels of the X-ray detectors in terms of electron microbursts. Maral (1970b), with a higher energy resolution system has reported the analysis of late morning microbursts in terms of a 20 keV exponential spectrum with time delays appropriate to a source at 2500 to 7500 km.

(d) *Synoptic Studies*

The main points in the extensive studies carried out in northern Europe with multiple balloon flights have been summarized by Trefall (1971).

In brief, X-ray emission accompanies substorms in the midnight sector and the emission tends to follow the electrojet region. There may be a build-up at the southern limit during the growth phase. In the morning sector there appears to be two kinds of emissions (i) along the auroral *zone* (i.e. at constant *L*) probably due to energetic trapped particles drifting eastwards from the midnight sector* and (ii) for sufficiently large substorms emissions along the auroral *oval* due to the extension of the substorm into the morning sector.

A systematic rise from ~30 keV to ~50 keV in characteristic energy from midnight to afternoon was reported by Bewersdorff *et al.* (1966) while Jentsch and Kremser (1972) have found that there are X-ray events with characteristic energies below 50 keV and others with much higher energies. The softer kind appears to be associated with the electrojet and substorms while the harder variety bears no such simple relationship. Recent studies of the behavior of X-rays in relation to substorms are by Bjordahl *et al.* (1971) and Pytte and Trefall (1972). An excellent account of this topic has been given by Akasofu (1968).

5.7. Radio Emission from Aurora

There are a number of mechanisms by which radio waves can be emitted by auroral particles or plasmas. Radio emissions in the ELF and VLF regions and less frequently in the VHF and UHF regions have been observed more or less in association with aurora or the auroral oval. A brief summary only of the topic will be given here since it appears likely that such emissions do not originate within the ionosphere as a consequence of auroral precipitation but belong rather to the class of magnetospheric phenomena. The generation mechanisms most widely discussed will be considered first.

5.7.1. Theory

(a) *Radiation from Plasma Oscillations*

The oscillation frequency of electrons against the ions in a plasma is $(1/2\pi)\sqrt{n_e e^2/\varepsilon_0 m_e}$. In fact critical reflection occurs when the incident wave frequency equals the plasma frequency so that Equation (5.4a) gives the electron density necessary for a given frequency of oscillation. Chamberlain (1961) gives references to the theory of radiation from oscillating plasmas.

* The theory of this is discussed in Section 6.2.4.

(b) *Cerenkov Radiation*

This arises when an electron moves through an ionized medium at a velocity which exceeds the phase velocity of electromagnetic radiation in the medium. Again a description of the theory is given by Chamberlain (1961). For emission to be possible the local refractive index must exceed unity which is possible only for certain wave modes and frequency ranges. The most important case is for the extraordinary mode for $\omega < \Omega_e \simeq 2 \times 10^6$ near the Earth (see Table 5.2). This corresponds to frequencies below 300 kHz. Jorgensen (1968) and Lim and Laaspere (1972) have calculated the emission intensities to be expected from auroral electrons with energies down to 100 eV.

(c) *Cyclotron Radiation*

This is due to circular motions at the gyrofrequency of energetic electrons in the magnetic field. Chamberlain (1961) discusses the nature of the radiation in free space. However in the last decade the question of the interaction between VLF waves generated by trapped particles and the energy and pitch angle distribution of these has assumed great importance since the generation of such waves constitutes an instability which limits the stably trapped energetic particle density. This question is discussed in Sections 6.2.4 and 5.

(d) *Synchrotron Radiation*

This is associated with the circular motion of relativistic electrons in the magnetic field. In effect the orbiting electron emits short pulses of emission repeating at the gyrofrequency and consequently having a virtually continuous VHF spectrum. The theory has been applied to the case of trapped electrons by Egan and Peterson (1960) and Peterson and Hower (1963).

5.7.2. Comparison with Observations

The radio emissions having some association with aurora and their probable causes are as follows.

(a) *Chorus*

This consists of multiple short duration rising tone emissions resembling birds waking at dawn. The frequencies range from a few hundred Hz to a few kHz. It is observed at middle and high latitudes. It appears likely to be due to electron cyclotron emission and to be generated in the equatorial region of the magnetosphere. The evidence for its origin and source has been reviewed by Jorgensen (1973) and Helliwell (1973). It appears to be mostly a daytime phenomenon and shows a good conjugate point correlation.

(b) *Auroral Hiss*

This is a band limited white noise peaked around 10 kHz covering the ELF and

VLF regions. Its origin has been a subject of some discussion but the evidence appears to favor Cerenkov radiation generated by electrons in the 100 eV to several keV energy range. The region of generation is still uncertain. Jorgensen (1973) has reviewed the evidence on the mechanism and origin. Helliwell (1973) reported that the conjugate point correlation is not high which suggests generation regions well off the equator although well above the ionosphere.

(c) *VHF and UHF Auroral Emissions*

Emissions at these frequencies have been reported many times over the past 20 yr usually in association with very intense auroral activity. The topic has been reviewed by Ellyett (1969). The evidence is that such emissions do occur but require exceptional conditions. Harang (1969) favored plasma oscillations as an explanation of a 28 MHz emission which he observed just before the sharp onset of auroral absorption in an auroral break-up. The only possible explanation for the 3000 MHz emission appears to be synchroton radiation. As shown by Egan and Peterson (1960) this may also explain the emissions in the 28 MHz range.

5.8. Infrasonic Emission from Aurora

Over the past decade auroral infrasonic waves of 10 to 100 s period have been studied following the first observations by Chrzanowski *et al.* (1961) and Campbell and Young (1963). A recent report on the topic has been published by Wilson (1972). The experimental method of detecting the direction of arrival of the infrasonic (IS) wave front has been described in Section 2.6.4.

5.8.1. Observations

The principal observations are as follows.

(i) The waves are associated with auroral forms moving *equatorwards* at *supersonic* velocity.

(ii) The generating forms are associated with a western electrojet current.

(iii) The direction of arrival of the waves corresponds to sources in the auroral oval. They arrive from the east before magnetic midnight, from the north near midnight and from the west after midnight. These directions correlate with the westward traveling surges, equatorward moving forms and eastward drifting loops observed in substorms in these sectors.

(iv) The waves can travel great distances and the variation in wave direction has been observed as far away as Boulder (geomagnetic latitude $\sim 49°$).

5.8.2. Theory of propagation

The theory of acoustic and gravity waves at low frequencies has been developed by Hines (1960) from whose work it may be concluded that waves with periods of several minutes or less can propagate in the acoustic mode. At the high frequency limit waves with periods of down to about 1 s can be transmitted fairly well ($\geq 10\%$) from the 110 km region (Procunier, 1971). It is clear that the 10 to 100 s period range provides favorable conditions for transmission to the surface. Procunier has reported observations of auroral signals in the 1 to 16 s range.

5.8.3. Theory of Generation

The interpretation of IS signals is still being developed and consequently it is not possible to give a completely quantitative theory of the effect. Wilson and Nichparenko (1967) first showed that the emitting auroral form must move with a supersonic velocity. The moving region of electron precipitation might generate a disturbance by (i) local particle heating, (ii) heating by an associated electric current, or (iii) Lorentz force interaction on the neutral atmosphere. The theory for these processes has been developed by Chimonas and Peltier (1970) in a paper which analyzes the effect of a moving linear heat source and also of a moving region in which a body force (the Lorentz force) is exerted on the neutral gas. The calculations show that the observed IS intensity levels could be produced by quite modest currents and heat sources. The moving heat source explanation should produce IS waves equally effectively for north- or southward motion contrary to observation. The Lorentz force model predicts a negative pulse for poleward moving forms. In fact Wilson's (1972) observations indicate that the sign of the observed pulses for equatorward motion can be positive or negative while nothing is detected in the opposite case. Very recently Swift (1973) has reexamined the question of IS wave generation, propagation and detection. Dispersive transmission and band-pass effects in the detecting system can produce a variety of types of observed IS signals. The Lorentz force and Joule heating effects appear both to be of importance and to be additive for forms moving equatorwards. There may also be other effects such as winds acting to favor wave generation by equatorward moving forms.

CHAPTER 6

MECHANISMS OF PRECIPITATION OF AURORAL PARTICLES

6.0. Introduction

The visible aurora and other ionospheric effects, as well as direct measurements, have shown that substantial particle fluxes with energies of 1 to 100 keV are incident upon the upper atmosphere. As has been noted in Section 1.3 these particles probably originate in the solar wind. The solar wind velocity changes from the supersonic values of Table 3.3 to subsonic at the collision free shock II in Figure 1.1. Here the ordered flow velocity of the protons is largely converted to random motions so that protons and electrons in the magnetosheath region EE have thermal energies of the order of 500 eV and 100 eV respectively. In the case of electrons this energy is largely thermal even in the solar wind (see Table 3.3). In order to achieve the energies of nighttime auroral particles a 50 × or larger increase in energy must be produced. The midday red aurora discussed in Section 3.3.6b may arise from magnetosheath plasma penetrating directly into the dayside cusp or clefts F of Figure 1.1. This aurora alone may be explained without the need for an acceleration mechanism. Thanks to direct satellite explorations of the magnetosphere and the solar wind the framework within which this acceleration process must take place is fairly well defined. It is difficult by direct observation to determine the details of dynamic processes in the magnetosphere-ionosphere system, consequently optical observations of aurora constitute a powerful tool for testing ideas about the processes occurring in magnetospheric substorms. The most challenging task is to understand not only how particles are able to enter the magnetosphere and be accelerated but also to know how the fine structure and motions of auroral forms may be explained.

In order to be able to outline current views on the mechanisms of precipitation of aurora particles it is necessary first to introduce the fundamental ideas on the structure and dynamics of the magnetosphere. We will first consider the steady state magnetosphere and then summarize ideas as to how transient effects could give rise to auroral substorms and the production of auroral forms.

6.1. The Steady State Magnetosphere

The concept of a steadily flowing supersonic solar wind continuously confining the terrestrial magnetic field as shown in Figure 1.1 superseded earlier ideas according to which the sun emitted streams of particles only at the times of geomagnetic disturbances. While it is clear that substorms correspond to perturbations of the steady state it is helpful to consider the latter first.

6.1.1. Shape and Structure

Chapman and Ferraro (see Chapman, 1964) considered the problem of the interaction of a rigid plasma with a plane surface sheet (idealized as a perfect conductor) approaching the Earth from the direction of the Sun. The solution, depicted schematically in Figure 6.1, is instructive. Induced currents flow on the surface in such a way

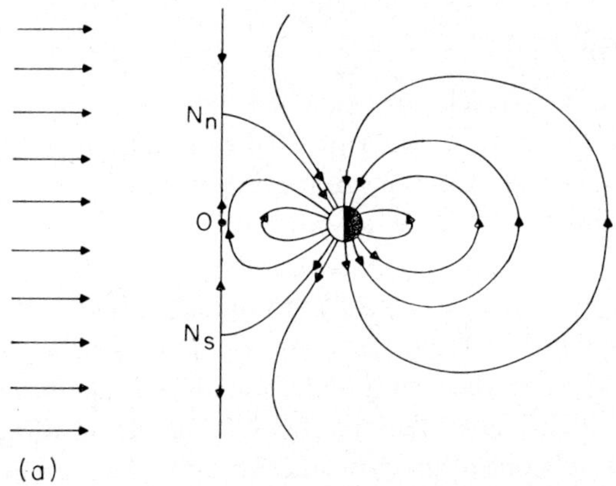

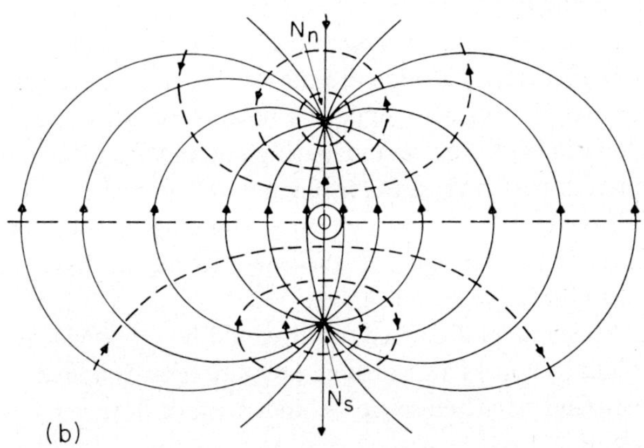

Fig. 6.1. Interaction of rigid plane surface of a conducting plasma with a dipole field: (a) section of field in plane of dipole showing neutral points N_n and N_s. (b) section in plane of 'magnetopause'. The solid lines (although not flux lines) show the direction of **B** in the interface while the dashed lines show the Chapman-Ferraro currents implied by the discontinuity in **B**.

as to confine the field to the right hand side of the surface where the field becomes the sum of that of the original dipole and its image in the plane of the surface. The field at the interface is everywhere tangential to the surface and there are two neutral points around which the induced eddy currents flow. This solution is topologically identical with more realistic ones.

Chapman and Ferraro examined various other exactly soluble cases involving plasma sheets approaching magnetic field regions but were unable to find an analytical solution for the case of plasma streaming past a dipole. However this problem can be solved numerically by means of a plausible assumption sometimes referred to as the Chapman-Ferraro approximation. This postulates that the solar wind particles striking the boundary at an angle of incidence ψ exert a pressure on the boundary given by the relation

$$p = knmv^2 \cos^2 \psi, \tag{6.1a}$$

where n, m and v are the particle concentration, mass and velocity, respectively. k is a constant between 1 and 2. A value of 2 corresponds to the kinetic pressure of particles specularly reflected from the surface. $k = 1$, corresponding to a thermalizing interaction may be more realistic. In any case the angular factor of Equation (6.1a) is the most important. This pressure is balanced by the magnetic pressure $B^2/2\mu_0$ of the tangential field on the other side of the boundary. The surface currents in the boundary, the Chapman-Ferraro currents, are related to $\mathbf{B}$ since curl $\mathbf{B} = \mu_0\mathbf{j}$. On this basis numerical solutions to the steady-state boundary shape have been obtained as reviewed by Mead (1967). These solutions closely resemble the observed configuration in the region (Figure 1.1) out to the nose and for a few Earth radii into the tail. The corresponding two-dimensional problem can be solved exactly (see e.g., Unti and Atkinson, 1968) and the solutions are very instructive. The calculated magnetospheric magnetic field does not match the observed field further back in the tail unless a current sheet (G in Figure 1.1) is added to the model. In fact a current system of the form GH is just what is needed to achieve a balance between the down-wind component of the solar wind pressure on the flairing tail and the attractive force between the dipole and the tail current system. It has not yet proved possible to include the current G in a self-consistent manner in a three dimensional solution of the shape of the magnetosphere. The dynamics of the magnetosphere have been extensively discussed by Siscoe (1966, 1972) and others.

The magnetospheric model discussed so far would be topologically closed. The tail would extend to infinity unless an isotropic pressure term is added to the solar wind in which case the wind could close up behind. In fact however a more important element is missing.

6.1.2. Convective circulation

The existence of an interplanetary magnetic field embedded in the solar wind has very important consequences if, as suggested by Dungey (1961), the merging process sketched in Figure 1.2 can thereby take place. In order to discuss fully the implications of this hypothesis it is convenient to introduce the 'frozen' field approximation. This is valid for low energy plasma for which (if parallel electric fields are negligible) the drift velocity is given by Equation (5.2m). It follows that the drift velocity $\mathbf{v}$, $\mathbf{B}$ and $\mathbf{E}$ are related by the equation

$$\mathbf{v} \times \mathbf{B} + \mathbf{E} = 0. \tag{6.1b}$$

In this case it can be shown (e.g., Hines, 1964), that if a certain quantity of plasma is bounded by a tube of magnetic flux enclosing a certain amount of flux that this amount will remain constant as the plasma drifts and that all plasma initially within a given tube of flux will remain within a common flux tube. In the steady state magnetosphere there will be a steady magnetic field which could be represented in the usual way by a configuration of lines of flux. Then if there are also electric fields perpendicular to the flux lines, the low energy plasma will drift as if the magnetic flux lines threading the plasma were 'frozen' into it with the lines constrained to move in such a way as to leave the field configuration unchanged, i.e., in a flow pattern which involves an interchange in position of the field lines in some circulatory pattern. Several interesting consequences arise as follows:

(i) If the low energy plasma motion is known in one region, then all the plasma on the same tube of force must move with it as mapped along the field lines.

(ii) This is true throughout the region where Equation (5.2m) is satisfied, i.e., for ions down to about 120 km and for electrons down to about 80 km.

(iii) Particles of energy above 1 keV are strongly influenced by additional drifts in inhomogeneous magnetic fields according to the equation

$$\mathbf{v}_\mathrm{d} = \frac{\mathbf{E} \times \mathbf{B}}{B^2} + \frac{\varepsilon_\perp \mathbf{B} \times \nabla B}{qB^3} + \frac{2\varepsilon_\parallel \mathbf{B} \times \mathbf{R}}{qB^2R^2}, \tag{6.1c}$$

where $\varepsilon_\perp$ and $\varepsilon_\parallel$ are the energy associated with the perpendicular and parallel velocities of particle motion, q is the particle charge and $\mathbf{R}$ is the vector radius of curvature of the field line directed from the line to the center of curvature. In a dipole-like field the two right hand terms of Equation (6.1c) give a westward azimuthal drift for protons and an eastward drift for electrons whereas for the low energy drift given by the $\mathbf{E} \times \mathbf{B}/B^2$ term, the velocity is independent of the sign of the charge.

(iv) From Equation (6.1b) it may be seen that the drift paths for energies $\ll 1$ keV are electric equipotential lines.

The transport of ionization within the magnetosphere, in accordance with these principles, is called equipotential *magnetospheric convection*. (The qualifier equipotential implies that the magnetic field lines are assumed to be equipotentials.) The application of these ideas provides a simple way of visualizing the consequences of the interaction between the solar wind and the magnetosphere. Originally Axford and Hines (1961) suggested that there might be a kind of 'viscous' interaction between the solar wind blowing past the magnetopause and the plasma inside. This could set up a plasma circulation which would be away from the sun on both flanks with a sunward return flow down the middle of the tail. This would be something like the pattern of Figure 1.3 except that there would be a region of antisunward flow at the edges. Such a flow, driven at the magnetopause would then map along the field lines right down to the ionosphere according to the principles of plasma convection. It is now generally believed that the circulation is driven by the process of merging of interplanetary and magnetospheric field lines as depicted in Figure 1.2. It appears that about 10% of the interplanetary field lines intercepted by the front of the magnetosphere merge as shown and are dragged back into the tail where they eventually reconnect and flow back toward the nose. Such a magnetosphere has its field lines open to the interplanetary medium for positions between (1) and (6) in Figure 1.2 until the north and south halves of the line reconnect again at (6). In this case the equatorial convection pattern should look strictly like Figure 1.3 since the lines are open and do not intersect the equatorial plane during their tailward flow.

The convective circulation is driven most readily by merging and reconnection if the interplanetary field has a southward component; (the rate of merging is proportional to $\sin\theta/2$ where θ is the angle between the directions of the merging flux lines). The correlation discussed in Section 3.5.3 between the southward component of the interplanetary field and geomagnetic disturbance (thought to be related to enhanced convection) is evidence in favor of the model of Figure 1.2. There are also observations of Anderson and Lin (1969) and Morfill and Scholer (1972) which show a simultaneity in the arrival of fast solar cosmic ray electrons from the tail direction along the open lines connected to the interplanetary field and their appearance in interplanetary space.

6.1.3. The effect of the ionosphere

If there were no interaction between the solar wind and the magnetospheric plasma there would still be a convective circulation (in the frame of the magnetosphere) driven by the Earth's rotation. The neutral atmosphere rotates with the Earth up to levels well above the lower limit for plasma convection (~ 120 km) and consequently according to Section 6.1.2(i) we would expect this motion to be impressed upon the whole magnetosphere since the flow of ionization would be driven by collisions with neutral molecules. Wind systems in the 100 to 200 km range could also impress their motions into the magnetosphere (and vice versa).

In the presence of a solar wind driven convection the two driving forces compete. In this case we must take a more fundamental viewpoint and remember that the convective drifts are caused by a perpendicular electric field which appears according to Equation (6.1b) as soon as any part of the plasma is set in motion. In the steady state case this field can be derived from a potential of which the equipotential surfaces coincide with the flow stream lines and the magnetic field lines. Axford and Hines (1961) assumed that one can then sum in the equatorial plane the potentials which would have been set up by the solar wind driven flow and the rotational flow and thus obtain new equipotential lines which define the resultant flow stream lines. Figure 1.3 was derived in this way. In the region inside the dashed line, the flow is closed and approximates corotation with the Earth while outside the solar wind driven flow from the tail dominates. The lines inside the dotted region are never open, leading to the formation of the plasmasphere as stressed by Nishida (1966) and Brice (1967).

The ionosphere can also affect the magnetospheric convection by shorting out the electric field which is the driving force for convection. As long as the ionospheric conductance is uniform the effect would depend on poorly understood factors influencing the internal resistance of the magnetopause region across which the emf appears. The main effect is likely to be on the shape of the magnetosphere since the convection rate is determined by the rate of merging at the nose (if we assume merging to be the principal driving mechanism). If the ionospheric conductance is not uniform then very striking perturbations of the convection pattern can take place (see e.g., Vasyliunas, 1970). In general one would expect the flow of perpendicular current in the ionosphere to create a stress at the feet of the convecting tubes of force in such a direction as to oppose their motion. This has been called line-tying. Thus one might expect the convective drifts to avoid regions of high conductivity and for perpendicular electric fields to be low in these regions.

6.1.4. The effect of merging on form

The form of the magnetosphere determined in the Chapman-Ferraro approximation may be further modified by merging at the nose since the rate, $\dot{F}_R$, at which the lines

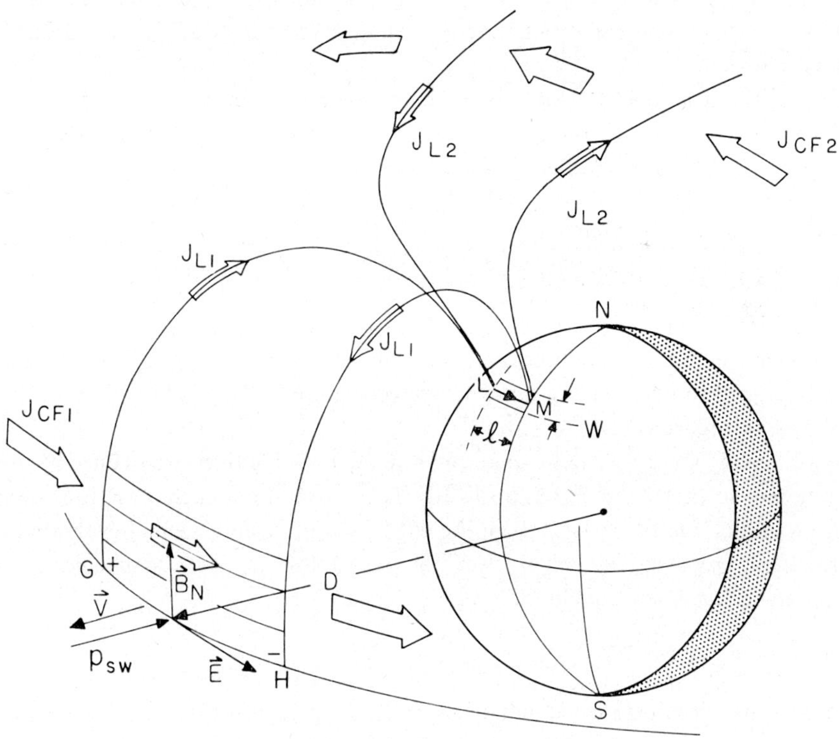

Fig. 6.2. Line-tying currents flowing through dayside cleft, LM, to limit convective return of flux peeled off from the nose, GH, by merging with the southward interplanetary field. The line-tying currents J_L result from the distortion of the magnetopause by merging and enable this distortion to persist by slowing convection.

flow back to the nose from the tail will influence the steady state configuration. Coroniti and Kennel (1973) showed how $\dot{F}_R$ may be estimated.

As they point out, the occurrence of merging at the nose would be expected to affect the radial distance, D, from the Earth's center to the magnetopause even at constant solar wind pressure. The geometry is illustrated in Figure 6.2. It has been observed experimentally that the onset of southward component of the interplanetary field can lead to a decrease in D without any change in the jump in magnetic field, ΔB_N, across the magnetopause at the nose. This implies that Chapman-Ferraro surface current density which is given by $\Delta B_N/2\mu_0$ remains constant even though the nose radius is reduced. The latter change, illustrated in Figure 6.4a and b, implies that the meridinal distance between the dayside cusps decreases so that the total E-ward Chapman-Ferraro current, J_{CF}, flowing between the cusps is decreased. At the same time the W-ward Chapman-Ferraro current at higher latitudes probably increases. Since these currents are closed in the suıface in the non-convecting case, the perturbations caused by the onset of merging are consistent with a diversion of the excess W-ward current down through the dayside cusp region. The steady-state situation would be somewhat as illustrated in Figure 6.2 which shows flux convecting out towards the nose at a velocity, **v**, under the influence of the perpendicular field **E** which must be consistent with the line tying current $J_L = J_{L1} + J_{L2}$ closing across the cusp region.

Coroniti and Kennel (1973) showed by elementary theory that the rate at which flux would flow back to the nose, $\dot{F}_R$, can be expressed in terms of the fractional decrease, r, in the nose radius by the relation

$$\dot{F}_R = -\left(\frac{B_N l D}{4\mu_0 \Sigma_p W}\right) r = -\dot{F}_* r, \tag{6.1d}$$

where B_N is the field just inside the nose, D is the nose radius, Σ_p is the height integrated atmospheric Pederson conductance in the dayside cusp region, l is the length of the cusp (assumed to limit the region of reconnection) and W is the NS width of the cusp as shown in Figure 6.2.

In the steady state $\dot{F}_R$ would equal the rate $\dot{F}_0$ at which lines are opened at the nose by merging so that from Equation (6.1d) $\dot{F}_R/\dot{F}_*$ would equal the fractional decrease in nose radius due to the line stripping effect. Coroniti and Kennel evaluated $\dot{F}_*$ by substituting reasonable values for the variables in the bracketed term of Equation (6.1d). They concluded that

$$\dot{F}_* \simeq 6 \times 10^6/\Sigma_p \text{ weber } s^{-1}.$$

The Pederson conductance is enhanced in the cusp region through the precipitation of electrons as discussed in Section 3.8. A reasonable estimate of the average value across the cusp is 5 mho, which gives a value of $\dot{F}_*$ of about 10^6 Wb $s^{-1} = 10^6$ V.

The merging rate across the nose can be roughly (Levy *et al.*, 1964) estimated by assuming that about 10% of the interplanetary lines intercepted by the nose of the magnetosphere are merged. Assuming a width of 15 R_E an interplanetary field of 20 γ, a solar wind velocity of 500 km s^{-1} one obtains a flux cutting rate of 1×10^5 Wb s^{-1}. This gives $r = 0.1$ or an inward displacement of 1 R_E.

The flux removed from the nose in this steady state is added to the tail. Coroniti and Kennel (1972b) developed a quantitative model of the tail based on this idea. Their theory predicts an increase in tail field as a function of the increased tail flux and other parameters. Unfortunately a completely self-consistent theory of the configuration of even an idealized magnetosphere is still lacking. There are many additional complications which may arise in the real magnetosphere. These include the effect of merging at the sides, of merging of N-ward interplanetary lines beyond the clefts and of the tilt of the Earth's axis relative to the solar wind direction and most important of all, the interaction of plasma within the magnetosphere with convective flow.

6.1.5. Particle Entry and Energization

Although magnetospheric and auroral substorms are obviously deviations from a steady-state condition, it is instructive as a first step to consider particle energization processes which are probably effective whenever convection proceeds. Variations in the NS component of the interplanetary field might indeed alone produce substorms as the convection process starts and stops although it is likely that some additional internal instabilities are involved because of the large energy release in so short a time.

6.1.5.1. *Origin of Auroral Particles*

In principle the particles of the plasma sheet C (Figure 1.1) could originate either from the terrestrial atmosphere or from the solar wind K (Axford, 1970). In the former case the immediate origin could be via the polar caps from which a steady stream of plasma (the polar wind) is lost along the lines of the tail lobes DD. It is conceivable that ions from the polar wind could be trapped into the plasma sheet by reconnection of the tail lobe lines. Solar wind flux is known from direct measurements to penetrate down into the dayside clefts, *F* and could then enter the tail lobes as a result of the nose reconnection process of Figure 1.2. Penetration may also be possible along the outer edges of the plasma sheet where the tail field is weak.

Measurements of the He^{++}/He^{+} ratio by Whalen *et al.* (1971a) favored the solar wind origin of auroral particles. There are now some measurements however of O^{+} ions (Shelley *et al.*, 1972) and of He^{+} ions (Johnson *et al.*, 1974) in the trapping region B and these could indicate at least a component of terrestrial origin.

6.1.5.2. *Adiabatic Energization*

Once a convective flow is set up in the plasma sheet, the drift motion described by Equation (6.1c) can lead to an increase in energy. As pointed out originally by Axford and Hines (1961) this may be described in two equivalent ways. From one viewpoint the energization can be regarded as a kind of adiabatic compression in that a given quantity of plasma confined to a particular tube of magnetic flux is compressed as the tube is convected in from the tail as in Figure 1.3. In the absence of exchange between the velocity components parallel and perpendicular to the field ($v_{\parallel}$ and $v_{\perp}$), there are actually two separate acceleration processes, namely Fermi acceleration of $v_{\parallel}$ due to the decrease in distance between the mirror points of the trapped particles on the flux tube and betatron acceleration of $v_{\perp}$ due to the increase in the field strength. The first gives an increase in energy proportional approximately to l^{-2} where l is the distance between mirror points in the flux tube. For a dipole field this would be proportional to R^{-2} where R is the equatorial distance to the flux tube and for distorted lines in the tail this would be a rough approximation. Betatron acceleration gives an increase proportional to the increase in magnetic field intensity which for a dipole like field is proportional to R^{-3}. In the presence of interchange between the two translational velocity modes, that is of pitch angle scattering, the adiabatic law $TV^{\gamma-1}$ = constant applies with $\gamma = 5/3$, where V is the flux tube volume. Since the particle energy, $E \propto T$ and $V \propto R^4$ the increase in energy is proportional to $R^{-8/3}$.

The second (and equivalent) viewpoint of the adiabatic energization is that it arises from energy gained in crossing equipotential lines of the magnetospheric electric field. As discussed above zero-energy plasma drifts according to the first term of Equation (6.1c) and follows the flow stream lines of Figure 1.2 which are also equipotential lines. Particles with some energy deviate from these stream lines in a dipole-like field

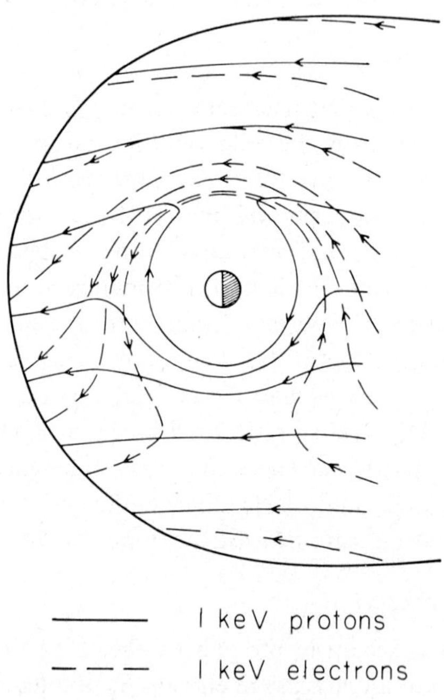

Fig. 6.3. Drift paths of 1 keV electrons (dashed lines) and protons (solid lines) under the influence of magnetospheric magnetic and electric fields after the calculations of Kavanagh *et al.* (1968).

according to the second and third terms of Equation (6.1c) and cross equipotential lines in such a sense as to gain energy. Since the electric field across the magnetosphere is of the order of 30 kV (according to the rate of cutting of interplanetary field flux at the nose as discussed above) the energies gained in the adiabatic energization process are normally limited to 10 to 20 keV.

Figure 6.3 shows drift paths determined by Equation (6.1c) according to a simple model computed by Kavanagh *et al.* (1968). A constant dawn-dusk electric field was assumed together with a corotational field $\sim 1/R^2$. The computed drift paths for ~ 1 keV protons and electrons are shown. The proton trajectories penetrate closer than do those of the electrons. It is clear that the protons and electron trajectories constitute an asymmetric ring current. (The magnetic field adopted is assumed not to be affected by the plasma so that the solution is not a self-consistent one.)

A steady state adiabatic particle energization would produce a constant diffuse band of aurora. Under the assumption of strong pitch angle scattering, the inner edge of the plasma sheet (and thus the equatorial edge of the oval) would occur where the flux tubes are depleted by loss in the atmosphere faster than they are replenished by the convective flow. Such a mechanism might explain the diffuse aurora or night-time mantle aurora discussed in Section 3.2(i) and 3.3.6c. An intermittent convective flow might even explain some aspects of substorms but it seems likely that additional processes are needed to account for the fine structure of auroral arc systems.

6.1.5.3. *Neutral Sheet Energization Processes*

Speiser (1967a, b) has described a mechanism by which protons and electrons can be accelerated by the electric field across the tail of the magnetosphere in the vicinity of a neutral point. It is believed (e.g., Axford, 1969) that this mechanism may be responsible for certain intense electron 'spikes' observed by satellites just above the auroral oval.

6.2. Transient Effects and Substorms

It is clear that auroral particles are precipitated intermittently in association with substorms. Since the interplanetary magnetic field direction appears to control the rate of reconnection and hence of magnetospheric convection it is conceivable that substorms are simply the direct response of the magnetosphere to periods of southward directed interplanetary field. If this were the case one might expect certain transient effects which we will discuss qualitatively, with references to detailed treatments.

6.2.1. Stable Transient Tail Growth

Coroniti and Kennel (1973) have developed a relation giving the response of the magnetosphere to a southward turning of the field leading to a step function in the nose reconnection rate, $\dot{F}_0$, i.e., an increase from 0 to $\dot{F}_0$ at time $t = 0$. The fractional decrease in nose radius, r, and the increase in tail flux ΔF_T are given by the equations

$$r(t) = \dot{F}_0/\dot{F}_*(1 - e^{-t/\tau}) \tag{6.2a}$$

$$\Delta\dot{F}_T = \dot{F}_0\tau(2 - e^{-t/\tau}), \tag{6.2b}$$

where the time constant

$$\tau = (\gamma + 2)M\pi/\dot{F}_*D. \tag{6.2c}$$

The symbols have the same meaning as in Equations (6.1d) to (f). γ is the ratio of total to dayside closed field lines and should be between 1 and 2; it would be 2 if the nightside lines were closed up to as high a latitude as that of the dayside cusp. By using the value of $\dot{F}_*$ of Section 6.1.4, one gets $\tau = 4\Sigma_p$ min if γ is taken to be 1.5. This indicates

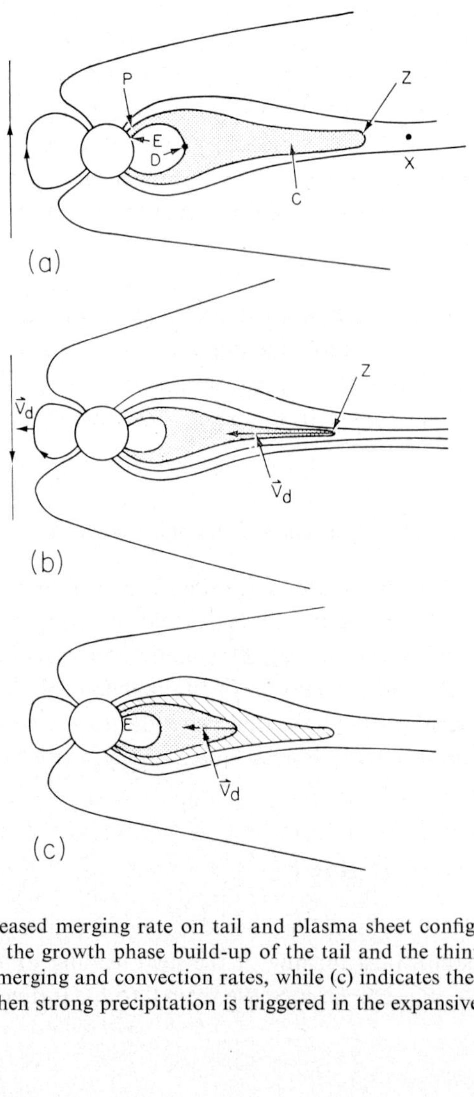

Fig. 6.4. Effect of increased merging rate on tail and plasma sheet configuration: (a) is the initial configuration. (b) shows the growth phase build-up of the tail and the thinning of the plasma sheet caused by the increased merging and convection rates, while (c) indicates the changes which probably occur when strong precipitation is triggered in the expansive phase.

that the step function leads to the erosion of the nose and an increase of flux in the tail with a time constant of about 20 min. If the rate at which flux returns to the nose, $\dot{F}_R$, is assumed to control the rate of convection in the tail then the latter also will increase with same time constant. (This assumption is debatable.) The situation which may arise is illustrated very schematically in Figure 6.4a. The loop in the last closed field line is probably very far down the tail. Reconnection at X is proceeding and there is a steady earthward convection between Z and the inner edge of the plasma sheet at D. If pitch angle scattering maintains isotropy in the plasma in the sheet C there would be a diffuse zone of precipitation between P and E. The effect of the onset of more rapid merging and convection in Figure 6.4b would be to move this zone equatorwards and to increase both the flux and energy of the precipitated particles. Since even in the limit of strong pitch angle scattering the loss of plasma is limited the collapse of the distorted lines in the plasma sheet tends to be blocked by a buildup of a plasma pressure gradient along the plasma sheet. The convective flow in the inner magnetosphere can no doubt still proceed even though the field lines are distorted.

The situation described in the previous paragraph could lead to a quiet sort of substorm with a slow intensification of a broad region of precipitation and an equatorward shift of this region. If the interplanetary field turned north again the situation should revert to the original condition with a similar time constant. The buildup process leading from Figures 6.4a to b has been widely considered to correspond to the growth phase of the magnetospheric substorm.

6.2.2. Instability in Tail Growth

The essential point of most recent auroral theories is that in some way the situation of Figure 6.4b becomes unstable so that a rapid precipitation of plasma occurs. This precipitation probably starts at the equatorial edge of the oval at E in Figure 6.4c and spreads polewards. The decrease in plasma pressure allows reconnection to proceed back at the neutral point in the tail and higher latitude lines ‘spring’ back in towards the Earth compressing and heating the plasma on higher latitude newly closed lines. The instability can continue to spread poleward while at the same time the field becomes more dipolar. This general view of the substorm process is due to Axford who has reviewed its development (Axford, 1969). Russell and McPherron (1973) have reviewed the latest ideas on this subject.

Recently there have been several proposals to explain the nature of the triggering process. Figure 6.5, from a paper by Atkinson (1974) illustrates some basic concepts proposed by Atkinson and by Coroniti and Kennel (1972a). (a) and (b) represent polar and equatorial cross sections during the growth phase and so correspond to Figure 6.4b. As noted earlier the presence of tail-like field lines implies the presence of the cross tail current. This is mostly carried by energetic protons which drift westward according to Equation (6.1c). At the same time as discussed in the previous section there should be a steady precipitation of electrons along the midnight oval region through pitch angle scattering. This precipitation leads to the appearance of a conducting path and hence to an EW Pederson current loop flowing from the magnetosphere down field lines on the down side along the oval and up along field lines on the evening side. Such a current would be driven by the cross tail electric field; along the field lines it would be mostly carried by electrons because of their higher mobility. Coroniti and Kennel (1972a) have discussed this situation and have

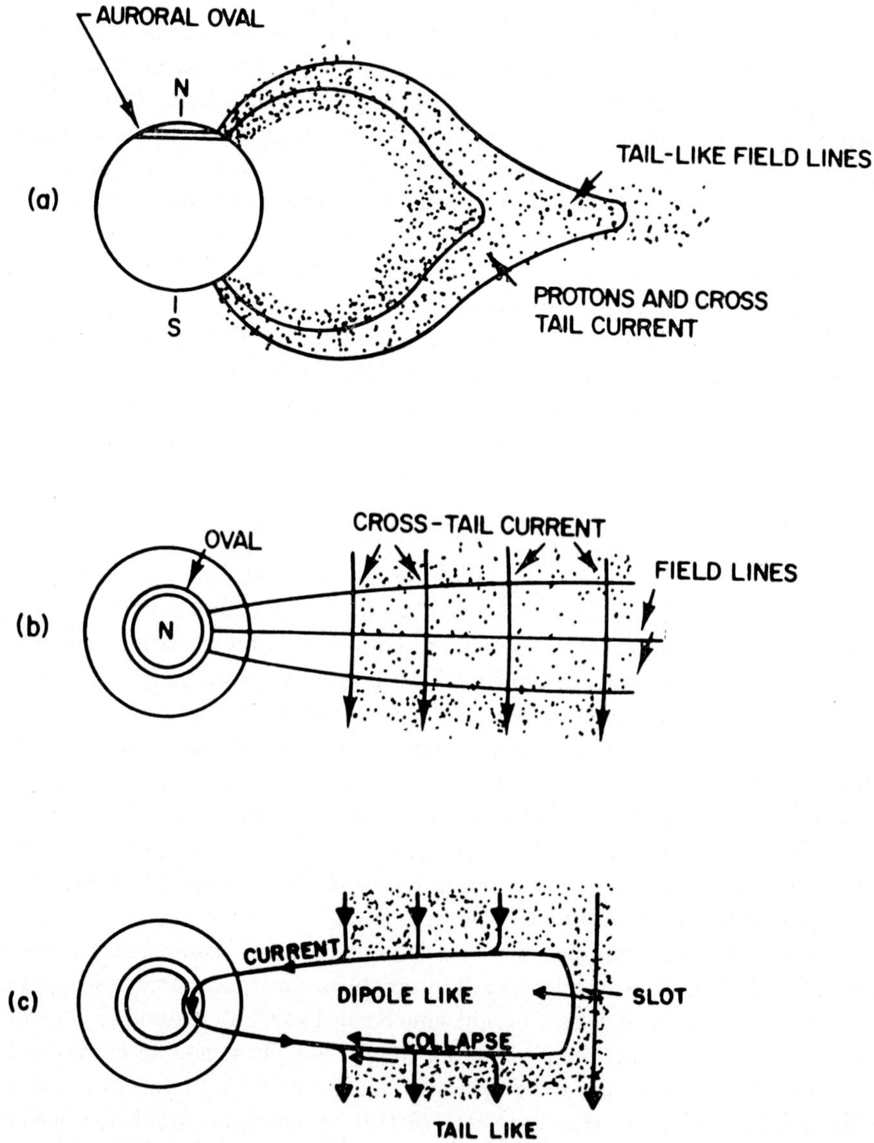

Fig. 6.5. Mechanism for expansive phase tail collapse after Atkinson (1974). (a) and (b) correspond to conditions at the end of the growth phase, and (c) shows the changes occurring with the expansive phase.

suggested that the electric field along the oval would also drive a poleward Hall current across the oval. Because the high conductivity is confined to the precipitation region, this current cannot close completely in the ionosphere but would either (i) close in the magnetosphere and hence give rise to a pair of sheet currents on either side of the precipitation region or (ii) lead to an equatorward polarization field across the 'arc' if the resistance of the magnetosphere loop is too high. Coroniti and Kennel supposed that the upward electron flow of the Hall current density on the equatorward edge might eventually exceed the limit for the electrostatic ion-cyclotron instability leading to the development of a high resistance in the upper ionosphere and thus to condition (ii). In this case the resulting polarization field itself would drive a Hall current along the oval in such a direction as to reinforce the original current. This enhanced (Cowling) current would be much larger than the original Pederson current. This might happen first for a limited longitude range near midnight where the conductivity is largest. McPherron *et al.* (1973), Atkinson (1971) and others suggest that this increased current could be regarded as closing in the current sheet in opposition to the original cross tail current. In other words the cross tail current might be considered to be 'shorted' down through the ionosphere as the enhanced western electrojet of the break-up phase of the substorm. Qualitatively this should lead to the situation in Figure 6.5c in which the field within the 'slot' region of reduced tail current collapses back into dipolar form. According to this viewpoint the plasma on the collapsing field lines is energized and largely lost by precipitation into the auroral oval in the case of electrons or by escape down the tail or westward drift in the case of protons. The slot region mapped along the field lines corresponds to the auroral substorm bulge of Figure 3.4c and the W end whence flows the upward branch of the shorted tail current is identified with the westward traveling surge. Atkinson (1974) suggests that this upward current may be carried by the auroral electrons or alternatively that the auroral electron flux may be precipitated from the slot region as a function of the decrease in tail current. The second alternative is consistent with the hypothesis of Kennel that the precipitating flux is produced by rapid pitch angle diffusion so that its rate is controlled by decrease in bounce time and by adiabatic energization as the flux tubes change from tail-like to dipolar form.

Parks *et al.* (1972) have suggested that the triggering is due to an increase in the ratio of $E_{\perp}/E_{\|}$ for trapped electrons due to betatron acceleration. As shown in the next section this can lead to whistler-mode precipitation (c.f., Equation (6.2i)) which in turn leads to further tail collapse and thus to a positive feedback situation. In this case pitch angle scattering would be weak until at some stage the convective increase in plasma density and energy reaches the stable trapping limit given by Equation (6.2k) when the onset of strong diffusion could initiate the collapse. The fate of the proton component of the plasma may constitute a problem.

It seems very likely that the auroral particle fluxes during the break-up phase are subject to further energization. Sharp *et al.* (1971), in a comparison of electron energy spectra at a synchronous satellite in the equatorial plane and a low level satellite on the same field line, find relatively smooth spectra at the former compared to peaked spectra (e.g., Figure 3.21b) at the latter. This provides direct evidence for acceleration processes operating between the equatorial plane and the ionosphere. Furthermore, the development of auroral forms, particularly the rapid transition of quiet arc to active type-b form with strong fluxes of electrons in the 50 to 100 keV range as attested by the appearance of X-rays and lower borders below 90 km, involves energies higher than can reasonably be obtained by adiabatic energization and suggests the operation of other acceleration processes.

6.2.3. Additional acceleration processes

Some of the mechanisms which probably energize auroral electrons and protons have been considered including Fermi and betatron acceleration and neutral sheet acceleration. It has been suggested that the adiabatic processes provide all the necessary energization and that the details of auroral forms can be explained by mechanisms which change pitch angle distributions. The latter will be discussed in Section 6.2.4. Nevertheless the observations noted at the end of the preceding section as well as the observation of field aligned electron fluxes above some auroral forms (see Section 3.8) argues in favor of some other acceleration process. As pointed out by Haskell and Southwood (1972) it is unlikely that even Fermi acceleration can produce field aligned fluxes just above auroral levels since this implies an unreasonable degree of parallelism in the flux at the equator.

The most plausible additional acceleration process and one which could produce field aligned fluxes involves some form of parallel electric field. The physical situation relating to the existence of such fields has been discussed in Section 5.2.2 where it is noted that the conductance along field lines should be so high that the largest fields to be expected should not exceed 20 μV m^{-1} i.e. $\sim$120 V R_E^{-1}. As implied by Equation (5.1n) the conductance is controlled by the electron collision frequency, i.e., the rate at which the electron can transfer momentum to the ambient plasma. The ion-cyclotron and ion-acoustic wave instabilities provide additional modes of momentum transfer. Hasegawa (1971) has discussed this problem and also suggested that the kink instability of a field aligned current might have an even lower threshold. Once a potential drop of several kV was produced, acceleration of higher energy electrons would be possible since the electron-ion momentum transfer cross section decreases with increasing energy as indicated by the expression for ν_{ei} (Equation (5.2g)). Coroniti and Kennel (1972a) suggest that this may happen with the upward electron Hall current flows discussed in Section 6.2.2. In this case the acceleration would be upward and the electrons would precipitate in the opposite hemisphere assuming they avoided the conjugate field region. The theory of parallel fields arising from parallel current instabilities is complex and not fully understood. It is clearly a topic of great importance.

The reality of accelerations by parallel fields has been questioned by O'Brien (1970) and Sharber and Heikkila (1972), largely on the grounds that (i) mixed proton and electron fluxes are common (ii) that 100 eV and several keV electrons are observed together in the particle streams and that (iii) plasma sheet flux spectra resemble those of auroral particles. These arguments appear to be persuasive as far as concerns the more diffuse or less active types of aurora but do not seem conclusive for the highly structured forms of the expansive phase. There are moreover some particle measurements noted in Section 3.6.3b which show anti-correlations between characteristics of electron and proton fluxes.

Besides quasi-steady electric fields there is also a possibility of excitation by transient fields associated with various types of plasma waves. Such processes have been suggested by Chamberlain (1963, 1969), Swift (1970), Perkins (1968), and D'Angelo (1969).

6.2.4. Particle Trapping and Pitch Angle Scattering

The trajectories of particles trapped in the geomagnetic field are discussed in Appendix 6A. Particles in orbits for which the mirror points are above 500 km have little chance of being scattered by the atmosphere and thus remain stably trapped while those which mirror below 200 km are likely to be lost. The magnetic field of a dipole of magnetic moment, M, is given by

$$B(R,\Lambda) = \frac{M}{R^3}(1 + 3\sin^2\Lambda)^{\frac{1}{2}}, \tag{6.2d}$$

where Λ is the geomagnetic latitude and R is the radial distance to the field point. Consequently the pitch angle at the equator, θ_{eq} where the field is B_{eq} for a particle which mirrors near the surface at radius R_E is from Equation (6A.2)

$$\theta_{eq} = \arcsin\sqrt{\frac{R_E^3}{R_{eq}^3}(1 + 3\sin^2\Lambda)}. \tag{6.2e}$$

For the auroral oval $\theta_{eq} \simeq 4°$ and consequently one might expect that the effect of atmospheric loss would be to remove only those particles in the range $\theta < 4°$ from the distribution measured in the equatorial plane. The loss rate would then become small.

However if some additional process were to operate which could redistribute pitch angles then there could be a steady loss. Fermi and parallel electric field acceleration are examples of such processes on the macroscopic scale. Random small changes in pitch angle may occur as a result of various interactions leading to so-called *pitch angle diffusion* by analogy with translational diffusion. If the diffusion is fast enough to redistribute the pitch angle distribution to isotropy in a time less than the particle bounce period one has the condition of *strong pitch angle diffusion*. In this case the particle loss rate is easily calculated since the probability of loss per bounce becomes $\Omega_{loss}/2\pi$ where Ω_{loss} is the solid angle of the loss cone.

An important mechanism for changing particle distribution functions is interaction with plasma waves. This mechanism was originally proposed by Dungey (1963).

Here we follow the classic paper of Kennel and Petschek (1966) which should be consulted for detailed references and theory. Wave particle interactions are of very great interest for the interpretation of a wide range of phenomena including VLF wave generation, magnetic pulsations and the stability of trapping of electrons and protons in the radiation and ring current belts. This topic was briefly discussed in Section 5.7.1c in connection with the generation of VLF waves by so-called cyclotron radiation. From a physical viewpoint it can be seen that an interaction is effective for a particle with a parallel velocity component v_R when the Doppler shifted frequency of a circularly polarized transverse plasma wave (as seen by the particle) is indentical to the gyrofrequency of the particle. In this case the gyrofrequency, Ω, of the particle is related to the angular frequency and propagation constant of the wave, ω and k respectively, by the equation

$$\omega = \Omega - kv_R\,. \tag{6.2f}$$

Normally electrons interact with right-hand circularly polarized* waves and protons with the left-hand polarization. ω/k is the phase velocity of the wave. For electrons the circularly polarized transverse waves in the frequency region of interest are often called electron cyclotron waves. From Equation (6.2f) and the expression for the wave velocity the energy for interaction, E_R, is found to be

$$E_R = (B^2/2\mu_0 n_e)\,(\Omega_e/\omega)\,(1 - \omega/\Omega_e)^3, \tag{6.2g}$$

where n_e is the plasma density which controls the wave velocity.

Equation (6.2g) gives the necessary energy for interaction with a wave of given frequency but does not indicate whether the wave grows at the expense of particle energy or vice versa. The wave amplitude can be shown to increase as $\exp(\gamma t)$ where

$$\gamma = \pi\Omega_e\left(1 - \frac{\omega}{\Omega_e}\right)^2 \eta(v_R)\left(A - \frac{1}{\dfrac{\Omega_e}{\omega} - 1}\right) \tag{6.2h}$$

where $\eta(v_R)$ is approximately the fraction of the electrons per unit parallel velocity interval in the neighborhood of the resonant velocity. A is a measure of the anisotropy of the velocity distribution and is given by the expression

$$A = \left\langle \frac{\tan\theta}{f}\frac{\partial f}{\partial\theta} \right\rangle_{v_\parallel = v_R}$$

where the average is computed over the range of perpendicular velocities and f is the one-electron velocity distribution function. In the simple case of an anisotropic Maxwellian distribution $f \sim e^{-mv_\perp^2/T_\perp}\, e^{-mv_\parallel^2/T_\parallel}$, $A = (T_\perp - T_\parallel)/T_\parallel$.

For γ to be positive

* In this context a right-hand (extraordinary) polarization is clockwise looking along **B** direction, i.e., rotating in the same sense as the electron cyclotron motion of Figure 5.9.

$$A > (\Omega_e/\omega - 1)^{-1}$$

or

$$E_R > \frac{B^2}{2\mu_0 n_e A(1+A)^2}. \tag{6.2i}$$

Thus for the waves to grow at the expense of electron energy the latter must have a minimum energy given by Equation (6.2i).

The effect of wave energy growth was shown by Brice (1964) to lead to a decrease in pitch angle for the resonant electrons and consequently to their diffusion towards the loss cone. The actual wave growth in the magnetosphere depends on the path length along which the growth rate given by Equation (6.2h) is maintained and the fraction R of the wave amplitude which is reflected back from the foot of the field lines. (The waves are the same type as whistler waves and follow the field lines approximately and are reflected back from the surface.) As shown by Kennel and Petschek (1966) it is reasonable to suppose that the growth is maintained over a magnetic field line segment of length of the order of LR_E centered on the equatorial plane (where L is the McIlwain coordinate). Thus the wave amplitude will grow if

$$e^{\gamma LR_E/v_g} R > 1, \tag{6.2j}$$

where v_g is the wave group velocity since in this case the growth per hop exceeds the reflection loss. Thus if $\eta(v_R)$ should be large enough in Equation (6.2h) to satisfy Equation (6.2j) wave growth will occur with the result that particles of resonant energy will be lost until $\eta(v_R)$ is reduced so that the left hand side of Equation (6.2j) is unity. Thus the maximum stable value of $\eta(v_R)$ is from Equations (6.2j) and (6.2h)

$$\eta_{max}(v_R) = \frac{2v_R\omega \ln(1/R)}{\pi(\Omega_e - \omega)^2 LR_E \left\{A - \dfrac{1}{\Omega_e/\omega - 1}\right\}}. \tag{6.2k}$$

Further applications and extensions of this equation are given by Kennel and Petschek (1966) and Brice and Lucas (1971). Once the stable trapping limit is exceeded for electrons of a given energy, whistler type waves will grow and the electrons will be precipitated. Brice (1971) suggested that this effect could be produced artificially by injecting 'cold' plasma into a flux tube in the auroral oval region so as to increase n_e in Equation (6.2i) and so lower the minimum value for which the resonance can occur. Calculations by Lucas and Brice (1973) suggest that the effect should occur with observed flux distributions.

Brice and Lucas (1971) suggested that the post midnight mantle type of auroral precipitation could be associated with this type of precipitation. There seems to be increasing experimental evidence for the occurrence of this process in causing energetic precipitation in the post midnight hours according to detailed balloon X-ray studies (Foster and Rosenberg, 1973), direct satellite observations (Tsurutani and Smith, 1973), and computer simulation (Matsumoto and Kimura, 1973).

When a strong injection source, such as powerful adiabatic compression maintains a flux level exceeding the Kennel-Petschek limit, then it would be expected that plasma wave resonance would maintain the condition of strong pitch angle diffusion discussed at the beginning of this section.

6.2.5. Proton precipitation

6.2.5.1. *Plasma-Wave-Proton Interactions*

As in the case of electrons a resonance is possible between the cyclotron motion of protons and the electric field of circularly polarized magnetohydrodynamic waves. The most important case is between waves with left-hand polarization (Doppler-shifted up to the proton gyrofrequency). Equations similar to Equations (6.2f) to (6.2k) were given by Kennel and Petschek. The theory of pitch angle diffusion was extended to lower energies by Gendrin (1968). Eather and Carovillano (1971) have discussed the application of these ideas to the precipitation of protons in the $L \simeq 3-7$ region where proton aurora is observed.

The analogue of Equation (6.2i) is

$$E_R > \frac{B^2}{2\mu_0 n_e A^2(1+A)} \tag{6.2l}$$

so that the minimum energy for the interaction rises rapidly in passing outwards from the plasmapause and then falls off at larger L values as B decreases. As with electrons it appears likely that protons in the energy range for proton aurora (10 to 100 keV) will generate ion-cyclotron waves and be precipitated at a rate sufficient to give rise to the diffuse type of proton aurora. At the moment both the theory of the proton wave interactions and observations of magnetospheric ULF waves are in a rapid state of development so that it is difficult to summarize the situation adequately. Some of the factors influencing the precipitation rate have been analyzed by Perraut and Roux (1973). Observations have been reviewed by McPherron *et al.* (1972).

6.2.5.2. *Energization of Protons and the Ring Current*

The protons which are precipitated as described above are presumably convected and energized as described in Section 6.1.5. Figure 6.3 illustrates the convection paths in the absence of wave-particle interactions. This pattern would be somewhat modified in the presence of the latter because of the energy losses associated with ion-cyclotron wave resonance. As pointed out earlier even with strong pitch angle scattering protons reach lower L values than those of electrons because of the longer bounce times of the former and the equatorward location of the zone of proton aurora may be a natural consequence of this effect.

The flux of westward drifting protons makes a major contribution to the ring current or DR current system discussed in Section 3.6.1. It is generally believed that protons are injected during substorms into the ring current belt in the $L = 3$ to 7 region and that they remain trapped for some time, drifting completely round the Earth to give the symmetrical D_{st} depression of the magnetic field. Methods of calculating the effects of such a current in a self-consistent manner were discussed, for example by Kendall *et al.* (1966). The build-up of the ring current by injections proportional to the AE index vs its decay rate has been discussed by Davis (1969b). The fluxes of ring current particles have been measured directly by Frank (1967, 1970) who found that about 75% of the current is carried by protons.

In recent years the ring current has been shown to be asymmetric during substorms. This was measured directly by Frank (1970) who found that the 5 to 50 keV protons detected at $L = 3.5$ to 5.0 on the nightside did not reach the midday sector. Kamide and Fukushima (1972) and Crooker and McPherron (1972) have suggested that the dusk-evening sector ring current closes down through the ionosphere to give the eastern electrojet current discussed in Section 3.6.1. Sugiura (1972) has argued that the quiet time ring current is a continuous extension of the neutral sheet current with 0.1 to 1 MeV protons carrying the current inside the plasmasphere and 10 to 50 keV protons outside. Thus the Earth is surrounded by an equatorial disc of current which extends to large distances in the anti-solar direction to form the current sheet. As shown by Kennel (1969) the 10 keV inner boundary for proton convection should be near $L = 6$ in quiet times and move closer when the magnetospheric electric field increases. Swift (1971) has considered a number of convection type models to account for the formation of the ring current. Cahill and Lee (1973) report from satellite magnetometer observations that the effects of the symmetrical ring current could be observed for as long as nine days after a major magnetic storm.

6.2.5.3. *Other Acceleration Processes for Protons*

Some of the strong proton fluxes reported at break-up (see Section 3.4 and Figure 3.11) and also the rapid increases in proton fluxes observed by synchronous satellite at 6.6 R_E suggest that other acceleration processes may be occurring. Coroniti (1973) has suggested that the downward electric fields postulated in the Coroniti-Kennel theory (Section 6.2.3) of arc polarization could be responsible for accelerating protons. Acceleration by AC electric fields associated with instabilities as discussed in Section 6.2.3 is a further possibility.

6.2.6. SAR- or M-arcs

The excitation of the $\lambda\lambda$6300–64 arcs described in Section 3.3.6f although very different from that of most types of aurora, bears a very interesting relationship to some of the magnetospheric phenomena discussed above. The theory has been fully presented by Rees and Roble (1974) and the following account is a brief outline only of recent theoretical ideas.

The location of SAR-arcs has been found to correspond with the instantaneous location of the plasmapause as shown in Figures 1.1 and 1.2. The currently accepted excitation mechanism is that of thermal excitation of $O(^1D)$ by high electron temperatures conducted in from the magnetosphere. The electrons are believed to be heated in the Landau damping of ion-cyclotron waves generated by the energetic protons of the ring current as discussed in the preceding section. According to this theory as developed by Cornwall *et al.* (1971) the ion-cyclotron waves are generated when the cold plasma of the plasmasphere expands outward into the ring current in the recovery phase of a magnetic storm. The threshold of Equation (6.21) is exceeded and about 50% of the protons energy is lost in generating ion-cyclotron waves. These lose their energy in heating electrons as their wave normal departs from the magnetic field direction in the regions of the field lines off the equatorial plane. The thermal energy so produced is conducted on the field line to the region where the O atomic density is sufficient for excitation according to the appropriate equation of Table 5.3. Chandra *et al.* (1972) have suggested that mixing processes in the lower thermosphere affect the threshold for arc formation. The region above SAR-arcs has been observed to be characterized by high values of $T_e(\simeq 3000\,\mathrm{K})$ but low n_e. Earlier hypotheses of excitation by electric field heating or very soft electron impact excitation do not appear to be valid. A comprehensive experimental test of these ideas is reported in the paper of Nagy *et al.* (1972).

6.3. Fine Structure of Auroral Precipitation

The ideas considered above either do not account at all for the detailed structure of auroral forms or have not been developed sufficiently to provide an explanation. Even the gross structure of quiet aurora, such as multiple arcs is not an obvious consequence of otherwise satisfactory theories. In this section we briefly outline some recent ideas.

6.3.1. Formation of Auroral Arcs

The tendency for particle precipitation to occur in thin arcs even during the quiet phases of substorms is characteristic. If the flux passing through 1 m long slice of an auroral arc 2 km in width is projected back into the tail its radial extent can be estimated from conservation of flux. The flux through such a slice would be about $2000 \times 5.5 \times 10^{-4}$ weber. In the tail the slice might project to a dimension of 10 m in a region near the neutral sheet where the field might not exceed 10 nanotesla (γ). In this case the radial extent must be $\sim 10^4$ km or $\sim 1\ R_E$. Even for an undistorted dipole the width in the tail would be of the order of 100 km at $\Lambda = 70°$. It is thus conceivable that single and multiple arc structures correspond to structure in the plasma of the neutral sheet as was suggested by Akasofu *et al.* (1967). An internal structure of the neutral sheet has been reported recently by Schindler and Ness (1972) but it remains to be seen whether this is related to auroral structures. In any case, as reviewed by Hasegawa (1971) there are possibilities of instabilities in the neutral sheet although it has not yet been shown how these can produce auroral arcs.

A second class of theory invokes interactions with the ionosphere to produce arc forms. Theories of this sort have been proposed by Atkinson (1970) and Ogawa and Sato (1971). In both cases a feedback mechanism is involved whereby perpendicular electric field variations due to ionospheric electron density fluctuations lead to parallel currents. It is assumed that the upward currents flow as downward fluxes of energetic electrons; positive feedback occurs when these precipitate with the right spatial phase to enhance the original electron density fluctuations. Parallel current instabilities may be involved as the acceleration mechanism. The Atkinson and Ogawa-Sato theories differ in assuming that the reactance of the magnetosphere is capacitive and inductive respectively. Both theories predict multiple arc systems with separations which depend on various theoretical parameters.

6.3.2. Deformation of Auroral Arcs

The distortion of regular homogeneous arcs into small scale curls and large scale spirals has been considered by Hallinan and Davis (1970) and Webster and Hallinan (1973). We may consider an arc initially to lie at the foot of a thin 'shell' of magnetic flux lines from which electrons are being precipitated and possibly accelerated by some mechanism such as pitch angle scattering or a parallel electric field. Thus from the upper-F region and perhaps well beyond the precipitating particles constitute a flat beam of downward flowing electrons. This electron beam is assumed to have a small net negative charge. This charge and the beam current give rise to two instabilities.

6.3.2.1. *Charge-Sheet Instability*

A uniformly charged plasma sheet with a uniform magnetic field parallel to its plane is unstable if the charge density exceeds a certain critical value. The following description of the instability is adapted from the paper of Webster and Hallinan (1973). Let Figure 6.6a represent a small random oscillation in a plane current sheet initially parallel to the xz plane. If we assume the neutralizing positive charges lie symmetrically on either side of the sheet, the minimum equipotential line will take a course qualitatively like the dashed line, i.e., it will be a little closer to the inside of the curve at A because of the net effect of the charge at larger values of y in the region of B and C. An extreme case for which there is a simple analytical solution is that of a cylindrical sheet as shown in Figure 6.6b. Here the minimum equipotential line lies at the inside edge. In this case for $b>r>c$ there is a radial inward electric field with $E_r = \rho(r^2 - b^2)/2\varepsilon_0 r$ where ρ is the charge density. According to Equation (5.2m) with **B** in the z direction this leads to a clockwise azimuthal drift of magnitude

$$v_\phi = (\rho/B\varepsilon_0)\,[(r^2 - b^2)/2r]. \tag{6.3a}$$

In the less extreme case of Figure 6.6a one would expect a clockwise drift about the center of curvature, D, above the minimum equipotential line and a smaller anticlockwise drift below. Thus there would be a net flow a negative charge towards C since the drift velocity is independent of charge sign and we are assuming a small excess negative charge density, ρ. The situation at E is symmetrical about C so that there will be a net accumulation of negative charge at C. Then, as originally suggested by Alfvén (1950), the radial inward electric field from this charge center causes a clockwise drift of the plasma sheet which winds itself up into a vortex about C. This

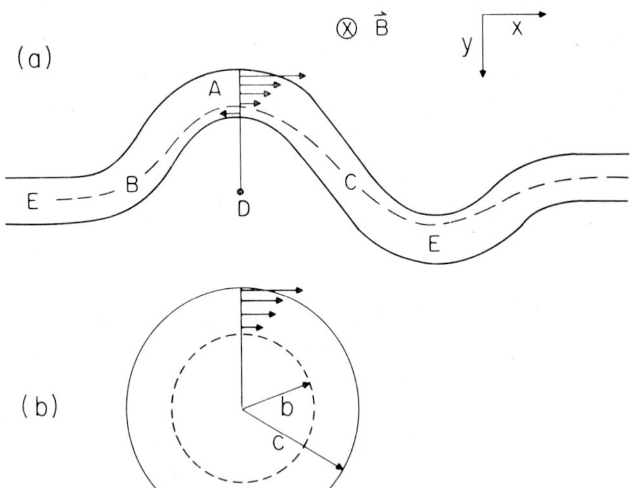

Fig. 6.6. Geometry for change-sheet instability according to Webster and Hallinan (1973).

process may be produced in laboratory electron beams. It seems to be certainly the process causing curls (and hence rays) in auroral arcs. Moreover the reverse curvature of the sheet at B should cause a flow of charge towards E leading to a new center of charge there and a second curl. Thus a series of curls should develop along the arc, as is observed. According to Webster and Hallinan the growth time for a curl is

$$t_k = (2B\varepsilon_0/\rho)(\lambda/d), \tag{6.3b}$$

where d is sheet thickness and λ is the lateral spacing. They point out that the theoretically most probable value of λ/d is 7.8 so that $t_k = 7.8(2B\varepsilon_0/\rho)$ while the time for the sheet to wind up 360° around its center is $2\pi(2B\varepsilon_0/\rho)$. Since the observed time constant for curl formation is of the order of 1/4 s (see Section 3.3.2) the net negative charge density is of the order of 10^{-2} e cm^{-3}. As pointed out by Hallinan and Davis, for arc element thicknesses of 200 m this implies electrostatic fields of several hundred mV m^{-1} towards the outer edge of the charge sheet. A similar value ($\sim$1000 mV m^{-1}) for perpendicular electric fields in the structure may be deduced if the observed 20 km s^{-1} curl velocities are attributed to $\mathbf{E} \times \mathbf{B}$ drifts.

As noted Section 3.3.2 the motions of curls in arc elements on either side of the center line of an arc system may be in opposite directions (E for the equatorial side).

This is a natural consequence of an $\mathbf{E} \times \mathbf{B}$ drift in the electrostatic field and excess negative charges in the arc system.

The charge sheet instability in auroral arcs is thought by Hallinan and Davis to be an *absolute instability*, i.e., the evolution of the structure is independent of the coordinate parallel to the magnetic field.

6.3.2.2. *The Current Sheet Instability*

A 'drift' velocity analogous to the $\mathbf{E} \times \mathbf{B}$ drift in Figure 6.6b appears as a result of the perturbation magnetic field associated with an electric current sheet flowing parallel to the original magnetic field. As an example, suppose that Figure 6.6b represents a cylindrical sheet of current flowing parallel to $\mathbf{B}$, i.e., into the paper. Then this current of density j, produces a azimuthal magnetic field of magnitude

$$B_\phi = \mu_0 j(r^2 - b^2)/2r \tag{6.3c}$$

and consequently, if the particles carrying the current have a velocity v_0 and are guided along the direction of the resultant field, they will have an azimuthal velocity v_ϕ given by

$$v_\phi = \frac{v_0}{B} B_\phi = \frac{\mu_0 j}{B} [(r^2 - b^2)/2r]. \tag{6.3d}$$

This effect was originally discovered by Hallinan and has been discussed by Webster and Hallinan (1973). They give reasons for believing that the magnetic drift should produce a vortex-forming instability analogous to the charge-sheet instability. Thus the characteristic time for the current sheet instability should be

$$t_\mathrm{j} = 2\pi(2B/v\mu_0 j). \tag{6.3e}$$

If values of the velocity and electron current density corresponding to 1 erg cm^{-2} of 10 keV electrons ($\simeq$ IBC3 aurora) are used, t_j comes out to about 10^2 s. Webster and Hallinan thus suggest that the *spirals* described in Section 3.3.2 are due to the current beam instability. Recently Davis and Hallinan (1973) have concluded that the westward traveling surge of the auroral substorm is in fact a such spiral. They conclude that a critical current density of 2.5×10^{-6} A m^{-2} is necessary.

The rotational sense of the current-sheet instability is clockwise for electrons when viewed from below in the northern hemisphere and this agrees with the observed sense for spirals.

6.3.3. Temporal Structure – Pulsing Aurora

The observations of pulsing aurora have been summarized earlier in Sections 3.3.7 and 3.1.3. Possible mechanisms for producing pulsing include:

(i) 'bouncing' of a 'blob' of energetic trapped plasma between conjugate mirroring regions,

(ii) the modulation of pitch angle scattering by ULF hydromagnetic waves, and
(iii) periodic acceleration by electrostatic waves as suggested in Section 6.2.3.

The bounce hypothesis would lead to a definite periodicity and probably to an anticorrelation between conjugate pulsations. Although bounce times of the order of 3 s could explain some observations of pulsations this cannot be so in general. In particular the coincidence in phase between conjugate points noted in Section 3.1.3 would not be expected. Moreover the particles would have to be nearly monoenergetic to maintain the pulsation since the bounce period is energy dependent (Equation (6A.5)).

The observed associations between magnetic pulsations, Earth current fluctuations and auroral pulsations strongly suggest that hydromagnetic waves are capable of modulating the rate of precipitation of energetic trapped electrons. Coroniti and Kennel (1970) have discussed this problem and concluded that direct perturbations of the magnetic field are unlikely to be strong or fast enough to produce precipitation. Instead they have suggested that the ULF disturbance could produce a strong effect on energetic plasma which already exceeded the threshold for whistler-mode instability. Thus the effect of the magnetic pulsation could be to induce periodic episodes of strong pitch angle scattering. The intermittant scattering should occur near the equatorial plane where whistler-wave growth can be rapid. The effect would be in phase at conjugate points. It would be most effective when the energetic plasma was marginally unstable initially so that a small perturbation could produce a large effect. Thus one would expect a considerable accompanying background precipitation.

Experimentally several investigators have sought to find the source region for pulsations by making measurements of the time delay between the arrival of particles of different energy in pulsations. This has been done directly as in the rocket results of Evans (1967) and Bryant *et al.* (1971) or indirectly as in the case of the study of flaming aurora by Carleton (1960) and Cresswell (1969) or the comparison of X-ray and optical pulsations of Rosenberg *et al.* (1971). For the rocket measurements there is in principle a problem in being sure that the observed variations are temporal and not spatial as a result of the motion of the detector perpendicular to the field lines. The measurement of Evans (1967) in what was apparently an event of 10 Hz flickering aurora showed that 16 keV and 120 keV pulsings were in phase and according to the most reasonable interpretation suggest an acceleration or modulation mechanism in the upper ionosphere whereas the direct measurements of Bryant *et al.* (1971) show conclusively that the source of an 8 s periodic event of pulsating aurora was close to the equatorial plane. Flaming aurora can be interpreted as being due to the pulses from a distant source from which the particles arrive with time delays increasing with decreasing energy. Carleton (1960) found the inferred time delay vs energy relation corresponded to an equatorial plane source but Cresswell (1969) concluded that this was not generally true. The X-ray-optical correlation results are likewise not conclusive as to source location although earlier results of Omholt and Pettersen (1967) on the variation of pulsation modulation with frequency favored a distant source. On the whole the evidence points to an equatorial plane or other distant modulation for pulsating aurora but a closer source for the higher frequency effects such as flickering.

The *streaming* or *fast auroral wave* phenomenon discussed in Section 3.3.7 has been tentatively explained by Boyd *et al.* (1972) as being due to the interaction between transverse hydromagnetic waves in the equatorial plane and trapped particles injected in an earlier substorm. They suggest some kind of precipitation mechanism involving the stimulation of reconnection and adiabatic acceleration or by the kind of process discussed by Swift (1970) and considered in Section 6.2.3. It would seem that the Coroniti-Kennel auroral pulsation mechanism would be an attractive possibility.

6.4. Conclusion

Those who profess to investigate the aurora are often asked the status of understanding of its origin. Two decades ago one could answer only very generally in terms of solar electrons and protons being guided by the Earth's magnetic field so as to precipitate in the auroral zones on the night-side. The reasons for the location of the auroral zones and the necessary increase in electronic energy over that inferred from the transit velocity remained obscure. The details of auroral forms were even more mysterious. Today, very reasonable theories to account for general nature of the phenomenon have been advanced and tested while much progress is being made in understanding the details. It is clear that a comprehensive theory of aurora depends on a full understanding of the fascinating processes of the magnetosphere. The time has arrived where comprehensive observations of aurora, particularly from high level satellites, can be interpreted as indications of the instantaneous state of the magnetosphere.

APPENDIX 2A

USE OF FILTER PHOTOMETERS FOR MEASUREMENT OF ABSOLUTE TOTAL INTENSITY OF EMISSION BANDS

The problem frequently arises of using a photometer to measure the total intensity of a spectral feature such as an electronic vibration-rotation band or an atomic multiplet. The relative intensity distribution of the feature to be measured must be known. Suppose $B(\lambda)\,\mathrm{d}\lambda$ is the absolute brightness (e.g., in rayleighs) in the wavelength interval λ to $\lambda+\mathrm{d}\lambda$. ($B(\lambda)$ might be in R Å^{-1} if $\mathrm{d}\lambda$ is in Å). For convenience we define the relative brightness,

$$b(\lambda) = B(\lambda)/\int_0^\infty B(\lambda)\,\mathrm{d}\lambda = B(\lambda)/B_{\mathrm{tot}}. \tag{2A.1}$$

The photometer output, V, (e.g., in mm of chart deflection) is given by

$$V = S\int s(\lambda)\,B(\lambda)\,\mathrm{d}\lambda, \tag{2A.2}$$

where $s(\lambda)$ is the relative photometer instrumental function normalized to unity at the wavelength of maximum sensitivity, λ_{m}, and S is then the absolute sensitivity at λ_{m} (in our example in units of mm R^{-1}).

The instrumental function, $s(\lambda)$ may be conveniently determined by means of a continuous source followed by a monochromator arranged to illuminate a screen large enough to fill the photometer field. Then the spectral brightness of this screen at wavelength λ' is

$$B_{\mathrm{M}}(\lambda,\lambda') = R(\lambda)\,m(\lambda',\lambda) \tag{2A.3}$$

where λ is the wavelength setting (wavelength of maximum output) of the monochromator, $m(\lambda',\lambda)$ is its instrumental function normalized so that $\int_0^\infty m(\lambda',\lambda)\,\mathrm{d}\lambda' = 1$, and $R(\lambda)$ is a factor which varies slowly with wavelength (mostly because of the wavelength variation of monochromator peak transmission and monochromator illuminator brightness). Then the total photometer output from the screen

$$V_{\mathrm{M}}(\lambda) = S\int_0^\infty s(\lambda')R(\lambda)m(\lambda',\lambda)\,\mathrm{d}\lambda'$$

$$\simeq Ss(\lambda)R(\lambda) \tag{2A.4}$$

if the monochromator instrumental function is sufficiently narrow that $s(\lambda')$ may be taken to have a constant value $s(\lambda)$ over the region where $m(\lambda, \lambda')$ differs significantly from zero. At λ_m

$$V_M(\lambda_m) = S\,R(\lambda_m)$$

since $s(\lambda_m) = 1$ and so

$$s(\lambda) = \frac{V_M(\lambda)}{V_M(\lambda_m)} \frac{R(\lambda_m)}{R(\lambda)}. \tag{2A.5}$$

In this way $s(\lambda)$ may be determined from a measurement of the relative response curve to the monochromator-illuminator-screen combination provided that the variation of $R(\lambda)/R(\lambda_m)$ can be estimated. $R(\lambda)$ should be nearly constant for narrow band-pass photometers.

S, the absolute sensitivity may be conveniently determined by measuring the response, V_c, to a calibrated-continuous source (such as a quartz-iodine lamp) illuminating a MgO screen. In this case

$$V_c = S \int_0^\infty s(\lambda) B(\lambda)\,d\lambda$$

$$\simeq SB(\lambda)\,\Delta\lambda_S \tag{2A.6}$$

if the known screen brightness $B(\lambda)$ may be taken as constant across the region where $s(\lambda)$ is non-zero. $\Delta\lambda_S = \int_0^\infty s(\lambda)\,d\lambda$ is the photometer equivalent band-pass. Since V_c and $B(\lambda)$ are known, S may be calculated.

Finally if the photometer is used to measure an auroral emission feature the output will be

$$V = SB_{tot} \int_0^\infty s(\lambda) b(\lambda)\,d\lambda$$

$$= SB_{tot}\varepsilon. \tag{2A.7}$$

The integral term can be regarded as an efficiency factor, ε, for measuring the emission feature. Clearly if $b(\lambda)$ is narrow compared to $s(\lambda)$ and located at the wavelength of maximum sensitivity, then $\varepsilon = 1$.

If two different emission features of total brightness B_1 and B_2 overlap, then B_1 and B_2 can in principle be determined by the outputs of two photometers since then

$$V_1 = S_1 B_1 \varepsilon_{11} + S_1 B_2 \varepsilon_{12}$$

$$V_2 = S_2 B_1 \varepsilon_{21} + S_2 B_2 \varepsilon_{22} \tag{2A.8}$$

where ε_{12} is the efficiency of photometer 1 for emission feature 2, etc.

The reliability of emission intensities derived from filter photometer observations is dependent on accurate determinations of the $s(\lambda)$ and $b(\lambda)$ functions. For the latter a reliable *a priori* knowledge of the spectral distribution is required. In general one photometer channel is required for each independently variable parameter of the spectrum. For example three channels would be needed to measure the intensity of a band of unknown rotational temperature superposed on a continuum of known spectral shape but variable intensity.

APPENDIX 3A

GEOMAGNETIC LATITUDE AND TIME

The magnetic field of the Earth may be represented as the sum of a dipole field plus terms corresponding to poles of higher order.

Details and further references are given by Chamberlain (1961) and Hess (1968). The dipole term gives a fairly good representation of the field well above the surface. Magnetic dipole latitude, Λ_d, and longitude are referred to a coordinate system with its axis along the dipole axis and zero longitude at the geographical north pole.

The real terrestrial field deviates significantly from the dipole and may be represented by various analytical models, such as the International Geomagnetic Reference Field 1965.0 (IAGA, 1969) which uses 80 spherical harmonic coefficients and their time derivatives.

Several more realistic types of geomagnetic coordinates have been used in recent years. The coordinates in these systems give more accurate information about the intersection with the equatorial plane of field lines originating from a point in the atmosphere. (This is not true for higher latitude lines above geomagnetic latitude 55–60° since as shown in Figure 1.1 these are distorted in an unpredictable manner.) Among the improved systems are *eccentric dipole coordinates* (Cole, 1963), equatorial ring coordinates (Mayaud, 1960) and the *corrected geomagnetic* coordinates, Λ_c, of Hakura (1965). Other systems are based on McIlwain's (1961) L parameter. This is precisely defined in the original paper (see also Hess, 1968) but can be taken roughly to give the distance in earth radii at which the real field line with a particular value of L crosses the equatorial plane. The *invariant latitude* defined by the equation,

$$\Lambda_i = \cos^{-1} \sqrt{1/L},$$

would give the latitude from which the dipole-like line would originate to intersect the equatorial plane at a distance of L Earth radii. Since invariant latitude is constant (very nearly) along field lines it must be specified for a particular height in the atmosphere in transforming from geographical coordinates to invariant latitude. *Invariant longitude* is subject to an arbitrary choice of zero and indeed of definition, as may be seen in comparing the systems of Kilfoyle and Jacka (1968) and Evans *et al.* (1969). The latter paper discusses the question in some detail as well as giving extensive graphs and tables to facilitate the transformation.

Geomagnetic time is simple in principle when referred to the dipole field and is the hour angle between the magnetic meridian through the station and through the Sun. Analytical formulae for its calculation have been given by Simonow (1963) and a graphical method for finding magnetic midnight by Montbriand (1965). Formulae for eccentric dipole time have been published by Cole (1963) and Simonow, while Kilfoyle and Jacka (1968) have proposed a simple definition for invariant time. Montbriand (1970) has extended his graphical procedure to give a corrected geomagnetic midnight based on Hakura's corrected geomagnetic coordinates. A simple method of visualizing and determining an approximate geomagnetic time by means of polar projection overlays has been described by Davis (1962).

APPENDIX 3B

MAGNETIC DISTURBANCE INDICES

Several indices of geomagnetic disturbance are in use and detailed definitions are given by Lincoln (1967). The most important points will be outlined here.

3B.1. K and K_p

K is an index ranging from 0.0 to 9.0 and covering 3 h intervals. It is a logarithmic representation of the maximum range of variation among the three components (*D*, *H* and *Z* or *X*, *Y* and *Z*). The conversion is adjusted to give the same range of K independent of the location of the observatory. K_p is essentially an average value of the K's from twelve standard stations distributed in geomagnetic latitude from 48° to 63° and a fairly even distribution in longitude.

3B.2. a_p and A_p

a_p is a linear index ranging from 0 to 400 derived from K_p while A_p is a 24 h average of a_p values.

3B.3. Q and Q_p

Q is a 15 minute logarithmic index based not on the range of variation but rather the maximum deviation from the undisturbed value of the most disturbed of the two horizontal field components. The same scale, rather similar to the K-conversion scale for auroral region stations, is used to convert this deviation to Q values ranging from 1 to 11. Q was compiled during the IGY by a number of higher latitude stations. Q_p is an average value from suitably distributed observatories.

3B.4. AE, AU and AL

These are continuous quantities suggested by Davis and Sugiura (1966) to give a measure of auroral electrojet activity. The magnetometer tracings for the horizontal component from 7 observatories just outside or in the auroral zone are superposed and the envelopes of their positive and negative excursions drawn in. The positive envelope is then AU which gives a measure of eastward electrojet currents while the envelope of negative excursions is AL corresponding to westward electrojet currents. The interval between AU and AL gives AE an overall index of electrojet activity.

3B.5. Equatorial D_{st}

This is the value of the horizontal disturbance component averaged over stations distributed in longitude. By choosing stations at low geomagnetic latitude (but avoiding the region of the equatorial electrojet) D_{st} can be a good index of the strength of the ring current.

APPENDIX 6A

TRAJECTORIES OF PARTICLES TRAPPED IN A DIPOLE FIELD

The motion of charged particles in a dipole field is fundamental to magnetospheric theory. Störmer (1955) investigated and described the trajectory of particles incident toward a dipole in the first attempt at a quantitative theory of aurora. He also discovered that there are certain forbidden regions not accessible from outside and within which particles below certain energies are trapped. These orbits are simply described by the guiding center approximation of Alfvén. As discussed in Section 5.2.1 a charged particle in a uniform magnetic field follows the circular trajectory of Figure 5.9a and may also have a velocity component parallel to **B** which is independent of the field. The motion can be thus resolved into the circular motion and the motion of the center of the circle-the *guiding center*. In a non-uniform magnetic field, such as that of a magnetic dipole, the trajectories become more complicated and the guiding center concept becomes very useful. The motion may be described in terms of paths which leave three quantities invariant. These so-called *adiabatic invariants* may be introduced as follows. A derivation and full discussion are given by Hess (1968).

(1) The first is the magnetic moment, $E_\perp/B$, of the particle around its guiding center. Thus in a motion having also a parallel component $v_\parallel$, in a field which varies in strength in the parallel direction

$$\frac{\frac{1}{2}mv^2 \sin^2\theta}{B} = \text{const.}, \tag{6A.1}$$

where v is the total velocity and θ is the pitch angle. In the absence of parallel electric fields the total kinetic energy is constant so that

$$\frac{\sin^2\theta}{B} = \text{const.} = \frac{1}{B_{\text{max}}}, \tag{6A.2}$$

where B_{max} is the field strength for which the pitch angle becomes 90°. In fact a particle in such a converging field experiences a force tending to drive its guiding center in the direction of decreasing B so that if it initially has a velocity $v_\parallel$ in the direction of increasing B it will be reflected when the field strength reaches B_{max}. If the particle is confined to a flux tube such as AA′ in Figure 6.7 with a minimum field strength at C on the equatorial plane and increasing toward the ends (as for a dipole field) the guiding center of the particle will to a first approximation oscillate back and forth between the two reflection points.

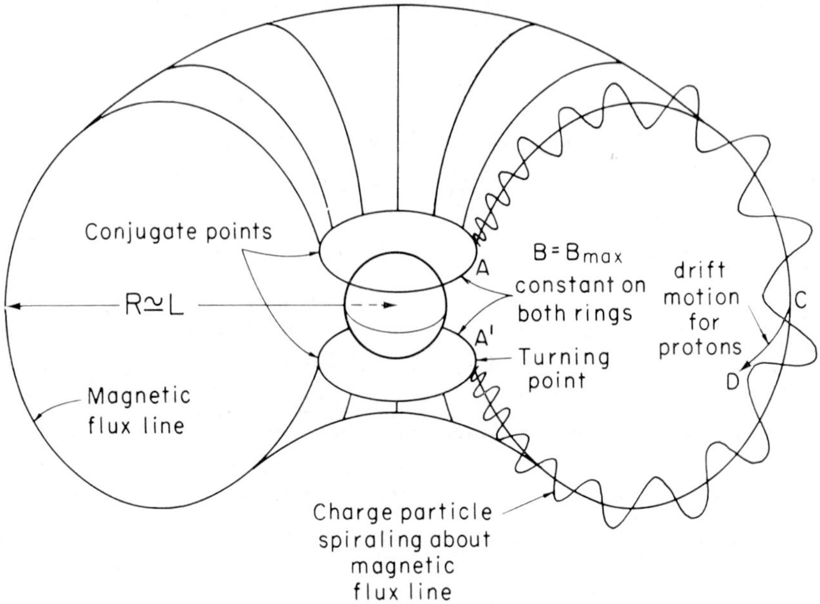

Fig. 6.7. Motion of trapped positive particles in a dipole-like field. A and A′ are conjugate mirror points. CD represents the azimuth drift motion. Adapted from Whalen and McDiarmid (1974).

(2) A second constant quantity is the integral, J of parallel particle momentum along the flux tube between the two mirror points, A, A′ just discussed in (1). Utilizing the first invariant this is given by

$$J = mv \int_{A}^{A'} \sqrt{1 - B/B_{\max}}\, ds \tag{6A.3}$$

If the magnetic field were to change slowly with time so that the distance between the mirror points becomes smaller, the integral would become smaller so mv would have to increase. This is the Fermi energization process which is analogous to the increase in temperature of a gas compressed in cylinder by a piston.

When the field B does not vary with time and in the absence of electric fields the particle momentum mv remains constant so that $I = J/mv$ remains constant. Usually

$$I = \int_{A}^{A'} \sqrt{1 - B/B_{\max}}\, ds \tag{6A.4}$$

is called the *second invariant*. For given values of $B_{\max}$ on a given field line it depends only on field geometry.

The utility of the second invariant arises when the longitudinal drifts due to line curvature and field intensity gradient in a direction perpendicular are taken into account. These effects lead to a drift of the guiding center as given by the two right hand terms of Equation (6.1c). In a quasidipole field a particle thus has a rapid helical motion back and forth between the mirror points of a flux tube and at the same time drifts in longitude perpendicular to the field direction. The drift for protons is shown by CD in Figure 6.7. The first invariant relation ensures that the reflection points remain on surfaces of constant B_{max} while the constancy of the second invariant determines the path of lateral drift. For a perfect dipole field I is obviously constant, by symmetry, along shells of constant equatorial radius. For a distorted dipolar field the shells of constant I will not have constant radius and can also have slightly different loci in the equatorial plane for different values of B_{max}, i.e., for different pitch angles. Non-relativistic protons and electrons of equal energy drift at the same rate on the same shells but in opposite directions, i.e., W for protons and E for electrons. The parameter L, introduced in Appendix 3A, as a coordinate for the distorted dipole field is constant for given particle drift shell (or L-shell) and was in fact defined to have this property.

The bounce and drift periods of auroral particles are of interest. Within 40% the bounce period is given by

$$\tau_B \simeq 4R/v \text{ s.} \tag{6A.5}$$

where R is the equatorial radius of the flux tube on which the particle is trapped and v its velocity. The drift period is given approximately by

$$\tau_D = 44/LE \text{ (min)}, \tag{6A.6}$$

where L is the L parameter ($\sim R/R_E$) for the drift shell and E is the particle energy in MeV. Substitution of $R \simeq 7\ R_E$ for the auroral zone gives for a 10 keV electron a bounce period of about 3 s and a drift period of about 10 h. For a proton of the same energy τ_D is unaltered but τ_B becomes about 2 min.

(3) The third adiabatic invariant is the total flux linked with the drift orbit. This remains constant for changes in B if they are slow compared to the drift period. It is obvious that this condition is easily violated and is consequently of little importance.

REFERENCES

Ackerson, K. L. and Frank, L. A.: 1972, 'Correlated Satellite Measurements of Low-Energy Electron Precipitation and Ground Based Observations of a Visible Auroral Arc', *J. Geophys. Res.* **77**, 1128.

Aggson, T. L.: 1969, 'Probe Measurements of Electric Fields in Space', in B. M. McCormac and A. Omholt (eds.), *Atmospheric Emissions*, Van Nostrand Reinhold, New York, p. 305.

Ahmed, M.: 1962, 'A Study of the Vibrational and Rotational Intensity Distributions in the $\Delta v = -1$ and -2 Sequences of the N_2^+ First Negative System in Aurora', M. Sc. Thesis University of Saskatchewan.

Ajello, J. M.: 1970, 'Emission Cross Sections of N_2 in the Vacuum UV by Electron Impact', *J. Chem. Phys.* **53**, 1156.

Akasofu, S.: 1963, 'The Dynamical Morphology of the Aurora Polaris', *J. Geophys. Res.* **68**, 1667.

Akasofu, S.: 1968, *Polar and Magnetosphere Substorms*, D. Reidel Publishing Co., Dordrecht, Holland.

Akasofu, S.: 1972, 'Midday Auroras and Magnetosphere Substorms,' *J. Geophys. Res.* **77**, 244.

Akasofu, S. and Chapman, S.: 1961, 'The Ring Current, Geomagnetic Disturbance, and the Van Allen Radiation Belts', *J. Geophys. Res.* **66**, 1321.

Akasofu, S. and Snyder, A. L.: 1972, 'Comments on the Growth Phase of Magnetospheric Substorms', *J. Geophys Res.* **77**, 6275.

Akasofu, S. and Yasuhara, F.: 1973, 'Red Auroras in the Morning Sector', *J. Geophys. Res.* **78**, 3027.

Akasofu, S., Chapman, S., and Meng, C. I.: 1965, 'The Polar Electrojet', *J. Atmospheric Terrest. Phys.* **27**, 1275.

Akasofu, S., Chapman, S., and Kendall, P. C.: 1967, 'The Significance of the Multiple Structure of the Auroral Arc', in B. M. McCormac (ed.), *Aurora and Airglow*, Reinhold Publishing Co., New York, p. 281.

Akasofu, S., Kimball, D. S., Buchau, J., and Gowell, A. W.: 1972, 'Alignment of Auroral Arcs', *J. Geophys. Res.* **77**, 4233.

Alcaydé, D.: 1968, 'Calcul des émissions lumineuses associées aux P.C.A. dans les bandes 3914 et 5577 Å', *Ann. Géophys.* **24**, 1031.

Alfvén, H.: 1950, *Cosmical Electrodynamics*, Clarendon Press, Oxford.

Allen, C. W.: 1963, *Astrophysical Quantities*, University of London, The Athlone Press, 2nd Ed., London.

Allen, J. H. and Bender, P. L.: 1972, 'Narrow Line Rb Magnetometer for High Accuracy Field Measurements', *J. Geomag. and Geoelectricity* **24**, 105.

Anderson, A. D. and Sharp, G. W.: 1972, 'Neutral Density Measurements Near 400 km by a Microphone Density Gauge on OGO 6 During July 12–15, 1969', *J. Geophys. Res.* **77**, 1878.

Anderson, K. A. and Enemark, D. C.: 1960, 'Balloon Observations of X-Rays in the Auroral Zone II', *J. Geophys. Res.* **65**, 3521.

Anderson, K. A. and Lin, R. D.: 1969, 'Observation of Interplanetary Field Lines in the Magnetotail', *J. Geophys. Res.* **74**, 3953.

Anger, C. D. and Lui, A. T. Y.: 1973, 'A Global View at the Polar Region on December 18, 1971', *Planetary Space Sci.* **21**, 873.

Anger, C. D., Lui, A. T. Y., and Akasofu, S. I.: 1973a, 'Observations of Auroral Oval and a Westward Traveling Surge from the ISIS 2 Satellite and the Alaskan Meridian All-Sky Cameras', *J. Geophys. Res.* **78**, 3020.

Anger, C. D., Fancott, T., McNally, J., and Kerr, H. S.: 1973b, 'ISIS-II Auroral Scanning Photometer', *Appl. Opt.* **12**, 1753.

Ansari, Z.A.: 1964, 'The Aurorally Associated Absorption of Cosmic Noise at College, Alaska', *J. Geophys. Res.* **69**, 4493.

Armstrong, E.B.: 1959, 'The Temperature in the Atmospheric Region Emitting the Nightlow O I 5577 Line and in Regions Above Faint Auroral Arcs', *J. Atmospheric Terrest. Phys.* **13**, 205.

Armstrong, J.C.: 1974, 'Field Aligned Currents in the Magnetosphere', in B.M. McCormac (ed.), *Magnetospheric Physics*, D. Reidel Publishing Co., Dordrecht, Holland, p. 155.

Armstrong, J.C. and Zmuda, A.J.: 1970, 'Field Aligned Current at 1100 km in the Auroral Region Measured by Satellite', *J. Geophys. Res.* **75**, 7122.

Arnoldy, R.L.: 1971, 'Signature in the Interplanetary Medium for Substorms', *J. Geophys. Res.* **76**, 5189.

Atkinson, G.: 1970, 'Auroral Arcs – Result of the Interaction of a Dynamic Magnetosphere with the Ionosphere', *J. Geophys. Res.* **75**, 4746.

Atkinson, G.: 1971, 'Magnetosphere Flows and Substorms', in *Proceedings of Advanced Study Inst. on Magnetosphere-Ionosphere Interactions*, Dalseter, Norway, p. 203.

Atkinson, G.: 1974, 'The Break-up Phase of the Magnetospheric Substorm', *J. Geophys. Res.*, submitted.

Atkinson, R. and Welge, K.H.: 1972, 'Temperature Dependence of $O(^1S)$ Deactivation by CO_2, O_2, N_2 and Ar', *J. Chem. Phys.* **57**, 3689.

Aubry, M.P. and McPherron, R.L.: 1971, 'Magnetotail Changes in Relation to Solar Wind Magnetic Field and Magnetospheric Substorms', *J. Geophys. Res.* **76**, 4381.

Aubry, M.P., Russell, C.T., and Kivelson, M.G.: 1970, 'Inward Motion of the Magnetopause before a Substorm', *J. Geophys. Res.* **75**, 7018.

Aubry, M.P., Kivelson, M.G., McPherron, R.L., and Russell, C.T.: 1972, 'Outer Magnetosphere Near Midnight at Quiet and Disturbed Times', *J. Geophys. Res.* **77**, 5487.

Axford, W.I.: 1968, 'Observations of the Interplanetary Plasma', *Space Sci. Rev.* **8**, 331.

Axford, W.I.: 1969, 'Magnetospheric Convection', *Rev. Geophys.* **7**, 421.

Axford, W.I.: 1970, 'On the Origin of Radiation Belt and Auroral Primary Ions', in B.M. McCormac (ed.), *Particles and Fields in the Magnetosphere*, D. Reidel Publishing Co., Dordrecht, Holland, p. 46.

Axford, W.I. and Hines, C.O.: 1961, 'A Unifying Theory of High Latitude Geophysical Phenomena and Geomagnetic Storms', *Can. J. Phys.* **39**, 1433.

Bailey, D.K., Brown, R.R., and Rees, M.H.: 1970, 'Simultaneous Forward-Scatter, Riometer and Bremsstrahlung Observations of a Daytime Electron Precipitation Event in the Auroral Zone', *J. Atmospheric Terrest. Phys.* **32**, 149.

Balsley, B.B. and Ecklund, W.L.: 1972, 'VHF Power Spectra of the Radar Aurora', *J. Geophys. Res.* **77**, 4746.

Balsley, B.B., Ecklund, W.L., and Greenwald, R.A.: 1973, 'VHF Doppler Spectra of Radar Echoes Associated With a Visual Auroral Form: Observations and Implications', *J. Geophys. Res.* **78**, 1681.

Banks, P.M.: 1966a, 'Collision Frequencies and Energy Transfer-Electrons', *Planetary Space Sci.* **14**, 1085.

Banks, P.M.: 1966b, 'Collision Frequencies and Energy Transfer-Ions', *Planetary Space Sci.* **14**, 1105

Banks, P.M.: 1966c, 'Charged Particle Temperatures and Electron Thermal Conductivity in the Upper Atmosphere', *Ann. Géophys.* **22**, 577.

Banks, P.M., Chappell, C.R., and Nagy, A.F.: 1974, 'A New Model for the Interaction of Auroral Electrons With the Atmosphere; Spectral Degradation, Backscatter, Optical Emission and Ionization', *J. Geophys. Res.* **79**, 1459.

Barcus, J.R. and Christensen, A.: 1965, 'A 75-Sec Periodicity in Auroral Zone X-Rays,' *J. Geophys. Res.* **70**, 5455.

Bardsley, J.N. and Biondi, M.A.: 1970, 'Dissociative Recombination,' *Adv. in Atomic and Mol. Phys.* **6**, 1.

Baron, M.J.: 1973, 'Electron Densities Within Aurorae and Other Auroral E-region Characteristics,' in *Symposium on Magnetospheric Substorms*, IAGA, Kyoto. (Also *Radio Sci.* **9**, 341 (1974).)

Barrington, R.E., Hartz, T.R., and Harvey, R.W.: 1971, 'Diurnal Distribution of ELF, VLF and LF Noise at High Latitudes as Observed by Alouette 2', *J. Geophys. Res.* **76**, 5278.

Bates, D.R.: 1949, 'The Intensity Distribution in the Nitrogen Band Systems Emitted from the Earth's Upper Atmosphere', *Proc. Roy. Soc.* (*London*), *Ser. A*, **196**, 217.

Bates, D. R.: 1960, *The Auroral Spectrum and its Interpretation* Ratcliffe, p. 321.
Bates, H. F., Belon, A. E., and Hunsucker, R. D.: 1973, 'Aurora and the Poleward Edge of the Main Ionosphere Trough', *J. Geophys. Res.* **78**, 648.
Bates, D. R., Massey, H. S. W., and Pearse, R. W. B.: 1948, in *Emission Spectra of the Night Sky and Aurora*, Reports of the Gassiot Committee, Physical Soc. London.
Bates, D. R., McDowell, M. R. C., and Omholt, A.: 1957, 'On the Energy Distribution of Secondary Auroral Electrons', *J. Atmospheric Terrest. Phys.* **10**, 51.
Bates, H. F., Sharp, R. D., Belon A. E., and Boyd, J. S.: 1969, 'Spatial Relationships Between HF Radar Aurora, Optical Aurora and Electron Precipitation', *Planetary Space Sci.* **17**, 83.
Beach, R., Cresswell, G. R., Davis, T.N., Hallinan, T.J., and Sweet, L.R.: 1968, 'Flickering, A 10 cps. Fluctuation Within Bright Auroras', *Planetary Space Sci.* **16**, 1525.
Belon, A. E. and Clark, K. C.: 1959, 'Spectroscopic Observations of the Great Aurora of 10 February 1958. II: Unusual Atomic Features', *J. Atmospheric Terrest. Phys.* **16**, 220.
Belon, A. E., Romick, G. J., and Rees, M.H.: 1966, 'Electron Energy Spectra from Luminosity Profiles of Aurora', *Planetary Space Sci.* **14**, 597.
Benesch, W., Vanderslice, J. T., Tilford, S. G., and Wilkinson, P. G.: 1966a, 'Franck-Condon Factors for Observed Transitions in N_2 above 6 eV', *Astrophys. J.* **143**, 236.
Benesch, W., Vanderslice, J. T., Tilford, S. G., and Wilkinson, P. G.: 1966b, 'Franck-Condon Factors for Permitted Transitions in N_2', *Astrophys. J.* **144**, 408.
Bens, A. R., Cogger, L.L., and Shepherd, G.G.: 1965, 'Upper Atmospheric Temperatures from Doppler Line Widths. III: Observation of the OI Dayglow Emission at 6300 Å', *Planetary Space Sci.* **13**, 551.
Berger, M. J. and Seltzer, S. M.: 1972, 'Bremsstrahlung in the Atmosphere', *J. Atmospheric Terrest. Phys.* **34**, 85.
Berger, M. J., Seltzer, S. M., and Maeda, K.: 1970, 'Energy Deposition by Auroral Electrons in the Atmosphere', *J. Atmospheric Terrest. Phys.* **32**, 1015.
Berger, M. J., Seltzer, S.M., and Maeda, K.: 1974, 'Some New Results on Electron Transport in the Atmosphere', *J. Atmospheric Terrest. Phys.* **36**, 591.
Bering, E. A., Kelley, M.C., and Mozer, F.S.: 1973, 'Split Langmuir Probe Measurements of Current. Density and Electric Fields in an Aurora', *J. Geophys. Res.* **78**, 2201.
Berkey, F. T.: 1968, 'Coordinated Measurements of Auroral Absorption and Luminosity Using the Narrow Beam Technique', *J. Geophys. Res.* **73**, 319.
Bernstein W., Inouye, G. T., Sanders, N. L., and Wax, R.L.: 1969, 'Measurements of Precipitated 1–20 keV Protons and Electrons During Breakup Aurora', *J. Geophys. Res.* **74**, 3601.
Berthelier, J.J. and Sturges, D.J.: 1967, 'Simultaneous Measurements of Electron Density and Temperature in the Northern Auroral Zone', *Planetary Space Sci.* **15**, 1049.
Bewersdorff, A., Dion, J., Kremser, G., Keppler, E., Legrand, J.P., and Riedler, W.: 1966, 'Diurnal Energy Variation of Auroral X-Rays', *Ann. Géophys.* **22**, 23.
Bird, G. R., Clark Jones, R., and Ames, A. E.: 1969, 'The Efficiency of Radiation Detection by Photographic Films; State-of-the-Art and Methods of Improvements', *Appl. Opt.* **8**, 2389.
Bjordal, J., Trefall, H., Ullaland, S., Bewersdorff, A., and Kangas, J.: 1971, 'On the Morphology of Auroral-Zone X-Ray Events. I: Dynamics of Midnight Events', *J. Atmospheric Terrest. Phys.* **33**, 605.
Blamont, J.-E. and Luton, J. M.: 1972, 'Geomagnetic Effect on the Neutral Temperature of the *F* Region During the Magnetic Storm of September 1969', *J. Geophys. Res.* **77**, 3534.
Bloom, A.L.: 1962, 'Rubidium Vapor Magnetometer', *Appl. Opt.* **1**, 61.
Bond, F.R. and Jacka, F.: 1962, 'Distribution of Auroras in the Southern Hemisphere. II: Nightly Probability of Overhead Aurora', *Australian J. Phys.* **15**, 261.
Bond, F.R. and Thomas, I.L.: 1971, 'The Southern Auroral Oval', *Australian J. Phys.* **24**, 97.
Booker, H. G.: 1956, 'Theory of Radar Scattering from Aurora', *J. Atmospheric Terrest. Phys.* **8**, 204.
Borst, W.L.: 1972, 'Excitation of Several Important Metastable States of N_2 by Electron Impact', *Phys. Rev.* **A5**, 648.
Borst, W.L. and Zipf, E.C.: 1970a, 'Cross Section for ē Impact Excitation of the 0,0 1N N_2^+ Band from Threshold to 3 keV', *Phys. Rev*: **A1**, 834.
Borst, W.L. and Zipf, E.C.: 1970b, 'Excitation of O_2^+ 1N Bands by Electron Impact on O_2', *Phys. Rev.* **A5**, 1410.
Boström, R.: 1971, 'Polar Magnetic Substorms', in B. M. McCormac (ed.), *The Radiating Atmosphere*,

D. Reidel Publishing Co., Dordrecht, Holland. p. 357.

Bouchareine, P. and Connes, P.: 1963, 'Interféromètre à champ compensé pour spectroscopie par transformation de Fourier', *J. de Phys.* **24**, 134.

Boyd, J.S., Belon, A., and Romick, G.J.: 1971, 'Latitude and Time Variations in Precipitated Electron Energy Inferred from Measurements of Auroral Heights', *J. Geophys. Res.* **76**, 7694.

Boyd, J.S., Davis, T.N., Brown, N.B., Hallinan, T.J., and Wallis, D.D.: 1972, 'Observations of Fast Auroral Waves', *Planetary Space Sci.* **20**, 437.

Brace, L.H., Mayr, H.B., and Carignan, G.R.: 1969, 'Measurements of Electron Cooling Rates in the Midlatitude and Auroral Zone Thermosphere', *J. Geophys. Res.* **74**, 257.

Brandy, J.H.: 1964, 'Rotational Temperatures of the Auroral 3914 Å Nitrogen Bands vs Height at Churchill, Manitoba', *Can. J. Phys.* **42**, 1793.

Brandy, J.H. and Hill, J.E.: 1964, 'Rapid Determination of Auroral Heights', *Can. J. Phys.* **42**, 1813.

Brekke, A.: 1971, 'On the Correlation Between Pulsating Aurora and Cosmic Noise Absorption', *Planetary Space Sci.* **19**, 891.

Brekke, A.: 1972, 'The Frequency of Pulsating Aurora and its Relationship to Other Characteristic Parameters', *Planetary Space Sci.* **20**, 285.

Brekke, A. and Pettersen, H.: 1972, 'A Possible Method for Estimating any Indirect Process in the Production of the $O(^1S)$ Atoms in Aurora', *Planetary Space Sci.* **20**, 1569.

Brice, N.: 1964, 'Fundamentals of Very Low Frequency Emission Generation Mechanisms', *J. Geophys. Res.* **69**, 4515.

Brice, N.M.: 1967, 'Bulk Motion of the Magnetosphere', *J. Geophys. Res.* **72**, 5193.

Brice, N.: 1971, 'Harnessing the Energy in the Radiation Belts', *J. Geophys. Res.* **76**, 4698.

Brice, N. and Lucas, C.: 1971, 'Influence of Magnetospheric Convection and Polar Wind on Loss of Electrons from the Outer Radiation Belt', *J. Geophys. Res.* **76**, 900.

Brinkmann, R.T. and Trajmar, S.: 1970, 'Electron Impact Excitation of N_2', *Ann. Géophys.* **26**, 201.

Broadfoot, A.L.: 1967, 'Resonance Scattering by N_2^+, *Planetary Space Sci.* **15**, 1801.

Broadfoot, A.L.: 1968, 'Vibrational Development of the Nitrogen Second Positive System in Aurora', *J. Atmospheric Terrest. Phys.* **30**, 305.

Broadfoot, A.L. and Hunten, D.M.: 1964, 'Excitation of N_2 Band Systems in Aurora', *Can. J. Phys.* **42**, 1212.

Brönstad, K. and Trefall, H.: 1968, 'East-West Movements of Pulsating Auroral-Zone X-ray Events', *J. Atmospheric Terrest. Phys.* **30**, 205.

Brown, N.B.: 1971, private communication.

Brown, R.R.: 1964a, 'On the Influence of Composition and Spectra of Solar Cosmic Rays on the Intensity Ratio $I(\lambda 5577)/I(\lambda 3914)$ in Polar-glow Aurora', *Planetary Space Sci.* **12**, 665.

Brown, R.R.: 1964b, 'On the Relationship Between Polar-glow Aurora and Solar Cosmic Ray Fluxes', *J. Atmospheric Terrest. Phys.* **26**, 805.

Brown, R.R.: 1966, 'Electron Precipitation in Auroral Zone', *Space Sci. Rev.* **5**, 311.

Brown, R.R.: 1968, 'Auroral Enhancement of Atomic Nitrogen', *J. Atmospheric Terrest. Phys.* **30**, 55.

Brown, R.R.: 1969a, 'Atomic Nitrogen Ion Distribution and $\lambda 5755$ Emission in the Auroral Ionosphere', *J. Atmospheric Terrest. Phys.* **31**, 601.

Brown, R.R.: 1969b, 'On Green Line Suppression in Type-B Aurora', *J. Atmospheric Terrest. Phys.* **31**, 855.

Bryant, D.A., Courtier, G.M., Skovli, G., Lindalen, H.R., Aarsnes, K., and Måseide, K.: 1970, 'Electron Density and Electron Flux in a Glow Aurora', *J. Atmospheric Terrest. Phys.* **32**, 1695.

Bryant, D.A., Courtier, G.M., and Bennett, G.: 1971, 'Equatorial Modulation of Electrons in Pulsating Aurora', *J. Atmospheric Terrest. Phys.* **33**, 859.

Buchau, J., Gasmann, G.J., and Pike, C.P.: 1972, 'Precipitation Patterns in the Arctic Ionosphere Determined from Airborne Observations', *Ann. Géophys.* **28**, 443.

Bullen, J.M.: 1968, 'Electric Fields in the Aurora', *J. Atmospheric Terrest. Phys.* **30**, 527.

Burch, J.L.: 1972, 'Preconditions for the Triggering of Polar Magnetic Substorms by Storm Sudden Commencements', *J. Geophys. Res.* **77**, 5629.

Burlaga, L.F. and Ogilvie, K.W.: 1969, 'Causes of Sudden Commencements and Sudden Impulses'. *J. Geophys. Res.* **74**, 2815.

Burridge, R.: 1971, 'The Acoustics of Pipe Arrays', *Geophys. J. Roy. Astron. Soc.* **26**, 53.

Burrows, K., Stolarik, J.D., and Heppner, J.P.: 1971, 'Rocket Measurements of the Magnetic Fields Associated With Visual Aurorae', *Planetary Space Sci.* **19**, 877.

Cahill, L.J. and Lee, Y.C.: 1973, 'Structure of the Evening Quadrant Magnetosphere', in *Symposium on Magnetospheric Configuration*, IAGA, Kyoto.

Campbell, W.H.: 1970, 'Rapid Auroral Intensity Fluctuations and Geomagnetic Field Pulsations', *J. Geophys. Res.* **75**, 6182.

Campbell, W.H. and Rees, M.H.: 1961, 'A Study of Auroral Coruscations', *J. Geophys. Res.* **66**, 41.

Campbell, W.H. and Young, J.M.: 1963, 'Auroral-Zone Observations of Infrasonic Pressure Waves Related to Ionospheric Disturbances and Geomagnetic Activity', *J. Geophys. Res.* **68**, 5909.

Carleton, N.P.: 1960, 'Dynamics of Rapid Fluctuation in Auroral Light', Private Communication.

Cartwright, D.C.: 1970, 'Total Cross Sections for Excitation of the Triplet States in Molecular N_2', *Phys. Rev.* **A2**, 1331.

Cartwright, D.C.: 1973a, 'Transition Probabilities for the Meinel Band System of N_2^+', *J. Chem. Phys.* **58**, 178.

Cartwright, D.C.: 1973b, 'Excitation of N_2 and N_2^+ Band Emissions in Normal Auroras', in *Symposium on Aurora and Airglow*, IAGA, Kyoto.

Cartwright, D.C.: 1974, private communication.

Cartwright, D.C., Trajmar, S., and Williams, W.: 1971, 'Vibrational Population of the $A^3\Sigma_u^+$ and $B^3\Pi_g$ States of N_2 in Normal Auroras', *J. Geophys. Res.* **76**, 8368.

Cartwright, D.C., Williams, W., and Trajmar, S.: 1972a, 'The IR Emission Spectrum of N_2 Excited Under Auroral Conditions', *Ann. Géophys.* **28**, 403.

Cartwright, D.C., Trajmar, S., and Williams, W.: 1972b, 'The Excitation of O_2 in Auroras', *Ann Géophys.* **28**, 397.

Cartwright, D.C., Trajmar, S., and Williams, W.: 1973, 'Reply to Comment by Shemansky and Broadfoot', *J. Geophys. Res.* **78**, 2365.

Cartwright, D.C., Pendleton, W.R., and Weaver, L.D.: 1974, 'Auroral Emissions of the N_2^+ Meinel Bands; *J. Geophys. Res*, submitted.

Cauffman, D.P. and Gurnett, D.A.: 1971, 'Double Probe Measurements of Convection Electric Fields With the Injun 5 Satellite', *J. Geophys. Res.* **76**, 6014.

Chabbal, R.: 1953, 'Recherche des meilleures conditions d'utilization d'un spectromètre photoélectrique Fabry-Perot', *J. Rech. CNRS* **24**, 138.

Chamberlain, J.W.: 1961, *Physics of the Aurora and Airglow*, Academic Press, New York and London.

Chamberlain, J.W.: 1963, 'Plasma Instability as a Mechanism for Auroral Bombardment', *J. Geophys. Res.* **68**, 5667.

Chamberlain, J.W.: 1969, 'Electric Acceleration of Auroral Particles', *Rev. Geophys.* **7**, 461.

Chandra, S., Maier, E.J., and Stubbe, P.: 1972, 'The Upper Atmosphere as a Regulator of Subauroral Red Arcs', *Planetary Space Sci.* **20**, 461.

Chapman, S.: 1964, *Solar Plasma, Geomagnetism and Aurora*, Gordon and Breach, Inc. N.Y.

Chapman, S.: 1968, 'Historical Introduction to Aurora and Magnetic Storms', *Ann. Géophys.* **24**, 497,

Chapman, S.: 1969, 'Auroral Science, 1600–1965, Towards its Golden Age', in B.M. McCormac and A. Omholt (eds.), *Atmospheric Emissions*, Van Nostrand-Reinhold, New York. p. 11.

Chapman, S. and Bartels, J.: 1940, *Geomagnetism*, Clarendon Press, Oxford.

Chase, L.M.: 1968, 'Auroral Zone Particle Spectra', *J. Geophys. Res.* **73**, 3469.

Chase, L.M.: 1970, 'Energy Spectra of Auroral Zone Particles', *J. Geophys. Res.* **75**, 7128.

Chimonas, G. and Peltier, W.R.: 1970, 'Bow Wave Generated by an Auroral Arc in Supersonic Motion', *Planetary Space Sci.* **18**, 599.

Choy, L.W., Arnoldy, R.L., Potter, W., Kintner, P., and Cahill, L.J.: 1971, 'Field Aligned Particle Currents Near an Auroral Arc', *J. Geophys. Res.* **76**, 8279.

Chrzanowski, P., Greene, G., Lemmon, K.T., and Young, J.M.: 1961, 'Traveling Pressure Waves Associated With Geomagnetic Activity', *J. Geophys. Res.* **66**, 3727.

Chubb, T.P. and Hicks, G.T.: 1970, 'Observations of the Aurora in the Far Ultraviolet from OGO 4', *J. Geophys. Res.* **75**, 1290.

Chutjian, A., Cartwright, D.C., and Trajmar, S.: 1973, 'Excitation of the $W^3\Delta_u$, $w^1\Delta_u$, $B'^3\Sigma_u^-$ and $a'^1\Sigma_u^-$ of N_2 by Electron impact', *Phys. Rev. Letters* **30**, 195.

Clark, T.A. and Anger, C.D.: 1967, 'Morphology of Electron Precipitation During Auroral Substorms', *Planetary Space Sci.* **15**, 1287.

Clark, K.C. and Belon, A.E.: 1959, 'Spectroscopic Observations of the Great Aurora of 10 February 1958. I: Abnormal Vibration of N_2^+', *J. Atmospheric Terrest. Phys.* **16**, 205.

Clark, K.C. and Romick, G.J.: 1959, 'An Auroral Spectrograph', *J. Opt. Soc. Am.* **49**, 141.

Cole, K. D.: 1963, Eccentric Dipole Coordinates', *Australian J. Phys.* **16**, 423.

Cole, K. D.: 1971, 'Atmospheric Excitation and Ionization by Ions in Strong Auroral and Man-made Electric Fields', *J. Atmospheric Terrest. Phys.* **33**, 1241.

Connes, J. and Gush, H. P.: 1960, 'Étude du ciel nocturne dans le proche infrarouge', *J. de Phys. et Radium* **21**, 645.

Cornwall, J. M., Coroniti, F. V., and Thorne, R. M.: 1971, 'Unified Theory of SAR Arc Formation at the Plasmapause', *J. Geophys. Res.* **76**, 4428.

Coroniti, F. V.: 1973, 'Plasma Turbulence in the Night Side Ionosphere', in *Symposium on Magnetospheric Configuration*, IAGA, Kyoto. (See Program and Abstracts.)

Coroniti, F. V. and Kennel, C. F.: 1970, 'Electron Precipitation Pulsations', *J. Geophys. Res.* **75**, 1279.

Coroniti, F. V. and Kennel, C. F.: 1972a, 'Polarization of the Auroral Electroject', *J. Geophys. Res.* **77**, 2835.

Coroniti, F. V. and Kennel, C. F.: 1972b, 'Changes in Magnetospheric Configuration During the Substorm Growth Phase', *J. Geophys. Res.* **77**, 3361.

Coroniti, F. V. and Kennel, C. F.: 1973, 'Can the Ionosphere Regulate Magnetospheric Convection?', *J. Geophys. Res.* **78**, 2837.

Cresswell, G. R.: 1968, 'Fast Auroral Waves', *Planetary Space Sci.* **16**, 1453.

Cresswell, G. R.: 1969, 'Flaming Auroras', *J. Atmospheric Terrest. Phys.* **31**, 179.

Cresswell, G. R.: 1972, 'The Morphology of Displays of Pulsating Auroras', *J. Atmospheric Terrest. Phys.* **34**, 549.

Cresswell, G. R. and Belon, A. E.: 1966, 'Observations of Fast Auroral Waves', *Planetary Space Sci.* **14**, 299.

Crooker, N. U. and McPherron, R. L.: 1972, 'On the Distinction Between the Auroral Electrojet and Partial Ring Current Systems', *J. Geophys. Res.* **77**, 6886.

Crosswhite, H. M. and Fastie, W. G.: 1962, 'Far-Ultraviolet Auroral Spectra', *J. Opt. Soc. Am.* **52**, 643.

Culp, G. and Stair, A. T.: 1967, 'Effective rotational Temperatures of N_2^+ (3914 Å) Excited by Monoenergetic Electrons in a Crossed Beam, *J. de Chimie Physique* **64**, 57.

Currie, B. W.: 1947, 'The Polar Aurora', *J. Roy. Astron. Soc. Can.* **41**, 249.

Currie, B. W. and Weaver, J. T.: 1955, 'Vertical Extent of Auroral Arcs', *Can. J. Phys.* **33**, 611.

Czechowsky, P. and Lange-Hesse, G.: 1971, 'Optical and Radio Observations of Aurora', in B. M. McCormac (ed.), *The Radiating Atmosphere*, D. Reidel Publishing Co., Dordrecht, Holland, p. 314.

Dalgarno, A. and Degges, T. C.: 1968, 'Electron Cooling in the Upper Atmosphere', *Planetary Space Sci.* **16**, 125.

Dalgarno, A. and Henry, R. J. W.: 1965, 'Electron Temperature in the *D*-region', *Proc. Roy Soc.* **A288**, 521.

Dalgarno, A. and Khare, S. P.: 1967, 'The $\lambda 5577/\lambda 3914$ Intensity Ratio in Auroras', *Planetary Space Sci.* **15**, 938.

Dalgarno, A. and McElroy, M. B.: 1965, 'The Fluorescence of Solar Ionizing Radiation', *Planetary Space Sci.* **13**, 947.

Dalgarno, A., McElroy, M. B., and Moffett, R. J.: 1963, 'Electron Temperatures in the Ionosphere', *Planetary Space Sci.* **11**, 463.

Dalgarno, A., Latimer, I. D., and McConkey, J. W.: 1965, 'Corpuscular Bombardment and N_2^+ Radiation', *Planetary Space Sci.* **13**, 1008.

Dalgarno, A., McElroy, M. B., and Walker, J. C. G.: 1967, 'The Diurnal Variation of Ionospheric Temperatures', *Planetary Space Sci.* **15**, 331.

Dalgarno, A., McElroy, M. B., and Stewart, A. I.: 1969, 'Electron Impact Excitation of the Dayglow', *J. Atmospheric Sci.* **26**, 753.

D'Angelo, N.: 1969, 'Role of the Universal Instability in Auroral Phenomena', *J. Geophys. Res.* **74**, 909.

D'Angelo, N.: 1973, 'Type III Spectra of the Radio Aurora', *J. Geophys. Res.* **78**, 3987.

Davidson, G. T.: 1965, 'Expected Spatial Distribution of Low-Energy Protons Precipitated in the Auroral Zones', *J. Geophys. Res.* **70**, 1061.

Davis, T. N.: 1962, 'Morphology of Auroral Displays 1957–58', *J. Geophys. Res.* **67**, 59.

Davis, T. N.: 1966, 'The Application of Image Orthicon Techniques to Auroral Observation', *Space Sci. Rev.* **6**, 222.

Davis, T. N.: 1967a, 'Worldwide Auroral Morphology', in B. M. Mc Cormac (ed.), *Aurora and Air-*

glow, Reinhold Publishing Co., New York, p. 41.

Davis, T.N.: 1967b, 'Cinematographic Observations of Fast Auroral Variations', in B.M. McCormac (ed.), *Aurora and Airglow*, Reinhold Publishing Co., New York, p. 133.

Davis, T.N.: 1969a, 'Television Observations of Auroras', in B.M. McCormac and A. Omholt (eds.), *Atmospheric Emissions*, Van Nostrand Reinhold, New York, p. 27.

Davis, T.N.: 1969b, 'Temporal Behavior of Energy Injection into the Geomagnetic Ring Current', *J. Geophys. Res.* **74**, 6266.

Davis, T.N.: 1971, 'Magnetospheric Convection Pattern Inferred from Magnetic Disturbance and Auroral Motions', *J. Geophys. Res.* **76**, 5978.

Davis, T.N.: 1972, 'Auroral Morphology', *Planetary Space Sci.* **20**, 1369.

Davis, T.N. and Hallinan, T.J.: 1973, 'Formation of Auroral Spirals During the Substorm Expansive Phase', in *Symposium on Magnetospheric Substorms*, IAGA, Kyoto.

Davis, T.N. and Hicks, G.T.: 1964, 'Television Cinemaphotography of Auroras and Preliminary Measurements of Auroral Velocities', *J. Geophys. Res.* **69**, 1931.

Davis, T.N. and Sugiura, M.: 1966, 'Auroral Electrojet Activity Index AE and its Universal Time Variations', *J. Geophys. Res.* **71**, 785.

Deehr, C.S. and Egeland, A.: 1972, 'Auroral Morphology', *Ann. Géophys.* **28**, 415.

DeForest, S.E. and McIlwain, C.E.: 1971, 'Plasma Clouds in the Magnetosphere', *J. Geophys. Res.* **76**, 3587.

Derblom, H.: 1964, 'Observations of Sodium Emission in Aurora', *J. Atmospheric Terrest. Phys.* **26**, 791.

Derblom, H.: 1968, 'On the Location of Hydrogen Emission with Respect to Auroral Forms', *Ann. Géophys.* **24**, 163.

Despain, A.M., Baker, D.J., Steed, A.T., and Tohmatsu, T.: 1971, 'Extended-Field Large-Aperture Interferometer-Spectrometer for Airglow Surveys', *Appl. Opt.* **10**, 1870.

Devlin, J.J., Oliver, N.I., and Carrigan, A.: 1964, 'Auroral Spectrograph Data', *Annals. I.G.Y.* **25**, ix.

DeWitt, R.N.: 1962, 'The Occurrence of Aurora in Geomagnetically Conjugate Areas', *J. Geophys. Res.* **67**, 1347.

Dick, K.A.: 1970a, 'Tentative Identification of Several N_2 $c'^1\Sigma_u^+ - a^1\Pi_g$ Bands in Auroras', *J. Geophys. Res.* **75**, 5609.

Dick, K.A.: 1970b, 'Some Spectrometric Results from the NASA 1968 Airborne Auroral Expedition', *J. Geophys. Res.* **75**, 5605.

Dick, K.A. and Fastie, W.G.: 1969, 'Up-Down Photometers for Auroral Profile Studies', *Appl. Opt.* **8**, 2457.

Dick, K.A. and Fastie, W.G.: 1970, 'A Report on Some Results from the NASA 1968 Airborne Auroral Expedition', Johns Hopkins University, Technical Report No. 24, NASA NGR 21-001-001.

Dobrowolski, J.A.: 1959, 'Interference Filters with Transmission Bands of Very Narrow Half-Widths', *J. Optical Soc. Am.* **49**, 794.

Doering, J.P., Fastie, W.G., and Feldman, P.D.: 1970, 'Photoelectron Excitation of N_2 in the Day Airglow', *J. Geophys. Res.* **75**, 4787.

Donahue, T.M. and Strickland, D.J.: 1970, 'Excitation and Radiative Transport of O I 1304 Resonance Radiation. II: The Aurora', *Planetary Space Sci.* **18**, 691.

Donahue, T.M., Zipf, E.C., and Parkinson, T.D.: 1970, 'Ion Composition and Ion Chemistry in an Aurora', *Planetary Space Sci.* **18**, 171.

Donat, H., Rème, H. and Bosqued, J.M.: 1971, 'Interaction des électrons avec l'atmosphère, application au cas d'une aurore diffuse', *Ann. Géophys.* **27**, 1.

Dreyer, J.W. and Perner, D.: 1973, 'Deactivation of $N_2(A^3\Sigma_u^+\ v=0-7)$ by Ground State N_2, C_2H_6, C_2H_4', *J. Chem. Phys.* **58**, 1195.

Druetta, M. and Dufay, M.: 1966, 'Determination of O_2^+ First Negative System Relative Vibrational Transition Probabilities', *Phys. Rev. Letters* **23**, 239.

Dungey, J.W.: 1961, 'Interplanetary Magnetic Field and the Auroral Zones', *Phys. Rev. Letters* **6**, 47,

Dungey, J.W.: 1963, 'Loss of Van Allen Electrons due to Whistlers', *Planetary Space Sci.* **11**, 591.

Duysinx, R. and Monfils, A.: 1972, 'Auroral Spectra Recorded Between 2000 and 3000 Å with a Fast Scanning Spectrometer', *Ann. Geophys.* **28**, 109.

Eather, R.H.: 1966a, 'Radiation from Positive Particles Penetrating the Auroral Atmosphere', *J. Geophys. Res.* **71**, 4133.

Eather, R.H.: 1966b, 'Red Shift of Auroral Hydrogen Profiles', *J. Geophys. Res.* **71**, 5027.
Eather, R.H.: 1967a, 'Secondary Processes in Proton Auroras', *J. Geophys. Res.* **72**, 1481.
Eather, R.H.: 1967b, 'Auroral Proton Precipitation and Hydrogen Emissions', *Rev. Geophys.* **5**, 207.
Eather, R.H.: 1968a, 'Spectral Intensity Ratios in Proton-Induced Auroras', *J. Geophys. Res.* **73**, 119.
Eather, R.H.: 1968b, 'Hydrogen Emission in Pulsating Auroras', *Ann. Géophys.* **24**, 525.
Eather, R.H.: 1968c, 'Some Recent Results on the Measurement of Hydrogen and Helium Emissions in the Auroral Zone', *Ann. Géophys.* **24**, 185.
Eather, R.H.: 1969, 'Short-Period Auroral Pulsations in λ6300 O I', *J. Geophys. Res.* **74**, 4998.
Eather, R.H.: 1970, 'Ionization Produced by Auroral Proton Precipitation', *Ann. Géophys.* **26**, 609.
Eather, R.H. and Akasofu, S.I.: 1969, 'Characteristics of Polar Cap Auroras', *J. Geophys. Res.* **74**, 4794.
Eather, R.H. and Burrows, K.M.: 1966, 'Excitation and Ionization by Auroral Protons', *Australian J. Phys.* **19**, 309.
Eather, R.H. and Carovillano, R.L.: 1971, 'The Ring Current as the Source for Proton Auroras', *Cosmical Electrodynamics* **2**, 105.
Eather, R.H. and Jacka, F.: 1966a, 'Auroral Absorption of Cosmic Radio Noise', *Australian J. Phys.* **19**, 215.
Eather, R.H. and Jacka, F.: 1966b, 'Auroral Hydrogen Emission', *Australian J. Phys.* **19**, 241.
Eather, R.H. and Mende, S.B.: 1971a, 'Airborne Observations of Auroral Precipitation Patterns', *J. Geophys. Res.* **76**, 1746.
Eather, R.H. and Mende, S.B.: 1971b, 'Auroral Precipitation Patterns', in B.M. McCormac (ed.), *The Radiating Atmosphere*, D. Reidel Publishing Co., Dordrecht, Holland. p. 255.
Eather, R.H. and Mende, S.B.: 1972, 'Systematics in Auroral Energy Spectra', *J. Geophys. Res.* **77**, 660
Eather, R.H. and Reasoner, D.L.: 1969, 'Spectrophotometry of Faint Light Sources with a Tilting-Filter Photometer', *Appl. Opt.* **8**, 227.
Eather, R.H. and Sandford, B.P.: 1966, 'The Zone of Hydrogen Emission in the Night Sky', *Australian J. Phys.* **19**, 25.
Ecklund, W.L. and Hargreaves, J.K.: 1968, 'Some Measurements of Auroral Absorption Structure Over Distances of about 300 km and of Absorption Correlation Between Conjugate Regions', *J. Atmospheric Terrest. Phys.* **30**, 265.
Ecklund, W.L., Balsley, B.B., and Greenwald, R.A.: 1973, 'Doppler Spectra of Diffuse Radio Auroras', *J. Geophys. Res.* **78**, 4797.
Egan, R.D. and Peterson, A.M.: 1960, 'Auroral Noise at HF', *J. Geophys. Res.* **65**, 3830.
Egeland, A.: 1962, 'Studies of Auroral Reflections in the VHF Band. II: Comparison of Experimental Results with Theoretical Models', *Arkiv för Geofysik* **4**, 171.
Egeland, A. and Omholt, A.: 1967, 'A Statistical Study of Carl Störmer's Height Measurements of Aurora Borealis', in B.M. McCormac (ed.), *Aurora and Airglow*, D. Reinhold Publishing Co., New York, p. 143.
Egeland, A., Måseide, K., Lindalen, H., Søraas, F., and Evans, D.S.: 1973, 'Proton Precipitation and Hydrogen Emissions in Post Break-up Auroral Glow', in *Symposium on Aurora and Airglow*, Kyoto.
Elliott, A. and Dickson, J.H.: 1951, *Laboratory Instruments*, Chapman and Hall, London.
Ellison, M.A.: 1957, 'Report of Working Group on Flare Classification', *Trans. IAU* **9**, 146.
Ellyett, C.D.: 1969, 'Radio Noise of Auroral Origin', *J. Atmospheric Terrest. Phys.* **31**, 671.
Evans, D.S.: 1963, 'A Pulsating Auroral-Zone X-Ray Event in the 100-Second Period Range', *J. Geophys. Res.* **68**, 395.
Evans, D.S.: 1967, 'A 10-cps Periodicity in the Precipitation of Auroral-Zone Electrons', *J. Geophys. Res.* **72**, 4281.
Evans, D.S.: 1968, 'Observations of a Near Monoenergetic Flux of Auroral Electrons', *J. Geophys. Res.* **73**, 2315.
Evans, D.S.: 1969, 'Fine Structure in the Energy Spectrum of Low Energy Auroral Electrons', in B.M. McCormac and A. Omholt (eds.), *Atmospheric Emissions*, Van Nostrand Reinhold, New York, p. 107.
Evans, D.S.: 1971, 'Direct Observations of Temporal and Spatial Structure in Auroral Electrons', in B.M. McCormac (ed.), *The Radiating Atmosphere*, D. Reidel Publishing Co., Dordrecht, Holland, p. 267.

Evans, D.S.: 1972, 'Auroral Particles and Fields', *Ann. Géophys.* **28**, 639.

Evans, D.S., Jacobsen, T., Maehlum, B.N., Skovli, G., and Wedde, T.; 1972, 'Low Energy Electron Precipitation and the Ionospheric *F* Region in and North of the Auroral Zone', *Planetary Space Sci.***20**, 233.

Evans, J.E., Newkirk, L.L., and McCormac, B.M.: 1969, 'North Polar, South Polar, World Maps and Tables of Invariant Mag. Coords for Six Altitudes 0, 100, 300, 600, 1000, 3000 km', DASA 2347.

Evans, R.D.: 1955, *The Atomic Nucleus*, McGraw-Hill, New York.

Evans, W.F.J. and Vallance Jones, A.: 1965, 'Some Observations of Type-B Red Aurora with a Multi-Channel Photometer', *Can. J. Phys.* **43**, 697.

Eyring, H., Walter, J., and Kimball, G.E.: 1944, *Quantum Chemistry*, Wiley, New York.

Fahleson, U.: 1967, 'Theory of Electric Field Measurements Conducted in the Magnetosphere with Electric Probes', *Space Sci. Rev.* **7**, 238.

Fahleson, U.V., Kelley, M.C., and Mozer, F.S.: 1970, 'Investigation of the Operation of a dc Electric Field Detector', *Planetary Space Sci.* **18**, 1551.

Fairfield, D.H.: 1973, 'Magnetic Field Signatures of Substorms on High Latitude Field Lines in the Nighttime Magnetosphere', *J. Geophys. Res.* **78**, 1553.

Fairfield, D.H. and Ness, N.F.: 1970, 'Configuration of the Geomagnetic Tail During Substorms', *J. Geophys. Res.* **75**, 7032.

Fan, C.Y.: 1958, 'Laboratory Studies of Auroral Excitation', in E.B. Armstrong and A. Dalgarno (eds.), *The Airglow and Aurora*, Pergamon Press, London, p. 276.

Farley, D.T.: 1963, 'A Plasma Instability Resulting in Field Aligned Irregularities in the Ionosphere', *J. Geophys. Res.* **68**, 6083.

Fastie, W.G., Crosswhite, H.M., and Markham, T.P.: 1961, 'Far Ultra-Violet Auroral Spectra with a Rocket Ebert Spectrophotometer', *Ann. Géophys.* **17**, 109.

Fedorova, N.I.: 1961, 'Observations of the (0, 0) 1PG N_2 Bands at 10420 Å and (0, 0) Bands of the Meinel Negative System of N_2^+ at 11109 Å', *Geomag. and Aeronomy* **1**, 702.

Fedorova, N.I.: 1963, 'Determination of Relative Populations of Vibrational Levels and of Rotation Temperature of the $A^2\Pi$–$X^2\Sigma N_2^+$ System Bands', *Aurorae and Airglow, Results of the International Geophysical Year's Researches* **10**, 80.

Fehsenfeld, F.C., Dunkin, D.B., and Ferguson, E.E.: 1970, 'Rate Constants for the Reaction of CO_2^+ with O, O_2 and NO; N_2^+ with O and NO; and O_2^+ with NO', *Planetary Space Sci.* **18**, 1267.

Feibelman, W.A., Hake, R.D., Sipler, D.P., and Biondi, M.A.: 1972, 'Twilight and Nighttime Ionospheric Temperatures from Oxygen λ6300 and λ5577 Spectral Line Profiles', *J. Geophys. Res.* **77**, 1869.

Feldman, P.D., Doering, J.P., and Moore, J.H.: 1971, 'Rocket Measurement of the Secondary Electron Spectrum in an Aurora', *J. Geophys. Res.* **76**, 1738.

Feldstein, Y.I.: 1964, 'Auroral Morphology: I. Location of Auroral Zone', *Tellus* **16**, 252.

Feldstein, Y.I.: 1966, 'Peculiarities in the Auroral Distribution and Magnetic Disturbance Distribution in High Latitudes Caused by the Asymmetrical Form of the Magnetosphere', *Planetary Space Sci.* **14**, 121.

Feldstein, Y.I.: 1973, 'Auroral Oval', *J. Geophys. Res.* **78**, 1210.

Feldstein, Y.I. and Starkov, G.V.: 1967, 'Dynamics of Auroral Belt and Polar Geomagnetic Disturbances', *Planetary Space Sci.* **15**, 209.

Fellgett, P.: 1951, 'The Theory of Infrared Sensitivities and its Application to Investigations of Stellar Radiation in the Near Infrared', Ph. D. Thesis, Cambridge University.

Findlay, F.D., Fortin, C.J., and Snelling, D.R.: 1969, 'Deactivation of $O_2(^1\Delta_g)$', *Chem. Phys. Letters* **3**, 204.

Findlay, F.D. and Snelling, D.R.: 1971, 'Collisional Deactivation of $O_2(^1\Delta_g)$', *J. Chem. Phys.* **55**, 545.

Föppl, H., Haerendel, G., Haser, L., Lust, R., Melzner, F., Meyer, B., Neuss, H., Rabben, H.–H., Reiger, E., Stocker, J., and Stoffregen, W.: 1968, 'Preliminary Results of Electric Field Measurements in the Auroral Zone', *J. Geophys. Res.* **73**, 21.

Forsyth, P.A.: 1968, 'Radio Aurora; a Review of Research in Radio Aurora for the Period 1963–67' *Ann. Géophys.* **24**, 555.

Forsyth, P.A.: 1969, 'The Rate of Occurrence of VHF Radio Aurora Over an 11-year Period', *J. Atmospheric Terrest. Phys.* **31**, 313.

Foster, J.C. and Rosenberg, T.J.: 1973, 'Electron Precipitation and VLF Emissions from Gyroreso nant Interactions Near the Plasmapause', in *Symposium on Magnetospheric Substorms*, IAGA, Kyoto.

Frank, L.A.: 1967, 'On the Extraterrestrial Ring Current During Magnetic Storms', *J. Geophys. Res.* **72**, 3753.

Frank, L.A.: 1970, 'Direct Detection of Asymmetric Increases in Extraterrestrial Ring Current Proton Intensities in Outer Radiation Zone', *J. Geophys. Res.* **75**, 1263.

Frank, L.A. and Ackerson, K.L.: 1971, 'Observations of Charged Particle Precipitation into the Auroral Zone', *J. Geophys. Res.* **76**, 3612.

Frank, L.A. and Ackerson, K.L.: 1972, 'Local-Time Survey of Plasma at Low Altitudes Over the Auroral Zones', *J. Geophys. Res.* **77**, 4116.

Fraser, P.A.: 1954, 'A Method of Determining the Electronic Transition Moment for Diatomic Molecules', *Can. J. Phys.* **32**, 515.

Fritz, H.: 1881, *Das Polarlicht*, F.A. Brockhaus, Leipzig.

Fukushima, N.: 1969, 'Equivalence in Ground Geomagnetic Effect of Chapman-Vestine and Birkeland-Alfvén's Electric Current Systems for Polar Magnetic Storms', *Rep. Ionos. Space Res. Japan* **23**, 219.

Fukunishi, H.: 1974, 'Dynamic Relationship Between Proton and Electron Auroral Substorms', *J. Geophys. Res.*, submitted.

Galperin, G.I.: 1959, 'Hydrogen Emission and Two Types of Auroral Spectra', *Planetary Space Sci.* **1**, 57.

Gartlein, C.W.: 1947, 'Unlocking Secrets of the Northern Lights', *Nat. Geograph. Mag.* **Nov.**, 673.

Gattinger, R.L.: 1969, 'Interpretation of Airglow Emissions – OH Emissions', *Ann. Géophys.* **25**, 825.

Gattinger, R.L. and Vallance Jones, A.: 1972, 'The Intensity Ratios of Auroral Emission Features', *Ann. Géophys.* **28**, 91.

Gattinger, R.L. and Vallance Jones, A.: 1973a, 'N_2^+ Meinel Auroral Spectra in the 1.5 μ Region', *Can. J. Phys.* **51**, 287.

Gattinger, R.L. and Vallance Jones, A.: 1973b, 'Observation and Interpretation of Hydroxyl Airglow Emissions', in B.M. McCormac (ed.), *Physics and Chemistry of Atmospheres*, D. Reidel Publishing Co., Dordrecht, Holland, p. 184.

Gattinger, R.L. and Vallance Jones, A.: 1973c, 'Observation and Interpretation of O_2 1.27 μ Emission Enhancements in Aurora', *J. Geophys. Res.* **78**, 8305.

Gattinger, R.L. and Vallance Jones, A.: 1974, 'Quantitative Spectroscopy of the Aurora. II: The Spectrum of Medium Intensity Aurora Between 4500 and 8900 Å', *Can. J. Phys.*, submitted.

Gault, W.A.: 1970, 'Decay of the O I λ5577 Line in Meteor Waves', *Can. J. Phys.* **48**, 1017.

Gendrin, R.: 1968, 'Pitch Angle Diffusion of Low Energy Protons Due to Gyroresonant Interaction with Hydromagnetic Waves', *J. Atmospheric Terrest. Phys.* **30**, 1313.

Gérard, J.C.: 1970, 'Metastable Oxygen Ions Distribution and Related Optical Emissions in the Aurora', *Ann. Géophys.* **26**, 777.

Gérard, J.C. and Harang, O.E.: 1973, 'Observation of O(^{1}D) and N(^{2}D) Emission in Polar Aurora', in B.M. McCormac (ed.), *Physics and Chemistry of Atmospheres*, D. Reidel Publishing Co., Dordrecht-Holland. p. 241

Gilmore, F.R.: 1965, 'Potential Energy Curves for N_2, NO, O_2 and Corresponding Ions', *J. Quant. Spectrosc. Radiat. Transfer* **5**, 369.

Gold, T.: 1959, 'Motions in the Magnetosphere of the Earth', *J. Geophys. Res.* **64**, 1219.

Goldberg, L.: 1935, 'Relative Strengths of Multiplets in LS Coupling', *Astrophys. J.* **82**, 1.

Goldberg, L.: 1936, 'Note on Absolute Multiplet Strengths', *Astrophys. J.* **84**, 11.

Gowell, R.W. and Akasofu, S.I.: 1969, 'Irregular Pulsations of the Morning Sky Brightness', *Planetary Space Sci.* **17**, 289.

Grafe, A.: 1969, 'On the Dynamics of the DP Current System', *Planetary Space Sci.* **17**, 1395.

Green, A.E.S., Peterson, L.R., and Prasad, S.S.: 1969, 'Semi-Empirical Cross Sections and the Airglow and Aurora', in B.M. McCormac and A. Omholt (eds.), *Atmospheric Emissions*, Van Nostrand Reinhold, New York, p. 523.

Green, D.W. and Whalen, B.A.: 1974, 'Ionospheric Ion Flow Velocities from Measurements of Ions Distribution Function-Technique', *J. Geophys. Res.* **79**, 2829.

Greenwald, R.A., Ecklund, W.L., and Balsley, B.B.: 1973, 'Auroral Currents, Irregularities and

Luminosity', *J. Geophys. Res.* **78**, 8193.
Gregory, P.C.: 1969, 'Radio Emission from Auroral Electrons', *Nature* **221**, 350.
Grover, F.H.: 1971, 'Experimental Noise Reducers for an Active Microbarograph Array', *Geophys. J. Roy. Astron. Soc.* **26**, 41.
Grün, A.E.: 1957, 'Lumineszenz, photometrische Messungen der Energie Absorption in Strahlungsfeld von Elektronenquellen eindimensionaler Fall in Luft', *Z. Naturforsch.* **12a**, 89.
Gurnett, D.A., Pfeiffer, G.W., Anderson, R.R., Mosier, S.R., and Cauffman, D.P.: 1969, 'Initial Observations of VLF Electric and Magnetic Fields with the Injun 5 Satellite', *J. Geophys. Res.* **74**, 4631.
Gustaffson, G.: 1964, 'Ionization in the *D*-Region During Auroral Break-up Events', *Planetary Space Sci.* **12**, 195.
Gustafsson, G.: 1969, 'Spatial and Temporal Relations Between Auroral Emissions and Cosmic Noise Absorption', *Planetary Space Sci.* **17**, 1961.
Gustafsson, G., Feldstein, Y.I., and Shevnina, N.P.: 1969, 'The Auroral Orientation Curves for the IQSY', *Planetary Space Sci.* **17**, 1657.
Haerendel, G. and Lüst, R.: 1970, 'Electric Fields in the Ionosphere and Magnetosphere', in B.M. McCormac (ed.), *Particles and Fields in the Magnetosphere*, D. Reidel Publishing Co., Dordrecht, Holland, p. 213.
Haerendel, G., Lüst, R., and Rieger, E.: 1967, 'Motion of Artificial Ion Clouds in Upper Atmosphere', *Planetary Space Sci.* **15**, 1.
Haerendel, G., Hedgecock, P.C., and Akasofu, S.I.: 1971, 'Evidence for Magnetic Field Aligned Currents During the Substorms of March 18, 1969', *J. Geophys. Res.* **76**, 2382.
Hakura, Y.: 1965, 'Geomagnetic Coordinates Corrected for Higher Order Spherical Harmonic Terms', *Rep. Ionos. Space Res. Japan* **19**, 121.
Hallinan, T.J. and Davis, T.N.: 1970, 'Small Scale Auroral Arc Distortions', *Planetary Space Sci.* **18**, 1735.
Hanna, P.B. and Anger, C.D.: 1971, 'Auroral Colour Variations', *Planetary Space Sci.* **19**, 399.
Hanson, W.B.: 1963, 'Electron Temperatures in the Upper Atmosphere', *Space Res.* **3**, 282.
Harang, L.: 1969, 'Radio Noise from Aurora', *Planetary Space Sci.* **17**, 869.
Harang, O.E. and Gérard, J.C.: 1970, 'On Possible Vibrational Enhancement of N_2^+ First Negative Bands in Auroral Hydrogen Arcs', *Physica Norvegica* **4**, 217.
Harang, L. and Larsen, R.: 1965, 'Radio Wave Emissions in the VLF Band Observed Near the Auroral Zone. I: Occurrence of Emissions During Disturbances', *J. Atmospheric Terrest. Phys.* **27**, 481.
Harang, O. and Pettersen, H.: 1967, 'Variation in Width of the H_α Line in Aurora', *Planetary Space Sci.* **15**, 1599.
Harang, L. and Stoffregen, W.: 1938, 'Scattered Reflections of Radio Waves from a Height of More Than 1000 km', *Nature* **142**, 832.
Harang, L. and Tröim, J.: 1970, 'Polarization of Auroral Echoes', *Planetary Space Sci.* **18**, 655.
Hargreaves, J.K.: 1968, 'Auroral Motions Observed with Riometers: Latitudinal Movements and a Median Global Pattern', *J. Atmospheric Terrest. Phys.* **30**, 1461.
Hargreaves, J.K.: 1969a, 'Conjugate and Closely-Spaced Observations of Auroral Radio Absorption', *Planetary Space Sci.* **17**, 1459.
Hargreaves, J.K.: 1969b, 'Auroral Absorption of HF Radio Waves in the Ionosphere: A Review of Results from the First Decade of Riometry', *Proc. I.E.E.E.* **57**, 1348.
Hargreaves, J.K.: 1970, 'Conjugate and Closely-Spaced Observations of Auroral Radio Absorption. IV: The Movement of Simple Features', *Planetary Space Sci.* **18**, 1691.
Hargreaves, J. K.: 1971, 'Conjugate and Closely-Spaced Observations of Auroral Radio Absorption. V: Alternation of Auroral Particle Precipitation Between Hemispheres', *Planetary Space Sci.* **19**, 513.
Harris, F.R., Peterson, R.N., and Bradley, D.T.: 1970, 'The Relative Intensities of the O_2^+ 1NG and N_2 1PG Band Systems in Normal and Type-B Aurora', *Can. J. Phys.* **48**, 2283.
Harrison, A.W.: 1969, 'Recent Observations of the Auroral Spectrum Between 1.02 and 1.13 μ', *Can. J. Phys.* **47**, 599.
Harrison, A.W.: 1972, 'Spectrophotometry of the Aurora', *Can. J. Phys.* **50**, 500.
Harrison, A.W. and Vallance Jones, A.: 1959, 'Observations of Meinel N_2^+ Band Near 1.5 μ in the Auroral Spectrum', *J. Atmospheric Terrest. Phys.* **13**, 291.

Hartz, T.R.: 1971, 'Particle Precipitation Patterns', in B.M. McCormac (ed.), *The Radiating Atmosphere*, D. Reidel Publishing Co., Dordrecht, Holland p. 225.

Hartz, T.R. and Brice N.M.: 1967, 'The General Pattern of Auroral Particle Precipitation', *Planetary Space Sci.* **15**, 301.

Hartz, T.R. and McAlpine, J.L.: 1962, 'The Dependence of Ionospheric Disturbances on Large Solar Flares', *Radio Wave Absorption in the Ionosphere, AGARDograph* **53**, 13.

Hasegawa, A.: 1971, 'Dynamics of Auroral Formation', in B.M. McCormac (ed.), *The Radiating Atmosphere*, D. Reidel Publishing Co., Dordrecht, Holland, p. 384.

Haskell, G.P. and Southwood, D.J.: 1972, 'Comments on "Fermi Acceleration of Auroral Particles"', *J. Geophys. Res.* **77**, 6926.

Hays, P.B. and Roble, R.G.: 1971, 'A Technique for Recovering Doppler Line Profiles from Fabry-Perot Interferometer Fringes of Very Low Intensity', *Appl. Opt.* **10**, 193.

Hays, P.B., Jones, R.A., and Rees, M.H.: 1973, 'Auroral Heating and the Composition of the Neutral Atmosphere', *Planetary Space Sci.* **21**, 559.

Heikkila, W.J. and Winningham, J.D.: 1971, 'Penetration of Magnetosheath Plasma to Low Altitudes Through the Dayside Magnetospheric Cusps', *J. Geophys. Res.* **76**, 883.

Heikkila, W.J., Winningham, J.D., Eather, R.H., and Akasofu, S.I.: 1972, 'Auroral Emissions and Particle Precipitation in the Noon Sector', *J. Geophys. Res.* **77**, 4100.

Helliwell, R.A.: 1973, 'Conjugate VLF Experiments', in *Symposium on Magnetospheric Configuration*, IAGA, Kyoto.

Henriksen, K.: 1973, 'The Excitation of O(^{1}S) Atoms by Energy Transfer from $N_2(A^3\Sigma_u^+)$ Molecules in Aurora', *Planetary Space Sci.* **21**, 863.

Henry, R.J.W., Burke, P.G., and Sinfailam, A.L.: 1969, 'Scattering of Electrons by C, N, O, N^+, O^+ and O^{++}', *Phys. Rev.* **178**, 218.

Heppner, J.P.: 1967, 'High Latitude Magnetic Disturbances', in B.M. McCormac (ed.), *Aurora and Airglow*, Reinhold Publishing Co., New York, p. 75.

Heppner, J.P., Byrne, E.C., and Belon, A.E.: 1952, 'The Association of Absorption and Es Ionization with Aurora at High Latitudes' *J. Geophys. Res.* **57**, 121.

Heppner, J.P., Stolarik, J.D., and Wescott, E.M.: 1971, 'Electric-Field Measurements and Identification of Currents Causing Magnetic Disturbances in the Polar Cap', *J. Geophys. Res.* **76**, 6028.

Hernandez, G.: 1970, 'A High Luminosity Spectrometer for Night Airglow Studies', *Appl. Opt.* **9**, 1225.

Herzberg, G.: 1950, *Molecular Spectra and Molecular Structure I Diatomic Molecules*, Van Nostrand, New York.

Hess, W.N.: 1968, *The Radiation Belt and Magnetosphere*, Blaisdell, Waltham, Mass.

Hilliard, R.L. and Shepherd, G.G.: 1966a, 'Wide-Angle Michelson Interferometer for Measuring Doppler Line Widths', *J. Opt. Soc. Am.* **56**, 362.

Hilliard, R.L. and Shepherd, G.G.: 1966b, 'Upper Atmosphere Temperatures from Doppler Line Widths. IV: A Detailed Study Using the O I 5577 Å Auroral and Night Glow Emission', *Planetary Space Sci.* **14**, 383.

Hines, C.O.: 1960, 'Internal Atmospheric Gravity Waves at Ionospheric Heights', *Can. J. Phys.* **38**, 1441.

Hines, C.O.: 1964, 'Hydromagnetic Motions in the Magnetosphere', *Space Sci. Rev.* **3**, 342.

Hoffmann, D.J. and Green R.A.: 1972, 'Balloon Observations of Simultaneous X-Ray and Visible Bursts', *J. Geophys. Res.* **77**, 776.

Hofstee, J. and Forsyth, P.A.: 1969, 'Ion-Acoustic Waves in the Auroral Plasma', *Can. J. Phys.* **47**, 2797.

Hofstee, J. and Forsyth, P.A.: 1972, 'Ion-Acoustic Waves and Aspect Sensitivity in Radio Aurora', *J. Atmospheric Terrest. Phys.* **34**, 893.

Holm, C., Maehlum, B.N., and Narheim, B.T.: 1972, 'Rotating Wedge Filter Photometer for High Altitude Sounding Rocket Application', *Appl. Opt.* **11**, 421.

Holt, O. and Lerfald, G.M.: 1967, 'Results of an RF Capacity Probe Experiment in the Auroral Ionosphere', *Radio Sci.* **2**, 1283.

Holt, O. and Omholt, A.: 1962, 'Auroral Luminosity and Absorption of Cosmic Radio Noise', *J. Atmospheric Terrest. Phys.* **24**, 467.

Hones, E.W.: 1969, 'Observations in the Magnetotail and Their Interpretation', in B.M. McCormac and A. Omholt (eds.), *Atmospheric Emissions*, Van Nostrand Reinhold, New York, p. 351.

Hones, E.W., Asbridge, J.R., Bame, S.J., and Singer, S.: 1971a, 'Energy Spectra and Angular Distributions of Particles in the Plasma Sheet and Their Comparison with Rocket Measurements Over the Auroral zone', *J. Geophys. Res.* **76**, 63.

Hones, E.W., Karas, R.H., Lanzerotti, L.J., and Akasofu, S.I.: 1971b, 'Magnetospheric Substorms of September 14, 1968', *J. Geophys. Res.* **76**, 6765.

Hones, E.W., Singer, S., Lanzerotti, L.J., Pierson, J.D., and Rosenberg, T.J.: 1971c, 'Magnetospheric Substorm of August 25–26, 1967', *J. Geophys. Res.* **76**, 2977.

Hooper, E.B. and Walker, J.C.G.: 1971, 'The $E \times B$ Instability in Auroral Arcs', *J. Atmospheric Terrest. Phys.* **33**, 1441.

Hughes, A.R.W. and Kaiser, T.R.: 1971, 'VLF Radio Emissions and the Aurora', in B.M. McCormac (ed.), *The Radiating Atmosphere*, D. Reidel Publishing Co., Dordrecht, Holland. p. 336.

Hultqvist, B.: 1961, 'Circular Form of Auroral Zones', *Planetary Space Sci.* **8**, 142.

Hultqvist, B.: 1966, 'Ionospheric Absorption of Cosmic Radio Noise', *Space Sci. Rev.* **5**, 771.

Hultqvist, B.: 1969, 'Auroral and Polar Cap Absorption, in B.M. McCormac and A. Ambolt (eds.) *Atmospheric Emissions*, Van Nostrand Reinhold, New York, p. 73.

Hultqvist, B.: 1971, 'On the Production of a Magnetic-Field-Aligned Electric Field by the Interaction Between the Hot Magnetospheric Plasma and the Cold Ionosphere', *Planetary Space Sci.* **19**, 749.

Hultqvist, B. and Egeland, A.: 1964, 'Radio Aurora', *Space Sci. Rev.* **3**, 27.

Hultqvist, B., Borg, H., Riedler, W., and Christopherson, P.: 1971, 'Observations of Magnetic-Field-Aligned Anisotropy for 1 and 6 keV Positive Ions in the Upper Ionosphere', *Planetary Space Sci.* **19**, 279.

Hunsucker, R.D. and Owren, L.: 1962, 'Auroral Sporadic-*E* Ionization', *J. Res. N.B.S.* **66D**, 581

Hunten, D.M.: 1955, 'Some Photometric Observations of Auroral Spectra', *J. Atmospheric Terrest. Phys.* **7**, 141.

Hunten, D.M.: 1956, 'An Automatic Auroral Recorder', *J. Opt. Soc. Am.* **46**, 578.

Hunten, D.M.: 1961, 'Temperatures Deduced from Aurora and Airglow Spectra', *Ann. Géophys.* **17**, 249.

Hunten, D.M., Rawson, E.G., and Walker, J.K.: 1963, 'Rapid Measurement of N_2^+ Rotational Temperatures in Aurora', *Can. J. Phys.* **41**, 258.

Hunten, D.M., Roach, F.E., and Chamberlain, J.W.: 1956, 'A Photometric Unit for the Airglow and Aurora', *J. Atmospheric Terrest. Phys.* **8**, 345.

Hunten, D.M., Rundle, H.N., Shepherd, G.G., and Vallance Jones, A.: 1967, 'Optical Upper Atmospheric Investigations at the University of Saskatchewan', *Applied Optics* **6**, 1609.

Ignatyev, V.M., Yugov, V.A. Alekseev, V.N., and Atlasov, K.V.: 1971, 'Interferometrical Measurement of Doppler Temperature from λ6300 Line Width in Auroras', in *Symposium on Aurora and Airglow*, IAGA, Moscow.

IAGA Working Group 4: 1969, 'International Geomagnetic Reference Field 1965.0', *J. Geophys. Res.* **74**, 4407.

Iijima, T. and Nagata, T.: 1972, 'Signatures for Substorm Development of the Growth Phase and Expansion Phase', *Planetary Space Sci.* **20**, 1095.

International Union of Geodesy and Geophysics: 1963, *International Auroral Atlas*, Edinburgh University Press.

Intriligator, D.S.: 1973, 'Solar Wind Parameters and Magnetospheric Activity', in *Workshop on Geophysical Indices*, IAGA, Kyoto.

Iversen, I.B., Olesen, O.H., and Ungstrup, E.: 1968, 'Observation of VLF-Radio Noise in the Ionosphere During an Auroral Absorption Event', *Ann. Géophys.* **24**, 313.

Jacquinot, P.: 1958, 'Caractères communs aux nouvelles méthodes de spectroscopie interférentielles; facteur de mérite', in *Les Progrès Récents en Spectroscopie Interférentielle*, Editions du CNRS, Paris, p. 39.

Jeannet, J.C. and Vassy, A.: 1963, 'Mesures interférométriques sur la raie verte des aurores polaires', *Ann. Géophys.* **19**, 80.

Jelly, D.H.: 1968, 'Apparent Poleward Motion of Onsets of Auroral Absorption Events', *Can. J. Phys.* **46**, 33.

Jelly, D.H.: 1969, 'Substorm Aspects of Auroral Absorption and Other Effects of the Upper Part of the Auroral Spectrum; Substorm Events', *Proceedings ESRO Colloquium*, Kiruna-ESRO SP38, p. 45.

Jelly, D.H.: 1970, 'On the Morphology of Auroral Absorption During Substorms', *Can. J. Phys.* **48** 335.

Jelly, D.H., McDiarmid, I., and Burrows, J.: 1964, 'Correlation Between Intensities of Auroral Absorption and Precipitated Electrons', *Can. J. Phys.* **42**, 2411.

Jentsch, V. and Kremser, G.: 1972, 'On observations of Auroral-Zone X-Rays with Energies Up To and Greater Than 200 keV', *J. Atmospheric Terrest. Phys.* **34**, 499.

Jespersen, M., Kane, J.A., and Landmark, B.: 1968, 'Electron and Positive Ion Density Measurements During Conditions of Auroral Absorption', *J. Atmospheric Terrest. Phys.* **30**, 1955.

Jeunehomme, M.: 1966, 'Transition-Moment Variation in the γ System of NO', *J. Chem. Phys.* **45**, 4433.

Johansen, O.E.: 1965 'Variations in Energy Spectrum of Auroral Electrons Detected by Simultaneous Observation with Photometer and Riometer', *Planetary Space Sci.* **13**, 225.

Johansen, O.E. and Omholt, A.: 1966, 'A Study of Pulsating Aurora', *Planetary Space Sci.* **14**, 207

Johnson, R.G., Sharp, R.D., and Shelley, E.G.: 1974, 'The Discovery of Energetic He^+ Ions in the Magnetosphere', *J. Geophys. Res.*, submitted.

Johnstone, A.D.: 1971, 'Correlation Between Electron and Proton Fluxes in Post Breakup Aurora', *J. Geophys. Res.* **76**, 5259.

Johnstone, A.D.: 1972, 'The Spreading of a Proton Beam by the Atmosphere', *Planetary Space Sci.* **20**, 292.

Jones, R., Oliver, D.J., and Pike, E.R.: 1971, 'Experimental and Theoretical Comparison of Photon-Counting and Current Measurements of Light Intensity', *Appl. Opt.* **10**, 1673.

Jones, R.A. and Rees, M.H.: 1973, 'Time Dependent Studies of the Aurora. I: Ion Density and Composition', *Planetary Space Sci.* **21**, 537.

Jørgensen, T.S.: 1968, 'Interpretation of Auroral Hiss Measured on OGO 2 and at Byrd as Incoherent Cerenkov Radiation', *J. Geophys. Res.* **73**, 1055.

Jørgensen, T.S.: 1973, 'Diagnostics of the Magnetosphere by Means of Naturally Occurring ELF and VLF Waves', *Geophys. Papers Danske Met. Inst.*, R-34.

Jørgensen, T.S., Friis-Christensen, H., and Wilhjelm, J.: 1972, 'Interplanetary Magnetic Field Direction and High Latitude Ionospheric Currents', *J. Geophys. Res.* **77**, 1976.

Kamide, Y. and Fukushima, N.: 1972, 'Positive Geomagnetic Bays in Evening High Latitudes and Their Possible Connection with Partial Ring Current', *Rep. Ionos. Res. Space Res. Japan* **26**, 79.

Kaneda, E., Oguti, T., and Nagata, T.: 1968, 'Photographic Atlas of Auroral Forms Observed at Syowa Station', *Japanese Antarctic Res. Exped. Sci. Reports.* **A4**.

Karandikar, R.V.: 1968, 'Interferometric Measurements of λ5577 Emission Temperatures in Aurora', *Planetary Space Sci.* **16**, 539.

Karandikar, R.V.: 1970, Private Communication.

Kavanagh, L.D., Freeman, J.W., and Chen, A.J.: 1968, 'Plasma Flow in the Magnetosphere', *J. Geophys. Res.* **73**, 5511.

Kawasaki, K. and Akasofu, S-I.: 1971, 'The Computed Distribution of Geomagnetic Disturbance Vectors for Model Three-Dimensional Current Systems', *Planetary Space Sci.* **19**, 543.

Kawasaki, K., Akasofu, S-I., Yasuhara, F., and Meng, C-I.: 1971, 'Storm Sudden Commencements and Polar Magnetic Substorms', *J. Geophys. Res.* **76**, 6781.

Kelley, M.C., Mozer, F.S., and Fahleson, U.V.: 1971a, 'Electric Fields in the Nighttime and Daytime Auroral Zone', *J. Geophys. Res.* **76**, 6054.

Kelley, M.C., Stair, J.A., and Mozer, F.S.: 1971b, 'Relationship Between Magnetospheric Electric Fields and Motion of Auroral Forms', *J. Geophys. Res.* **76**, 5269.

Kendall, P.C., Chapman, S., Akasofu, S.I., and Swartztrauber, P.N.: 1966, 'The Computation of the Magnetic Field of any Axisymmetric Current Distribution', *Geophys. J.* **11**, 349.

Kennel, C.F.: 1969, 'Consequences of a Magnetospheric Plasma', *Rev. Geophys. Space Phys.* **7**, 379.

Kennel, C.F. and Petschek, H.E.: 1966, 'Limit on Stably Trapped Particle Fluxes', *J. Geophys. Res.* **71**, 1.

Kieffer, L.J. and Dunn. G.H.: 1966, 'Electron Impact Ionization Cross Section Data for Atoms, Atomic Ions and Diatomic Molecules. I: Experimental Data', *Rev. Mod. Phys.* **38**, 1.

Kilfoyle, B.P. and Jacka, F.: 1968, 'Geomagnetic *L* Coordinates', *Nature* **220**, 773.

Kindel, J.M. and Kennel, C.F.: 1971, 'Topside Current Instabilities', *J. Geophys. Res.* **76**, 3055.

King-Hele, D.G. and Walker, Doreen M.C.: 1971, 'Air Density at Heights Near 150 km in 1970 from the Orbit of Cosmos 316 (1969–108A)', *Planetary Space Sci.* **19**, 1637.

Kinsey, J.H.: 1965, 'Auroral Parallactic Measurements at Byrd Station, Antarctica During 1963', *J. Geophys. Res.* **70**, 579.

Kisabeth, J.L. and Rostoker, G.: 1971, 'Development of the Polar Electrojet During Polar Magnetic Substorms', *J. Geophys. Res.* **76**, 6815.

Knecht, R.W.: 1956, 'Aurora and Sporadic *E* Echoes At Barrow', *J. Geophys. Res.* **61**, 59.

Knox, F.B.: 1972, 'A Theory of Discrete Radio Aurora', *J. Atmospheric Terrest. Phys.* **34**, 747.

Koehler, J.A. and Kavadas, A.: 1969, 'A Rocket Released Magnetometer Probe for Upper Atmospheric Measurements', *Can. J. Phys.* **47**, 1529.

Kokubun, S.: 1971, 'Polar Substorm and Interplanetary Magnetic Field', *Planetary Space Sci.* **19**, 697.

Kokubun, S.: 1972, 'Relationship of Interplanetary Magnetic Field Structure with Development of Substorm and Substorm Main Phase', *Planetary Space Sci.* **20**, 1033.

Kvifte, G.J. and Pettersen, H.: 1969, 'Morphology of Pulsating Aurora', *Planetary Space Sci.* **17**, 1599.

Lane, N.F. and Dalgarno, A.: 1969, 'Electron Cooling by Vibrational Excitation of O_2', *J. Geophys. Res.* **74**, 3011.

Lange-Hesse, G.: 1967, 'Radio Aurora', in B.M. McCormac (ed.), *Aurora and Airglow*, Reinhold Publishing Co., New York, p. 519.

Lassen, K., Laursen, O.R., and Olsen, J.: 1964, *Danish Visual Aurora Observations*, **1957–60**, Publikationer fra det Danske Meteorologiske Institut Meddelelser Nr. 18.

Lassen, K.: 1969, 'Polar Cap Emissions', in B.M. McCormac and A. Omholt (eds.). *Atmospheric Emissions*, Van Nostrand Reinhold, New York, p. 63.

Lassen, K.: 1970, 'The Position of the Auroral Oval Over Greenland and Spitzbergen', *Danske Meteorologiske Institut Geophys. Papers*, R16.

Lassen, K.: 1972, 'On the Classification of High Latitude Auroras', *Danske Meteorologiske Institut Geophys. Papers*, R28.

Levy, R.H., Petchek, H.E., and Siscoe, G.L.: 1964, 'Aerodynamic Aspects of the Magnetospheric Flow', *AIAA Journal* **2** (December), 2065.

Lim, T.L. and Laaspere, T.: 1972, 'An Evaluation of the Intensity of Cerenkov Radiation from Auroral Electrons with Energies Down to 100 eV', *J. Geophys. Res.* **77**, 4145.

Lincoln, J.V.: 1967, 'Geomagnetic Indices', in S. Matsushita and W.H. Campbell (eds.), *Physics of Geomagnetic Phenomena*, Vol. I, Academic Press, New York, p. 67.

Little, C.G. and Leinbach, H.: 1959, 'The Riometer – A Device for the Continuous Measurement of Ionospheric Absorption', *Proc. IRE* **47**, 315.

Loughnan, C.J.: 1961, 'Longitudinal Dependence of Radiation-Belt Scattering, and Primary Auroral Particles', *Planetary Space Sci.* **8**, 13.

Lucas, C. and Brice, N.: 1973, 'Cyclotron Resonance Wave Amplification in the Magnetosphere and Energetic Particle Stability', *J. Geophys. Res.* **78**, 8338.

Lyon, G.F. and Kavadas, A.: 1958, 'Horizontal Motions in Radar Echoes from Aurora', *Can. J. Phys.* **36**, 1661.

Mack, J.E., McNutt, D.P., Roesler, F.L., and Chabbal, R.: 1963, 'The PEPSIOS Purely Interferometric High-Resolution Scanning Spectrometer. I: The Pilot Model', *Appl. Opt.* **2**, 873.

Maeda, K.: 1965, 'Diffusion of Low Energy Auroral Electrons in the Atmosphere', *J. Atmospheric Terrest. Phys.* **27**, 259.

Maeda, K. and Aikin, A.C.: 1968, 'Variations of Polar Mesospheric Oxygen and Ozone During Auroral Events', *Planetary Space Sci.* **16**, 371.

Maeda, H., Sakurai, K., Ondoh, T., and Yamamoto, M.: 1962, 'A Study of Solar-Terrestrial Relationships During the IGY and IGC', *Ann. Géophys.* **18**, 305.

Maehlum, B.N.: 1968, 'Universal-Time Control of the Low Energy Electron Fluxes in the Polar Regions', *J. Geophys. Res.* **73**, 3459.

Maggs, J.E. and Davis, T.N.: 1968, 'Measurement of the Thicknesses of Auroral Structures', *Planetary Space Sci.* **16**, 205.

Mahadevan, P. and Roach, F.E.: 1968, 'Mechanism for the Aurora Emission of O I(6300 Å)', *Nature* **220**, 150.

Maier, E.J. and Rao, B.C.N.: 1970, 'Observations of Suprathermal Electron Flux and the Electron Temperature at High Latitudes', *J. Geophys. Res.* **75**, 7168.

Maral, G.: 1970a, 'Mesure du rayonnement X auroral et application à l'étude des précipitations électroniques', Ph. D. Thesis, Univ. of Toulouse.

Maral, G.: 1970b, 'Détermination indirecte des caractères énergétiques des précipitations électroniques dans la zone aurorale pendant la journée', *Ann. Géophys.* **26**, 71.

Mariska, J.T. and Oster, L.: 1972, 'Solar Activity and the Variations of the Geomagnetic K_p-Index', *Solar Phys.* **26**, 241.

Matsumoto, H. and Kimura, I.: 1973, 'Computer Experiments of VLF Triggered Emissions', *Symposium on Magnetospheric Substorms*, IAGA, Kyoto.

Mayaud, P.N.: 1960, 'Un nouveau système de coordonnées magnétiques pour l'étude de la haute atmosphère: les coordonnées de l'anneau equatorial', *Ann. Géophys.* **16**, 278.

Mayr, H.G. and Volland, H.: 1972, 'Magnetic Storm Effects in the Neutral Composition', *Planetary Space Sci.* **20**, 379.

Mayr, H.G. and Volland, H.: 1973, 'Magnetic Storm Characteristics of the Thermosphere', *J. Geophys. Res.* **78**, 2151.

McDiarmid, D.R.: 1970, 'Ion-Acoustic Waves and Radio Aurora', *Can. J. Phys.* **48**, 1863.

McDiarmid, D.R.: 1972, 'On the Aspect Sensitivity of Radio Aurora', *Can. J. Phys.* **50**, 2557.

McDiarmid, D.R. and McNamara, A.G.: 1969, 'A Physical Model of a Radio Aurora Event', *Can. J. Phys.* **47**, 1271.

McDiarmid, I.B. and Hruška, A.: 1972, 'Anisotropy of High-Latitude Electron Fluxes during Substorms and Structure of the Magnetotail', *J. Geophys. Res.* **77**, 3377.

McDiarmid, I.B., Burrows, J.R., and Wilson, M.D.: 1972, 'Solar Particles and the Dayside Limit of Closed Field Lines', *J. Geophys. Res.* **77**, 1103.

McEwen, D.J. and Sivjee, G.G.: 1972, 'Rocket Measurements of Electron Influx during a Major Magnetic Storm with Type-A Aurora', *J. Geophys. Res.* **77**, 5523.

McIlwain, C.E.: 1961, 'Coordinates for Mapping the Distribution of Magnetically Trapped Particles', *J. Geophys. Res.* **66**, 3681.

McNamara, A.G.: 1969, 'Rocket Measurements of Plasma Densities and Temperatures in Visual Aurora', *Can. J. Phys.* **47**, 1913.

McNamara, A.G.: 1971, 'Survey of Radio Reflections from Aurora', in B.M. McCormac (ed.), Radiating Atmosphere, D. Reidel Publishing Co., Dordrecht-Holland, p. 301.

McNeal, R.J. and Birely, J.H.: 1973, 'Laboratory Studies of Collisions of Energetic H^+ and Hydrogen with Atmospheric Constituents', *Rev. Geophys. Space Phys.* **11**, 633.

McPherron, R.L.: 1970, 'Growth Phase of Magnetospheric Substorms', *J. Geophys. Res.* **75**, 5592.

McPherron, R.L.: 1974, 'Critical Problems in Establishing the Morphology of Substorms in Space', in B.M. McCormac (ed.), *Magnetospheric Physics*, D. Reidel Publishing Co., Dordrecht, Holland, p. 225.

McPherron, R.L., Russell, C.T., and Coleman, P.J.: 1972, 'Fluctuating Magnetic Fields in the Magnetosphere. II: ULF Waves', *Space Sci. Rev.* **13**, 411.

McPherron, R.L., Russell, C.T., and Aubrey, M.P.: 1973, 'Satellite Studies of the Magnetospheric Substroms on August 15, 1968, 9. Phenomenological Model for Substorms', *J. Geophys. Res.* **78**, 3131.

Mead, G.D.: 1967, 'Neutral Points on the Boundary of a Closed Magnetosphere', *Space Sci. Rev.* **7**, 158.

Meek, J.H.: 1953, 'Correlation of Magnetic, Auroral and Ionospheric Variations at Saskatoon', *J. Geophys. Res.* **58**, 445.

Meek, J.H.: 1954, 'Correlation of Magnetic, Auroral and Ionospheric Variations at Saskatoon – Part 2', *J. Geophys. Res.* **59**, 87.

Meek, J.H.: 1955, 'The Location and Shape of the Auroral Zone', *J. Atmospheric Terrest. Phys.* **6**, 313.

Meek, J.H. and Hector, F.S.: 1955, 'A Recording Magnetic Variometer', *Can. J. Phys.* **23**, 364.

Meinel, A.B.: 1951, 'Doppler Shifted Auroral Emission', *Astrophys. J.* **113**, 50.

Meinel, A.B. and Schulte, D.H.: 1953, 'Auroral Motions', *Astrophys. J.* **117**, 454.

Meinel, A.B., Negaard, B.J., and Chamberlain, J.W.: 1954, 'Variation of the 27-Day Recurrence Period for Low Latitude Aurorae', *J. Geophys. Res.* **59**, 407.

Mende, S.B., Sharp, R.D., Shelley, E.J., Haerendel, G., and Hones, E.W.: 1972, 'Coordinated Observations of the Magnetosphere: the Development of a Substorm', *J. Geophys. Res.* **77**, 4682.

Meng, C.I. and Akasofu, S.I.: 1969, 'A Study of Polar Magnetic Substorms. II: Three Dimensional Current System', *J. Geophys. Res.* **74**, 4035.

Meredith, H.T. and Montalbetti, R.: 1960, 'A Scanning Photometer for Monitoring H_β', *Defence Res. Northern Lab. Tech. Note* **8/60**.

Meriwether, J.W., Benesch, J.M., and Tilford, S.G.: 1970, 'Characteristics of a Visual Aurora Following a Solar Flare', *Planetary Space Sci.* **18**, 525.

Meyer, J.A., Setser, D.W., and Stedman, D.H.: 1969, 'Excitation of the Auroral Green Line of Atomic Oxygen ($^1S - {}^1D$)', *Astrophys. J.* **157**, 1023.

Miiller, J.R., and Shepherd, G.G.: 1968, 'Auroral Measurements Using Rocket-Borne Photometers', *Ann. Géophys.* **24**, 305.

Miller, J.R. and Shepherd, G.G.: 1969, 'Rocket Measurements of H Beta Production in a Hydrogen Aurora', *J. Geophys. Res.* **74**, 4987.

Miller, J.R. and Shepherd, G.G.: 1970, 'Photometric Rocket Measurements in Hydrogen Auroras', *Space Res.* **X**, 876.

Miller, R.E., Fastie, W.G., and Isler, R.C.: 'Rocket Studies of Far-UV Radiation in Aurora', *J. Geophys. Res.* **73**, 3353.

Miller, R.E. and Zeitz, W.M.: 1971, 'Observations on Pulsating Auroras', *Planetary Space Sci.* **19**, 693.

Millman, P.M.: 1968, 'The Canadian Visual Auroral Program', *Ann. Géophys.* **24**, 513.

Montalbetti, R. and McEwen, D.J.: 1962, 'Hydrogen Emissions and Sporadic *E* Layer Behaviour', *J. Phys. Soc. Japan* **17**, Supplement A-I, p. 212.

Montbriand, L.E. : 1965, 'A Simple Method for Determining the Local Time of Geomagnetic Midnight', *Can. J. Phys.* **43**, 1074.

Montbriand, L.E.: 1969, 'Morphology of Auroral Hydrogen Emissions during auroral Substorms', Ph. D. Thesis, Univ. of Saskatchewan.

Montbriand, L.E.: 1970, 'A Simple Method for Calculating the Local Time of Corrected Geomagnetic Midnight', *J. Geophys. Res.* **75**, 5634.

Montbriand, L.E.: 1971, 'The Proton Aurora and Auroral Substorm', in B.M. McCormac (ed.), *The Radiating Atmosphere*, D. Reidel Publishing Co. Dordrecht-Holland, p. 366.

Montbriand, L.E. and Vallance Jones, A.: 1962, 'Studies of Auroral Hydrogen Emissions in West-Central Canada. I: Time and Geographical Variations', *Can. J. Phys.* **40**, 1401.

Moorcroft, D.R.: 1961a, 'Models of Auroral Ionization. Part I: Auroral Ionization Models and their Radio-Reflection characteristics', *Can. J. Phys.* **39**, 677.

Moorcroft, D.R.: 1961b, 'Models of Auroral Ionization. Part II: Applications to Radio Observation of Aurora', *Can. J. Phys.* **39**, 695.

Moorcroft, D.R.: 1966, 'The Interpretation of the Frequency Dependence of Radio Aurora', *Planetary Space Sci.* **14**, 269.

Moorcroft, D.R.: 1972, 'Radio Aspect Sensitivity and the Two Stream Instability', *J. Geophys. Res.* **77**, 765.

Moore, C.E.: 1945, 'A Multiplet Table of Astrophysical Interest', *Contrib. Princeton Univ. Obs.*, No. 20.

Moore, J.H. and Doering J.P.: 1969a, 'Rotational Excitation in Ion-Molecule Collisions. II: H^+, D^+, Ne^+, H_2^+ and D_2^+, *Phys. Rev.* **182**, 176.

Moore, J.H. and Doering, J.P.: 1969b, 'Vibrational Excitation in Ion Molecule Collisions: H^+, H_2^+, He^+, N^+, Ne^+ and Electrons on N_2', *Phys. Rev.* **177**, 218.

Morfill, G. and Scholer, M.: 1972, 'Reconnection of the Geomagnetic Tail Deduced from Solar Particle Observations', *J. Geophys. Res.* **77**, 4021.

Mozer, F.S.: 1971, 'Origin and Effects of Electric Fields during Isolated Magnetospheric Substorms', *J. Geophys. Res.* **76**, 7595.

Mozer, F.S.: 1973, 'Analysis of Techniques for Measuring DC and AC Electric Fields in the Magnetosphere', *Space Sci. Rev.* **14**, 272.

Mozer, F.S. and Bruston, P.: 1966, 'Auroral Zone Electron Source and Electric Fields', *J. Geophys. Res.* **71**, 4451.

Mozer, F.S. and Bruston, P.: 1967, 'Electric Field Measurements in Aurora', *J. Geophys. Res.* **72**, 1109.

Mozer, F.S. and Fahleson, U.V.: 1970, 'Parallel and Perpendicular Electric Fields in an Aurora', *Planetary Space Sci.* **18**, 1563.

Mozer, F.S. and Manka, R.H.: 1971, 'Magnetospheric Electric Field Properties Deduced from Simultaneous Balloon Flights', *J. Geophys. Res.* **76**, 1697.

Mozer, F.S. and Serlin, R.: 1969, 'Magnetospheric Electric Field Measurements with Balloons', *J. Geophys. Res.* **74**, 4739.

Mumma, M.J. and Zipf, E.C.: 1971a, 'Calibration of Vacuum-UV Monochromators by the Molecular Branching Ratio Technique', *J. Opt. Soc. Am.* **61**, 83.

Mumma, M.J. and Zipf, E.C.: 1971b, 'Dissociative Excitation of Vacuum-Ultraviolet Emission Features by Electron Impact on Molecular Gases. II: N_2', *J. Chem. Phys.* **55**, 5582.

Murcray, W.B.: 1959, 'Some Properties of the Luminous Aurora as Measured by a Photo-Electric Photometer', *J. Geophysics. Res.* **64**, 955.

Murcray, W.B.: 1966, 'Height-Luminosity Distribution of Auroral H_β', *Planetary Space Sci.* **14**, 812.

Mylyarchik, T.M.: 1960, 'Interferometric Measurements of the Widths of the Emissions (O I) 6300 Å and 5198–5200 Å (N I) in the Polar Aurora', *Doklady Acad. Sci. USSR* **130**, 303, No.2.

Nagata, T. and Kokobun, S.: 1961, 'A Photoelectric Analyser of All-Sky Camera Auroral Photographs', *J. Geophys. Res.* **66**, 2259.

Nagy, A.F., Hanson, W.B., Hoch, R.J., and Aggson, T.L.: 1972, 'Satellite and Ground-Based Observations of a Red Arc', *J. Geophys. Res.* **77**, 3613.

Neo, Y.P. and Rundle, H.N.: 1969, 'An Identification of Twilight Helium 10 830 Å Emission with Increased Resolution and Preliminary Photometric Measurements', *Planetary Space Sci.* **17**, 715

Newton, G.P. and Pelz, D.T.: 1969, 'Latitudinal Variations in the Neutral Atmospheric Density', *J. Geophys. Res.* **74**, 4169.

Nicholls, R.W.: 1961, 'Franck-Condon Factors to High Vibrational Quantum Numbers. I: N_2 and N_2^+', *J. Res. Nat. Bur. Standards* **65A**, 451.

Nicholls, R.W.: 1962, 'Laboratory Astrophysics', *J. Quant. Spectrosc. Radiat. Transfer* **2**, 433.

Nicholls, R.W.: 1964a, 'Transition Probabilities of Aeronomically Important Spectra', *Ann. Geophys.* **20**, 144.

Nicholls, R.W.: 1964b, 'Franck-Condon Factors to High Vibrational Quantum Numbers IV NO Band Systems', *J. Res. Nat. Bur. Standards* **68A**, 536.

Nicholls R.W.: 1965a, 'Franck-Condon Factors to High Vibrational Quantum Numbers. V: O_2. Band Systems', *J. Res. Nat. Bur. Standards* **69A**, 369.

Nicholls, R.W.: 1965b, 'Franck-Condon Factors for O_2^+ First and Second Negative Band Systems', *Can. J. Phys.* **43**, 1390.

Nicholls, R.W.: 1969, 'Aeronomically Important Transition Probability Data', *Can. J. Chem.* **47**, 1847.

Nicholls R.W.: 1974, Private Communication.

Nishida, A.: 1966, 'Magnetosphere Theory', *J. Geophys. Res.* **71**, 5669.

Nishida, A.: 1968, 'Geomagnetic D_P2 Fluctuations and Associated Magnetospheric Phenomena', *J. Geophys. Res.* **73**, 1795.

Noxon, J.F.: 1961, 'Observation of the ($b^1\Sigma_g^+$–$a^1\Delta_g$) Transition in O_2', *Can. J. Phys.* **39**, 1110.

Noxon, J.F.: 1962, 'Active Nitrogen at High Pressure', *J. Chem. Phys.* **36**, 926.

Noxon, J.F.: 1970, 'Auroral Emission from $O_2(^1\Delta_g)$', *J. Geophys. Res.* **75**, 1875.

Noxon, J.F. and Evans, J.V.: 1974, in preparation.

Obayashi, T. and Nishida, A.: 1968, 'Large, Scale Electric Field in the Magnetosphere', *Space Sci. Rev.* **8**, 3.

O'Brien, B.J.: 1967, 'Satellite Observations of Particle Fluxes and Atmospheric Emissions', in B.M. McCormac (ed.), *Aurora and Airglow*, Reinhold Publishing Co., New York, p. 623.

O'Brien, B.J.: 1970, 'Considerations that the Source of Auroral Energetic Particles is not a Parallel Electrostatic Field', *Planetary Space Sci.* **18**, 1821.

Ogawa, T. and Sato, T.: 1971, 'New Mechanism of Auroral Arcs', *Planetary Space Sci.* **19**, 1393.

Ohman, Y.: 1958, 'Solar Flare Importance Scale', *IGY Annals* **5**, 264.

Oliven, M.N. and Gurnett, D.A.: 1968, 'Microburst Phenomena. III: Microbursts and VLF Chorus', *J. Geophys. Res.* **73**, 2355.

Oliven, M.N., Venkatesan, D., and McCracken, K.G.: 1968, 'Microburst Phenomena. II: Auroral Zone Electrons', *J. Geophys. Res.* **73**, 2345.

Omholt, A.: 1955, 'Auroral *E* Layer Ionization and Auroral Luminosity', *J. Atmospheric Terrest. Phys.* **7**, 73.

Omholt, A.: 1957, 'Red and Near IR Auroral Spectrum', *J. Atmospheric Terrest. Phys.* **10**, 320.

Omholt, A.: 1959, 'Studies of the Excitation of Aurora Borealis. I: The Hydrogen Lines', *Geophys. Pub.* **20**, No. 11.

Omholt, A.: 1967, 'The Spectrum and Excitation Mechanisms in Aurora', in B.M. McCormac (ed.), *Aurora and Airglow*, Reinhold Publishing Co., New York, p. 59.

Omholt, A.: 1971, *The Optical Aurora*, Springer-Verlag, Heidelberg.
Omholt, A. and Harang, L.: 1955, 'Measurements of the Mean Lifetime of the Metastable ^{1}S-State of O I in Aurora', *J. Atmospheric Terrest. Phys.* **7**, 247.
Omholt, A. and Pettersen, H.: 1967, 'Characteristics of High Frequency Auroral Pulsations', *Planetary Space Sci.* **15**, 347.
Omholt, A., Kvifte, G.J., and Pettersen, H.: 1969, 'Pulsating Aurora', in B.M. McCormac and A Omholt (eds.), Van Nostrand Reinhold, New York, p. 131.
O'Neill, R. and Davidson, G.: 1969, 'Luminous Efficiency of Electron Induced Emission in Nitrogen', in B.M. McCormac and A. Omholt (eds.), *Atmospheric Emissions*, Van Nostrand Reinhold, New York, p. 477.
Opal, C.S., Moos, H.W., and Fastie, W.G.: 1970, 'Far-Ultraviolet Altitude Profiles and Molecular Oxygen Densities in an Aurora', *J. Geophys. Res.* **75**, 788.
Opal, C.S., Peterson, W.K., and Beaty, E.C.: 1971, 'Measurements of Secondary Electron Spectra Produced by Electron Impact ionization of a Number of Simple Gases', *J. Chem. Phys.* **55**, 4100.
Ossakow, S.L.: 1968, 'Anomalous Resistivity along Lines of Force in the Magnetosphere', *J, Geophys. Res.* **73**, 6366.
Paresce, F., Lampton, M., and Holberg, J.: 1972, 'Extreme Ultraviolet Emissions from an Aurora'. *J. Geophys. Res.* **77**, 4773.
Park, R.J. and Cloutier, P.A.: 1971, 'Rocket-Based Measurement of Birkeland Currents Related to an Auroral Arc and Electrojet', *J. Geophys. Res.* **76**, 7714.
Parkinson, T.D.: 1971, 'A Phase and Amplitude Study of Auroral Pulsations', *Planetary Space Sci.* **19**, 251.
Parkinson, T.D. and Zipf, E.C.: 1970, 'Energy Transfer from N_2 ($A^3\Sigma_u^+$) as a Source of O(^{1}S) in the Aurora', *Planetary Space Sci.* **18**, 895.
Parkinson, T.D., Zipf, E.C., and Donahue, T.M.: 1970, 'Rocket Investigation of the Auroral Green Line', *Planetary Space Sci.* **18**, 187.
Parks, G.K., Coroniti, F.V., McPherron, R.L., and Anderson, K.A.: 1968, 'Studies of the Magnetospheric Substorm. I: Characteristics of Modulated Energetic Electron Precipitation Occurring During Auroral Substorms', *J. Geophys. Res.* **73**, 1685.
Parks, G.K., Laval, G., and Pellat, R.: 1972, 'Behaviour of Outer Radiation Zone and a New Model of Magnetospheric Substorm', *Planetary Space Sci.* **20**, 1391.
Parsons, L.W. and Wiatr, Z.M.: 1962, 'Rubidium Vapour Magnetometer', *J. Sci. Instr.* **39**, 292.
Paulson, K.V. and Shepherd, G.G.: 1966a, 'Fluctuations in Brightness from Quiet-Form Auroras', *Can. J. Phys.* **44**, 837.
Paulson, K.V. and Shepherd, G.G.: 1966b, 'Short-Lived Brightness Oscillations in Active Auroras', *Can. J. Phys.* **44**, 921.
Pecker, J.C. and Roberts, W.O.: 1955, 'Solar Corpuscles Responsible for Geomagnetic Disturbances', *J. Geophys. Res.* **60**, 33.
Peek, H.M.: 1970, 'Vacuum UV Emission from Auroras', *J. Geophys. Res.* **75**, 6209.
Pendrel, J.V., Koehler, J.A., and Kavadas, A.: 1973, 'Ionospheric Magnetic Field Measurements at Auroral Latitudes', *Can. J. Phys.* **51**, 1313.
Penn, W.D. and Currie, B.W.: 1949, 'A Recording Meter for Auroral Radiations', *Can. J. Phys.* **A27**, 45.
Perkins, F.W.: 1968, 'Plasma-Wave Instabilities in the Ionosphere over the Aurora', *J. Geophys. Res.* **73**, 6631.
Perraut, S. and Roux, A.: 1973, 'Effects of Cold and Hot Plasma Densities on the Generation of ULF Emissions and the Precipitation in ring Current Protons', in *Symposium on Micropulsations*, IAGA, Kyoto. (See also *J. Atmospheric Terrest. Phys.*, submitted (1974).)
Peterson, A.M. and Hower, G.L.: 1963, 'Synchrotron Radiation from High-Energy Electrons', *J. Geophys. Res.* **68**, 723.
Peterson, J.R. and Moseley, J.T.: 1973, 'Time of Flight Measurements of Lifetime of N_2^+ ($A^2\Pi_u$) – The Meinel Band System', *J. Chem. Phys.* **58**, 172.
Petrie, W.: 1963, '*Keoeeit – The Story of the Aurora Borealis*', Pergamon Press, London.
Petrie, W. and Small, R.: 1953, 'The Intensities of Atomic and Molecular Features in the Auroral Spectrum', *Can. J. Phys.* **31**, 911.
Pfister, W.: 1967, 'Auroral Investigations by Means of Rockets', *Space Sci. Rev.* **7**, 642.
Pilkington, G.R.: 1970, 'A Study of Auroral X-Rays and Aurora', Ph. D. Thesis, Univ. of Calgary.

Pilkington, G. R. and Anger, C. D.: 1971, 'A Monte-Carlo Analysis of the Passage of Auroral X-Rays Through the Atmosphere', *Planetary Space Sci.* **19**, 1069.

Pilkington, G. R., Anger, C. D., and Clark, T. A.: 1968, 'Auroral X-Rays and Their Association with Rapidly Changing Forms', *Planetary Space Sci.* **16**, 815.

Ponomarev, V. H.: 1972, 'Partial Effects of Collisions in the Polar Upper Atmosphere for Monoenergetic and Monodirectional Beams of Protons (Proton Aurora)', Accad. Sci. U.S.S.R. Inst. Space Res., Pr. 116.

Pope, J. H.: 1963, 'A High-Latitude Investigation of the Natural Very-Low-Frequency Electromagnetic Radiation Known as Chorus', *J. Geophys. Res.* **68**, 83.

Potter, W. E.: 1970, 'Rocket Measurements of Auroral Electric and Magnetic Fields', *J. Geophys. Res.* **75**, 5415.

Potter, W. E. and Cahill, L. J.: 1969, 'Electric and Magnetic Field Measurements near an Auroral Electrojet', *J. Geophys. Res.* **74**, 5159.

Prasad, S. S. and Green, A. E. S.: 1971, 'UV Emission of Atomic Nitrogen in Aurora', *J. Geophys. Res.* **76**, 2419.

Primdahl, F.: 1970, 'The Fluxgate Mechanism. I: Gating Curves of Parallel and Orthogonal Fluxgates', *IEEE Transaction on Magnetics* **MAG**, **6**, 376.

Primdahl, F.: 1971, 'Bibliography of Fluxgate Magnetometers', *Danish Meteorological Inst. Geophys. Papers*, R22.

Procunier, R. W.: 1971, 'Observations of Acoustic Aurora in the 1–16 Hz Range', *Geophys. J. Roy. Astron. Soc.* **26**, 185.

Pytte, T. and Trefall, H.: 1972, 'Auroral-Zone Electron Precipitation Events Observed Before and at the Onset of Negative Magnetic Bays', *J. Atmospheric Terrest. Phys.* **34**, 315.

Ratcliffe, J. A.: 1960, *Physics of the Upper Atmosphere*, Academic Press, London.

Rearwin, S.: 1971, 'Rocket Measurements of Low-Energy Auroral Electrons', *J. Geophys. Res.* **76**, 4505.

Reasoner, D. L., Eather, R. H., and O'Brien, B. J.: 1968, 'Detection of α-Particles in Auroral Phenomena', *J. Geophys. Res.* **73** (13), 4185.

Reddy, C. A., Rao, M. M., Matsushita, S., and Smith, L. G.: 1969, 'Rocket Observations of Electron Densities in the Night-Time Auroral *E*-Region at Fort Churchill, Canada', *Planetary Space Sci.* **17**, 617.

Reed, R. D., Shelley, E. G., Bakke, J. G., Sanders, T. C., and McDaniel, J. D.: 1969, 'Low-Energy Channel-Multiplier Spectrometer for ATS-E', *Trans. I.E.E.E.* **NS-16**, 359.

Rees, D.: 1971, 'Upper Atmosphere Neutral Temperature Profiles in the Auroral Zone 1968–1970', *Planetary Space Sci.* **19**, 233.

Rees, D.: 1973, 'Neutral Wind Structure in the Thermosphere During Quiet and Disturbed Geomagnetic Periods', in B. M. McCormac (ed.) *Physics and Chemistry of Atmospheres*, D. Reidel Publishing Co., Dordrecht-Holland, p. 11.

Rees, M. H.: 1963a, 'A Method for Determining the Height and Geographical Position of an Auroral Arc from one Observing Station', *J. Geophys. Res.* **68**, 175.

Rees, M. H.: 1963b, 'Auroral Ionization and Excitation by Incident Energetic Electrons', *Planetary Space Sci.* **11**, 1209.

Rees, M. H.: 1964, 'Ionization in the Earth's Atmosphere by Aurorally Associated Bremsstrahlung X-Rays', *Planetary Space Sci.* **12**, 1093.

Rees, M. H.: 1969, 'Auroral Electrons', *Space Sci. Rev.* **10**, 413.

Rees, M. H.: 1974, 'Observations of NaD in Aurora', in preparation.

Rees, M. H. and Benedict, P. C.: 1970, 'The Auroral Proton Oval', *J. Geophys. Res.* **75**, 4763.

Rees, M. H. and Jones, R. A.: 1973, 'Time Dependent Studies of the Aurora. II: Spectroscopic Morphology', *Planetary Space Sci.* **21**, 1213.

Rees, M. H. and Maeda, K.: 1973, 'Auroral Electron Spectra', *J. Geophys. Res.* **78**, 8391.

Rees, M. H. and Roble, R. G.: 1974, 'Observations and Theory of the Formation of Stable Aurora. Red Arcs', *Fundamentals of Cosmic Phys.*, Submitted.

Rees, M. H. and Walker, J. C. G.: 1968, 'Ion and Electron Heating by Auroral Electric Fields', *Ann. Géophys.* **24**, 193.

Rees, M. H., Belon, A. E., and Romick, G. J.: 1961, 'The Systematic Behavior of Hydrogen Emission in the Aurora – I', *Planetary Space Sci.* **5**, 87.

Rees, M. H., Jones, R. A., and Walker, J. C. G.: 1971, 'The Influence of Field-Aligned Currents on

Auroral Electron Temperatures', *Planetary Space Sci.* **19**, 313.
Rees, M.H., Stewart, A.I., and Walker, J.C.G.: 1969, 'Secondary Electrons in Aurora', *Planetary Space Sci.* **17**, 1997.
Rees, M.H., Walker, J.C.G., and Dalgarno, A.: 1967, 'Auroral Excitation of the Forbidden Lines of Atomic Oxygen', *Planetary Space Sci.* **15**, 1097.
Reid, G.C. and Collins, C.: 1959, 'Observations of Abnormal VHF Radio Wave Absorption at Medium and High Latitudes', *J. Atmospheric Terrest. Phys.* **14**, 63.
Reid, J.S. and Phillips, J.: 1971, 'Time Lags in the Auroral Zone Ionosphere', *Planetary Space Sci.* **19**, 959.
Rème, H. and Bosqued, J.M.: 1971, 'Evidence near the Auroral Ionosphere of a Parallel Electric Field Deduced from Energy and Angular Distributions of Low Energy Particles', *J. Geophys. Res.* **76**, 7683.
Ring, J. and Schofield, J.W.: 1972, 'Field Compensated Michelson Spectrometers', *Appl. Opt.* **11**, 507.
Robben, F.: 1971, 'Noise in the Measurement of Light with Photomultipliers', *Appl. Opt.* **10**, 777.
Robertson, I.W.H., Wallis, D.D., Anger, C.D., and Johnson, D.W.: 1974, 'A Rocket Instrument for Imaging the Aurora by Means of a High Speed Photometric Scanner', *Planetary Space Sci.* **22**, 1003.
Rogers, E.H., Morse, F.A., Nelson, D.F., and Savage, R.C.: 1973, 'Auroral Photography from Satellites', in *IAGA Symposium on Aurora and Airglow*, Kyoto. (See also *Science* **183**, 951 (1974).)
Romick, G.J. and Belon, A.E.: 1967a, 'The Spatial Variation of Auroral Luminosity. I: The Behaviour of Synthetic Model Auroras', *Planetary Space Sci.* **15**, 475.
Romick, G.J. and Belon, A.E.: 1967b, 'Spatial Variation of Auroral Luminosity. II: Determination of Volume Emission Rate Profiles', *Planetary Space Sci.* **15**, 1695.
Romick, G.J. and Brown, N.B.: 1971, 'Midday Auroral Observations in the Oval, Cusp Region and Polar Cap', *J. Geophys. Res.* **76**, 8420.
Romick, G.J. and Elvey, C.T.: 1958, 'Variations in the Intensity of the Hydrogen Emission Line H_β during Auroral Activity', *J. Atmospheric Terrest. Phys.* **12**, 283.
Rosen, B.: 1970, *Données spectroscopiques rélatives aux molécules diatomiques*, Pergamon Press.
Rosenberg, T.J.: 1968, 'Correlated Bursts of VLH Hiss, Auroral Light and X-Rays', *Planetary Space Sci.* **16**, 1419.
Rosenberg, T.J., Bjordal, J., Trefall, H., Kvifte, G.J., Omholt, A., and Egeland, A.: 1971, 'Correlation Study of Auroral Luminosity and X-Rays', *J. Geophys. Res.* **76**, 122.
Rostoker, G.: 1972, 'Polar Magnetic Substorms', *Rev. Geophys. Space Phys.* **10**, 157.
Rothwell, P. and Lynam, C.: 1969, 'The Plasmapause, The Plasma Sheet and Energetic Trapped Electrons in the Earth's Magnetosphere', *Planetary Space Sci.* **17**, 447.
Russell, C.T. and McPherron, R.L.: 1973, 'The Magnetotail and Substorms', *Space Sci. Rev.* **15**, 205.
Russell, H.N.: 1936, 'Table for Intensities of Lines', *Astrophys. J.* **83**, 129.
Ryan, K.R.: 1969, 'Ionic Collision Processes in Gaseous Mixtures of Oxygen and Nitrogen', *J. Chem.* Phys. **51**, 4136.
Sandford, B.P.: 1961, 'Enhancement of Night-Sky Molecular Nitrogen Emission in Polar Cap Absorption Events', *Nature* **190**, 245.
Sandford, B.P.: 1963, 'Optical Studies of Particle Bombardment in Polar Cap Absorption Events', *Planetary Space Sci.* **10**, 195.
Sandford, B.P.: 1964, 'Aurora and Airglow Intensity Variations With Time and Magnetic Activity at Southern High Latitudes', *J. Atmospheric Terrest. Phys.* **26**, 749.
Sandford, B.P.: 1967, 'Polar Glow Aurora', *Space Res.*, **7**, 836.
Sandford, B.P.: 1968, 'Variations of Auroral Emissions with Time Magnetic Activity and Solar Cycle', *J. Atmospheric Terrest. Phys.* **30**, 1921.
Sandford, B.P. and Heiser, P.: 1959, 'Colour Photography of the Aurora', *Nature* **184**, 540.
Sawchuk, W. and Anger, C.D.: 1972, 'A dual Wavelength Ground-Based Auroral Scanner', *Planetary Space Sci.* **20**, 1935.
Schaeffer, R.C., Feldman, P.D., and Zipf, E.C.: 1972, 'The Dayglow [O I] $\lambda\lambda$ 6300 and 5577 Å Lines in the Early Morning Ionosphere', *J. Geophys. Res.* **77**, 6828.
Schiff, H.I.: 1972, 'Laboratory Measurements of Reactions Related to Ozone Photochemistry', *Ann. Géophys.* **28**, 67.

Schiff, H.I.: 1973, 'Neutral Atmospheric Chemistry – Introduction and Review', in B.M. McCormac (ed.), *Physics and Chemistry of Atmospheres*, D. Reidel Publishing Co., Dordrecht-Holland, p. 85.

Schindler, K. and Ness, N.F.: 1972, 'Internal Structure of the Geomagnetic Neutral Sheet', *J. Geophys. Res.* **77**, 91.

Schunk, R.W. and Hays, P.B.: 1971, 'Photoelectron Energy Losses to Thermal Electrons', *Planetary Space Sci.* **19**, 113.

Schunk, R.W. and Walker, J.C.G.: 1970a, 'Thermal Diffusion in the F-2 Region of the Ionosphere', *Planetary Space Sci.* **18**, 535.

Schunk, R.W. and Walker, J.C.G.: 1970b, 'Minor Ion Diffusion in the *F*-2 Region of the Ionosphere', *Planetary Space Sci.* **18**, 1319.

Schunk, R.W. and Walker, J.C.G.: 1971, 'Transport Processes in the *E* region of the Ionosphere', *J. Geophys. Res.* **76**, 6159.

Scourfield, M.W.J. and Parsons, N.R.: 1969, 'Auroral Pulsations and Flaming – Some Initial Results of a Cinematographic Study Using an Image Intensifier', *Planetary Space Sci.* **17**, 1141.

Scourfield, M.W.J. and Parsons, N.R.: 1971, 'Television Imaging of Fast Auroral Waves', *Planetary Space Sci.* **19**, 437.

Scourfield, M.W.J., Cresswell, G.R., Pilkington, G.R., and Parsons, N.R.: 1970, 'Auroral Pulsations-Television Image and X-Ray Correlations', *Planetary Space Sci.* **18**, 495.

Scourfield, N.W.J., Innes, W.F., and Parsons, N.R.: 1972, 'Spatial Coherency of Pulsating Aurora', *Planetary Space Sci.* **20**, 1843.

Seaton, M.J.: 1958, 'Oxygen Red Lines in the Airglow. II: Collisional Deactivation Effects', *Astrophys. J.* **127**, 67.

Sen, H.K. and Wyller, A.A.: 1960, 'On the Generalization of the Appleton-Hartree Magnetoionic Formulas', *J. Geophys. Res.* **65**, 3931.

Sharber, J.R. and Heikkila, W.J.: 1972, 'Fermi Acceleration of Auroral Particles', *J. Geophys. Res.* **77**, 3397.

Sharp, R.D. and Johnson, R.G.: 1968, 'Some Average Properties of Auroral Electron Precipitation as Determined from Satellite Observations', *J. Geophys. Res.* **73**, 969.

Sharp, W.E.: 1971, 'Rocket Borne Spectroscopic Measurements in the Ultraviolet Aurora; Nitrogen Vegard-Kaplan Bands', *J. Geophys. Res.* **76**, 987.

Sharp, R.D., Carr, D.L., Johnson, R.G., and Shelley, E.G.: 1971, 'Coordinated Auroral-Electron Observations from a Synchronous and a Polar Satellite', *J. Geophys. Res.* **76**, 7669.

Sharp, W.E. and Rees, M.H.: 1972, 'The Auroral Spectrum Between 1200 and 4000 Å', *J. Geophys. Res.* **77**, 1810.

Shefov, N.N.: 1969, 'Hydroxyl Emission of the Upper Atmosphere. I: The Behaviour during a Solar Cycle, Seasons and Geomagnetic Disturbances', *Planetary Space Sci.* **17**, 797.

Shefov, N.N. and Yurchenko, O.T.: 1970, 'Absolute Intensities of the Auroral Emissions in Zvenigorod', Results of Researches on the International Geophys. Projects, *Aurora and Airglow* **18**, 50.

Shelley, E.G., Johnson, R.G., and Sharp, R.D.: 1972, 'Satellite Observations of Energetic Heavy Ions During a Geomagnetic Storm', *J. Geophys. Res.* **77**, 6104.

Shemansky, D.E.: 1969a, 'N_2 Vegard-Kaplan System in Absorption', *J. Chem. Phys.* **51**, 689.

Shemansky, D.E.: 1969b, 'Transition Probabilities and Collision Broadening Cross-Section of the N_2 Lyman-Birge-Hopfield System', *J. Chem. Phys.* **51**, 5487.

Shemansky, D.E. and Broadfoot, A.L.: 1971a, 'Excitation of N_2 and N_2^+ Systems by Electrons. II: Excitation Cross-Sections and N_2 1PG Low Pressure Afterglow', *J. Quant. Spectrosc. Radiat. Transfer* **11**, 1401.

Shemansky, D.E. and Broadfoot, A.L.: 1971b, 'Excitation of N_2 and N_2^+ Systems by Electrons. I: Absolute Transition Probabilities'. *J. Quant. Spectrosc. Radiat. Transfer* **11**, 1385.

Shemansky, D.E. and Broadfoot, A.L.: 1973, 'Comment on Paper by D.C. Cartwright, S. Trajmar and W. Williams on Vibrational Population of $A^3\Sigma_u^+$ and $B^3\Pi_g$ States of N_2 in Normal Auroras', *J. Geophys. Res.* **78**, 2357.

Shemansky, D.E. and Vallance Jones, A.: 1968, 'Type-B Red Aurora: The O_2^+ First Negative System and the N_2 First Positive System', *Planetary Space Sci.* **16**, 1115.

Shemansky, D.E., Donahue, T.M., and Zipf, E.C., Jr.: 1972, 'N_2 Positive and N_2^+ Band Systems and the Energy Spectra of Auroral Electrons', *Planetary Space Sci.* **20**, 905.

Shemansky, D.E., Zipf, E.C., and Donahue, T.M.: 1971, 'Deactivation of $N_2 A^3\Sigma_u^+$ Molecules in Aurora', *Planetary Space Sci.* **19**, 1669.

Shepherd, G.G.: 1971, 'Auroral Excitation and Ionization Intensity and Temperature Variations', in B.M. McCormac (ed.), *The Radiating Atmosphere*, D. Reidel Publishing Co. Dordrecht-Holland, p. 150.

Shepherd, G.G., Lake, C.W., Miiller, J.R., and Cogger, L.L.: 1965, 'A Spatial Spectral Scanning Technique for the Fabry-Perot Spectrometer', *Appl. Opt.* **4**, 267.

Shepherd, G.G., Anger, C.D., Brace, L.H., Burrows, J.R., Heikkila, W.J., Hoffman, J., Maier, E.J., and Whitteker, T.H.: 1973a, 'An Observation of Polar Aurora and Airglow from the ISIS-II Spacecraft', *Planetary Space Sci.* **21**, 819.

Shepherd, G.G., Fancott, T., McNally, J., and Kerr, S.: 1973b, 'ISIS-II Atomic Oxygen Red Line Photometer', *Appl. Opt.* **12**, 1767.

Shepherd, G.G. and Pemberton, E.V.: 1968, 'Characteristics of Auroral Brightness Fluctuations', *Radio Sci.* **3**, 650.

Shuiskaya, F.K.: 1961, 'Determination of Relative Rates of Population of the Various Vibrational Levels of 1NG N_2^+ and Rotational Temperatures in Aurorae', Results of Researches on the International Geophys. Projects Pub. House 'NAUK', Moscow, Articles, *Spectral, Electrophotometrical and Radar Researches of Aurora and Airglow* **5**, 49.

Simonow, G.V.: 1963, 'Geomagnetic Time', *Geophys. J.* **8**, 258.

Simpson, F.R. and McConkey, J.W.: 1969, 'Excitation of the $A^2\Pi_u$ State of N_2^+ by Electrons', *Planetary Space Sci.* **17**, 1941.

Siscoe, G.L.: 1966, 'A Unified Treatment of Magnetospheric Dynamics with Applications to Magnetic Storms', *Planetary Space Sci.* **14**, 947.

Siscoe, G.L.: 1972, 'On the Plasma Sheet Contribution to Force Balance Requirements in the Geomagnetic Tail', *J. Geophys. Res.* **77**, 6230.

Sivjee, G.G.: 1974, 'Coordinated Analysis of Airborne Spectrophotometric Measurements from Mid-Day Auroras', *J. Geophys. Res.*, Submitted.

Slanger, T.G., Wood, B.J., and Black, G.: 1971, 'Temperature Coefficients for $N(^2D)$ Quenching by O_2 and N_2O', *J. Geophys. Res.* **76**, 8430.

Smith, D.: 1972, 'The Application of Langmuir Probes to the Measurement of Very Low Electron Temperatures', *Planetary Space Sci.* **20**, 1715.

Snyder, A.L. and Akasofu, S.I.: 1972, 'Observations of the Auroral Oval by the Alaskan Meridian Chain of Stations', *J. Geophys. Res.* **77**, 3419.

Sofko, G.J.: 1974, Private Communication.

Sofko, G.J. and Kavadas, A.: 1969, 'Periodic Fading in 42-MHz Auroral Back Scatter', *J. Geophys. Res.* **74**, 3651.

Sofko, G.J. and Kavadas, A.: 1971, 'Polarization Characteristics of 42-MHz Auroral Back Scatter', *J. Geophys. Res.* **76**, 1778.

Speiser, T.W.: 1967a, 'Auroral Particles Accelerated in the Geomagnetic Tail', in B.M. McCormac (ed.), *Aurora and Airglow*, Reinhold Publishing Co., New York, p. 491.

Speiser, T.W.: 1967b, 'Particle Trajectories in Model Current Sheets. II: Applications to Auroras Using a Geomagnetic Tail Model', *J. Geophys. Res.* **72**, 3919.

Spitz, A.L. and Watson, M.D.: 1972, 'Auroral Photometric Observations at Geomagnetically Conjugate Points', *Planetary Space Sci.* **20**, 1337.

Stair, A.T., Huppi, E.R., Sandford, P.B., Murphy, R.E., O'Neil, R.R., Hart, A.M., Huppi, R.J., and Pendleton, W.R.: 1971, 'Infrared Investigations of Aurora and Airglow', in B.M. McCormac (ed.), *The Radiating Atmosphere*, D. Reidel Publishing Co. Dordrecht-Holland, p. 185.

Starkov, G.V. and Feldstein, Ya. I.: 1968, 'Auroral Oval and its Dependence on Magnetic Disturbance', Results of Researches on the International Geophys. Projects Pub. House 'NAUK', Moscow, Articles, *Polar Aurora* **17**, 22.

Stenbaek-Nielsen, H.C., Davis, T.N., and Glass, N.W.: 1972, 'Relative Motion of Auroral Conjugate Points during Substorms', *J. Geophys. Res.* **77**, 1844.

Stenbaek-Nielsen, H.C., Wescott, E.M., Davis, T.N., and Peterson, R.W.: 1973, 'Differences in Auroral Intensity at Conjugate Points', *J. Geophys. Res.* **78**, 659.

Stetson, H.: 1939, 'The Present State of Solar Activity and Associated Phenomena', *Science* **90**, 482.

Stoffregen, W.: 1968, 'On the Influence of the Moon on the Auroral Activity', *Ann. Géophys.* **24**, 159.

Stoffregen, W.: 1969, 'Transient Emissions on the Wavelength of He I, 5876 Å recorded during Auroral Break-up', *Planetary Space Sci.* **17**, 1927.

Stolarski, R.S.: 1968, 'Calculation of Auroral Emission Rates and Heating Effects', *Planetary Space*

Sci. **16**, 1265.
Stolarski, R.S. and Green, A.E.S.: 1967, 'Calculations of Auroral Intensities from Electron Impact', *J. Geophys. Res.* **72**, 3967.
Stolarski, R.S., Dulock, V.A., Watson, C.E., and Green, A.E.S.: 1967, 'Electron Impact Cross-Sections for Atmospheric Species. II: Molecular Nitrogen', *J. Geophys. Res.* **72**, 3953.
Stone, E.J. and Zipf, E.C.: 1973, 'Excitation of Atomic N by Electron Impact', *J. Chem. Phys.* **58**, 4278.
Stone, E.J. and Zipf, E.C.: 1974, 'Electron Impact Excitation of the $^3S^0$ and $^5S^0$ States of Atomic Oxygen', *J. Chem. Phys.* **60**, 4237.
Stone, J.M.: 1963, *Radiation and Optics*, McGraw-Hill, New York, p. 346.
Störmer C.: 1955, *The Polar Aurora*, Clarendon Press, Oxford.
Strickland, D.J.: 1970, 'Resonance Transport of N I 1200 Å Radiation in Aurora', *Trans. Am. Geophys. Union* **51**, 368.
Stringer, W.J. and Belon, A.E.: 1967, 'Statistical Auroral Zone for IQSY', *J. Geophys. Res.* **72**, 245.
Strobel, D.F.: 1971, 'Diurnal Variation of NO in the Upper Atmosphere', *J. Geophys. Res.* **76**, 2441.
Strobel, D.F. and McElroy, M.B.: 1970, 'The *F*-2 Layer at Middle Latitudes', *Planetary Space Sci.* **18**, 1181.
Struve, O.: 1951, 'Photography of the Counterglow', *Sky Telesc.* **10**, 215.
Stubbe, P.: 1968, 'Frictional Forces and Collision Frequencies Between Moving Ion and Neutral Gases', *J. Atmospheric Terrest. Phys.* **30**, 1965.
Stubbe, P.: 1970, 'Simultaneous Solution of the Time Dependent Coupled Continuity Equations, Heat Conduction Equations and Equations of Motion for a System Consisting of a Neutral Gas, an Electron Gas and a Four Component Ion Gas', *J. Atmospheric Terrest. Phys.* **32**, 865.
Stuhl, F. and Niki, H.: 1969, 'Formation of Metastable $N_2(A^3\Sigma_u^+)$ in Charge Transfer Reaction of N_2^+ with NO', *Chem. Phys. Letters* **4**, 417.
Sugiura, M.: 1972, 'Equatorial Current Sheet in the Magnetosphere', *J. Geophys. Res.* **77**, 6093.
Svalgaard, L.: 1973, 'Polar Cap Magnetic Variations and Their Relationship with Interplanetary Magnetic Sector Structure', *J. Geophys. Res.* **78**, 2064.
Swartz, W.E., Nisbet, J.S., and Green, A.E.S.: 1971, 'Analytic Expression for the Energy-Transfer Rate from Photoelectrons to Thermal Electrons', *J. Geophys. Res.* **76**, 8425.
Swider, W. and Narcisi, R.S.: 1970, 'On the Ionic Contribution of Class I Auroras', *Planetary Space Sci.* **18**, 379.
Swift, D.W.: 1965, 'A Mechanism for Energizing Electrons in Magnetosphere', *J. Geophys. Res.* **70**, 3061.
Swift, D W.: 1970, 'Particle Acceleration by Electrostatic Waves', *J. Geophys. Res.* **75**, 6324.
Swift, D.W.: 1971, 'Possible Mechanisms for Formation of the Ring Current Belt', *J. Geophys. Res.* **76**, 2276.
Swift, D.W.: 1973, 'The Generation of Infrasonic Waves by Auroral Electrojets', *J. Geophys. Res.* **78**, 8205.
Taeusch, D.R., Carignan, G.R., and Reber, C.A.: 1971, 'Neutral Composition Variation Above 400 km During a Magnetic Storm', *J. Geophys. Res.* **76**, 8318.
Takayanagi, K. and Itikawa, Y.: 1970, 'Elementary Processes Involving Electrons in the Ionosphere', *Space Sci. Rev.* **11**, 380.
Taylor, H.E. and Hones, E.W.: 1965, 'Adiabatic Motion of Auroral Particles in a Model of the Electric and Magnetic Fields Surrounding the Earth', *J. Geophys. Res.* **70**, 3605.
Title, A.M.: 1971, 'Fabry-Perot Interferometers as Narrow Band Optical Filters', in F. Labuhn and R. Lüst (eds.) 'New Techniques in Space Astronomy', *IAU Symp.* **41**, 325.
Trajmar, S., Cartwright, D.C., and Williams, W.: 1971, 'Differential and Integral Cross-Sections for the Electron-Impact Excitation of the $a^1\Delta_g$ and $b^1\Sigma_g$ States of O_2', *Phys. Rev.* **4**, 1482.
Trefall, H.: 1971, 'X-Rays Associated with Auroras', in B.M. McCormac (ed.), *The Radiating Atmosphere*, D. Reidel Publishing Co., Dordrecht-Holland, p. 288.
Truttse, Yu.L.: 1972, 'Oxygen Emission at 6300 Å', *Ann. Géophys.* **28**, 169.
Tsurutani, B.T. and Smith, E.J.: 1973, 'Midnight Sector Chorus', in *Symposium on Magnetospheric Substorms*, IAGA., Kyoto.
Turgeon, E.C. and Shepherd, G.G.: 1962, 'Upper Atmosphere Temperatures from Doppler Line

Widths. II: Measurements of the O I 5577 and O I 6300 Å Lines in Aurora', *Planetary Space Sci.* **9**, 295.

Ullaland, S. L., Trefall, H. Kremser, G., and Bewersdorff, A.: 1967, 'Large Scale Coherent Pulsa tion in Bremsstrahlung', *J. Atmospheric Terrest. Phys.* **29**, 395.

Unti, T. and Atkinson, G.: 1968, 'Two-Dimensional Chapman-Ferraro Problem with Neutral Sheet. I: The Boundary', *J. Geophys. Res.* **73**, 7319.

Unwin, R. S. and Baggaley, W. T.: 1972, 'The Radio Aurora', *Ann. Géophys.* **28**, 111.

Vaisberg, O. L.: 1962, 'Auroral Spectroelectrophotometry of the Emission N_2, N_2^+, O I, N II', Results of Researches of the Program of the IGY. Pub. House Acad. Sci. U.S.S.R., *Polar Aurora and Night Airglow*, **8**, 43.

Vaisberg, O. L.: 1967, 'Pitch-Angle and Energy Distribution of Protons in Aurora', Results of Researches on the International Geophys. Projects Pub. House 'NAUK', Moscow, Articles, *Aurora and Airglow*, **13**, 73.

Vallance Jones, A.: 1955, 'An Atlas of the Auroral Spectrum', Scientific Report AR 20, Contract A.F. 19 (122)-152.

Vallance Jones, A.: 1960, 'An Analysis of a Spectrogram of the Red Aurora of February 10/11, 1958, in the Wavelength Range 7300–8700 Å', *Can. J. Phys.* **38**, 453.

Vallance Jones, A.: 1967, 'Auroral Observations from West-Central Canada', in B. M. McCormac (ed.), *Aurora and Airglow*, Reinhold Pubblishing Co., New York, p. 465.

Vallance Jones, A.: 1971, 'Auroral Spectroscopy', *Space Sci. Rev.* **11**, 776.

Vallance Jones, A. and Gattinger, R. L.: 1971, 'Infrared Spectrum of Aurora', in B. M. McCormac (ed.), *The Radiating Atmosphere*, D. Reidel Publishing Co. Dordrecht, Holland, p. 177.

Vallance Jones, A. and Gattinger, R. L., 1972, 'Low Intensity Auroral Red Arcs', *Ann. Géophys.* **28**, 85.

Vallance Jones, A. and Gattinger, R. L.: 1973, 'Indirect Excitation Process in Aurora, in B. M. McCormac (ed.), *Physics and Chemistry of Atmospheres*, D. Reidel Publishing Co. Dordrecht, Holland, p. 232.

Vallance Jones, A. and Gattinger, R. L.: 1974, 'The $O_2\, b^1\Sigma_g^+ – X^3\Sigma_g^-$ System in Aurora', *J. Geophys. Res.* **79** (in press).

Vallance Jones, A. and Harrison, A.W.: 1955, 'Rotational Temperatures of the Auroral N_2^+ bands', *J. Atmospheric Terrest. Phys.* **6**, 336.

Vallance Jones, A. and Hunten, D.M.: 1960, 'Rotational and Vibrational Intensity Distribution of the First Negative N_2^+ Bands in Sunlit Auroral Rays', *Can. J. Phys.* **38**, 458.

Vasyliunas, V.M.: 1968, 'A Survey of Low-Energy Electrons in the Evening Sector of the Magnetosphere with OGO-1 and OGO-3', *J. Geophys. Res.* **73**, 2839.

Vasyliunas, V.M.: 1970, 'Mathematical Models of Magnetospheric Convection and its Coupling to the Ionosphere', in B.M. McCormac (ed.), *Particles and Fields in the Magnetosphere*, D. Reidel Publishing Co., Dordrecht-Holland, p. 60.

Vegard, L. and Tonsberg, E.: 1937, 'Variations of the Intensity Distribution within the Auroral Spectrum', *Geophys. Pub.* **11** (16), 1.

Venkatesan, D., Oliven, M.N., Edwards, P.J., McCracken, K.G., and Steinback, M.: 1968, 'Microburst Phenomena. I: Auroral Zone X-Rays', *J. Geophys. Res.* **73**, 2333.

Vestine, E.H.: 1944, 'Geographic Incidence of Aurora and Magnetic Disturbance, Northern Hemisphere', *Terr. Magn.* **49**, 77.

Vestine, E.H.: 1960, *The Upper Atmosphere and Geomagnetism*, Ratcliffe, Cambridge, p. 495.

Vreux, J.-M.: 1971, 'Enregistrement spectrophotographique d'émissions ultraviolettes aurorales durant un vol de fusée-sonde', *Ann. Géophys.* **27**, 493.

Vreux, J.-M. and Marette, G.: 1971, 'Étude de variations rapides de la lumière aurorale', *Bull. Soc. Roy. Sci. Liège* **40**, 451.

Wacks, M.E. and Krauss, M.: 1961, 'Franck-Condon Factors and Shape of Ionization Efficiency Curves', *J. Chem. Phys.* **35**, 1902.

Walker, J.C.G.: 1968, 'Electron and Nitrogen Vibrational Temperatures in the *E* Region of the Ionosphere ', *Planetary Space Sci.* **16**, 321.

Walker, J.C.G.: 1970, 'Electric Field Excitation of O_2 ($^1\Delta_g$) in Auroras', *Planetary Space Sci.* **18**, 1043.

Walker, J.C.G.: 1971, 'Ionosphere Effects of Auroral Electric Fields' in B.M. McCormac (ed.), *Physics and Chemistry of Atmospheres*, D. Reidel Publishing Co. Dordrecht-Holland, p. 214.

Walker, J.C.G.: 1973, 'Oxygen and Nitrogen Vibration in the Thermosphere', in B.M. McCormac (ed.), *Physics and Chemistry of Atmospheres*, D. Reidel Pub. Co. Dordrecht-Holland, p. 203.

Walker, J.C.G. and Rees, M.H.: 1968, 'Ionospheric Electron Densities and Temperatures in Aurora', *Planetary Space Sci.* **16**, 459.

Walker, J.C.G., Brace, L.H., and Rees, M.H.: 1968, 'Langmuir Probe Evidence for a Nocturnal Ionization Source at Fort Churchill', *J. Geophys. Res.* **73**, 7285.

Walker, J.C.G., Stolarski, R.S., and Nagy, A.F.: 1969, 'The Vibrational Temperature of Molecular Nitrogen in the Thermosphere', *Ann. Géophys.* **25**, 831.

Wallace, L.: 1959, 'An Analysis of a Spectrogram of the Aurora of 11 February 1958, in the Wavelength Range 3710–4420 Å, *J. Atmospheric Terrest. Phys.* **17**, 46.

Wallace, L. and Chamberlain, J.W.: 1960, 'Excitation of O_2 Atmospheric Bands in the Aurora', *Planetary Space Sci.* **2**, 60.

Walt, M.: 1967, 'Penetration of Electrons into the Atmosphere', in B.M. McCormac (ed.), *Aurora and Airglow*, Reinhold Publishing Co., New York, p. 287.

Warnecke, G., Reed, E.I., Fowler, W.B., Krems, E.R., Allison, L.J., and Blamont, J.E.: 1969, 'Meteorological Results from Multispectral Photometry in Airglow Bands by the OGO-4 Satellite', *J. Atmospheric Terrest. Phys.* **26**, 1329.

Warwick, C.S. and Hansen, R.T.: 1959, 'Geomagnetic Activity Following Large Solar Flares, *J. Atmospheric Terrest. Phys.* **14**, 287.

Watson, C.E., Dulock, V.A. Jr., Stolarski, R.S., and Green, A.E.S.: 1967, 'Electron Impact Cross-Sections for Atmospheric Species. III: Molecular Oxygen', *J. Geophys. Res.* **72**, 3961.

Wax, R.L. and Bernstein, W.: 1970, 'Rocket-Borne Measurements of H_β Emissions and Energetic Hydrogen Fluxes during an Auroral Breakup', *J. Geophys. Res.* **75**, 783.

Webster, H.F. and Hallinan, T.J.: 1973, 'Instabilities in Charge Sheets and Current Sheets and Their Possible Occurrence in the Aurora', *Radio Sci.* **8**, 475.

Weill, G.M.: 1969, 'NI($^4S-^2D$) Radiation in the Night Airglow and Low Latitude Aurora', in B.M. McCormac and A. Ombolt (eds.), *Atmospheric Emissions*, Van Nostrand Reinhold Co., New York, p. 449.

Weller, A.E.: 1961, 'Interferometric Study of the Green Line in the Spectrum of Some Auroral Forms', Results of Researches on the International Geophys. Projects Pub. House 'NAUK', Moscow, Articles, *Aurora and Airglow*, **7**, 65.

Wells, W.C., Borst, W.L., and Zipf, E.C.: 1971, 'Absolute Cross-Section for Production of O($^5S^0$) by Electron Impact Dissociation of O_2', *Chem. Phys. Letters* **12**, 288.

Wescott, E.M.: 1961, 'Magnetic Variations at Conjugate Points', *J. Geophys. Res.* **66**, 1789.

Wescott, E.M., Stolarik, J.D., and Heppner, J.P.: 1969, 'Electric Fields in the Vicinity of Auroral Forms from Motions of Ba Vapor Releases', *J. Geophys. Res.* **74**, 3469.

Westerlund, L.H.: 1969, 'The Auroral Electron Energy Spectrum Extended to 45 eV', *J. Geophys. Res.* **74**, 351.

Whalen, B.A. and McDiarmid, I.B.: 1969, 'Summary of Rocket Measurements of Auroral Particle Precipitation', in B.M. McCormac and A. Omholt (eds.), *Atmospheric Emissions* Van Nostrand-Reinhold, New York, p. 93.

Whalen, B.A. and McDiarmid, I.B.: 1972, 'Observations of Magnetic Field Aligned Auroral-Electron Precipitation', *J. Geophys. Res.* **77**, 191.

Whalen, B.A. and McDiarmid, I.B.: 1974, 'Auroral Particle Precipitation Observations', *Fundamentals of Cosmic Phys.*, Submitted.

Whalen, B.A., Miller, J.R., and McDiarmid, I.B.: 1971a, 'Evidence for Solar Wind Origin of Auroral Ions from Low-Energy Ion Measurements', *J. Geophys. Res.* **76**, 2406.

Whalen, B.A., Miller, J.R., and McDiarmid, I.B.: 1971b, 'Energetic Particle Measurements in a Pulsating Aurora', *J. Geophys. Res.* **76**, 978.

Whalen, J.A., Buchau, J., and Wagner, R.A.: 1971c, 'Airborne Ionospheric and Optical Measurements of Noontime Aurora', *J. Atmospheric Terrest. Phys.* **33**, 661.

Whalen, B.A., Green, D.W., and McDiarmid, I.B.: 1974, 'Observations of Ionospheric Ion Flow and Related Convective Electric Fields in and near an Auroral Arc', *J. Geophys. Res.* **79**, 2835.

Whitehead, J.P.: 1968, 'Low Frequency Plasma Instabilities in the Ionosphere', *J. Atmospheric Terrest. Phys.* **30**, 1563.

Wiens, R.H.: 1968, 'Average Time Latitude Distributions of Hydrogen and [O I] Green Line Emissions in Aurora', Ph. D. Thesis, Univ. of Saskatchewan.

Wiens, R.H. and Vallance Jones, A.: 1969, 'Studies of Auroral Hydrogen Emissions in West-Central Canada. III: Proton and Electron Auroral Ovals', *Can. J. Phys.* **47**, 1493.

Wilcox, J.M.: 1968, 'Interplanetary Magnetic Field', *Space Sci. Rev.* **8**, 258.

Wilcox, J.M. and Colburn, D.S.: 1972, 'Interplanetary Sector Structure at Solar Maximum', *J. Geophys. Res.* **77**, 751.

Wilson, C.R.: 1969, 'Two-Station Auroral Infrasonic Wave Observations', *Planetary Space Sci.* **17**, 1817.

Wilson, C.R.: 1972, 'Auroral Infrasonic Wave Generation Mechanism', *J. Geophys. Res.* **77**, 1820.

Wilson, C.R.: 1970, private communication.

Wilson, C.R. and Nichparenko, S.: 1967, 'Infrasonic Waves and Auroral Activity', *Nature* **214**, 1299.

Winters, H.F.: 1966, 'Ionic Absorption and Dissociation Cross-Section for Nitrogen', *J. Chem. Phys.* **44**, 1472.

Yevlashin, L.S.: 1970, 'Monochromatic Morphology of Polar Aurora According to the Data of the Patrol Spectrograph in Murmansk', Results of Researches on the International Geophys. Projects, Pub. House 'NAUK', Moscow, Articles, *Aurora* **19**, 12.

Young, R.A., Black, G., and Slanger, T.G.: 1969, 'Vacuum UV Photolysis of N_2O. II: Deactivation of $N_2(A^3\Sigma_u^+)$ and $N_2(B^3\Pi_g)$', *J. Chem. Phys.* **50**, 303.

Zipf, E.C.: 1969, 'The Collisional Deactivation of Metastable Atoms and Molecules in the Upper Atmosphere ', *Can. J. Chem.* **47**, 1863.

Zipf, E.C.: 1970, 'The Dissociative Recombination of O_2^+ Ions into Specifically Identified Final Atomic States', *Bull. Am. Phys. Soc.* **15**, 418.

Zipf, E.C.: 1973a, 'Comment on the Role of Photoionization in Auroral Arcs', *Trans AGU* **54**, 403.

Zipf, E.C.: 1973b, 'Photo Dissociation of O_2: A Major Source of the Aurocal Green Line', *Trans. AGU* **54**, 403.

Zipf, E.C.: 1973c, private communication.

Zipf, E.C.: 1974, private communication.

Zipf, E.C., Borst, W.L., and Donahue, T.M.: 1970, 'A Mass Spectrometer Observation of NO in an Auroral Arc', *J. Geophys. Res.* **75**, 6371.

Zwick, H.H. and Shepherd, G.G.: 1963, 'Some Observations of Hydrogen-Line Profiles in the Aurora', *J. Atmospheric Terrest. Phys.* **25**, 604.

Zwick, H.H. and Shepherd, G.G.: 1973, 'Upper Atmospheric Temperatures from Doppler Line Widths', *Planetary Space Sci.* **21**, 605.

SUBJECT INDEX

Boldface page numbers refer to major discussions of the topic; italic page numbers to figures or tables. Atoms and molecules are listed alphabetically by their chemical formulae.

GEOPHYSICS AND ASTROPHYSICS MONOGRAPHS

AN INTERNATIONAL SERIES OF FUNDAMENTAL TEXTBOOKS

1. R. Grant Athay, *Radiation Transport in Spectral Lines.* 1972, XIII + 263 pp.
2. J. Coulomb, *Sea Floor Spreading and Continental Drift.* 1972, X + 184 pp.
3. G.T. Csanady, *Turbulent Diffusion in the Environment.* 1973, XII + 248 pp.
4. F.E. Roach and Janet L. Gordon, *The Light of the Night Sky.* 1973, XII + 125 pp.
6. J. Iribarne and W. Godson, *Atmospheric Thermodynamics.* 1973, X + 222 pp.
7. Z. Kopal, *The Moon in the Post-Apollo Era*, 1974, VIII + 223 pp.
9. A. Vallance Jones, *Aurora*, 1974.
10. C.-J. Allègre and G. Michard, *Introduction to Geochemistry.* 1974, XI + 142 pp.
12. E. Tandberg-Hanssen, *Solar Prominences.* 1974, XVI + 155 pp.

Forthcoming:

5. R. Grant Athay, *The Solar Chromosphere and Corona*
8. Z. Švestka and L. De Feiter, *Solar High Energy Photon and Particle Emission*
11. G. Haerendel, *Magnetospheric Processes*